# REINFORCED CONCRETE DESIGN

# REINFORCED CONCRETE DESIGN

**Kenneth Leet**

*Professor of Civil Engineering*
*Northeastern University*

**McGraw-Hill Book Company**

New York   St. Louis   San Francisco   Auckland   Bogotá   Hamburg
Johannesburg   London   Madrid   Mexico   Montreal   New Delhi
Panama   Paris   São Paulo   Singapore   Sydney   Tokyo   Toronto

This book was set in Times Roman.
The editors were Julienne V. Brown and Madelaine Eichberg;
the production supervisor was Phil Galea.
The drawings were done by J & R Services, Inc.
The cover was designed by Charles A. Carson.
R. R. Donnelley & Sons Company was printer and binder.

**REINFORCED CONCRETE DESIGN**

Copyright © 1982 by McGraw-Hill, Inc. All rights reserved. Printed in the United States
of America. Except as permitted under the United States Copyright Act of 1976, no part
of this publication may be reproduced or distributed in any form or by any means, or
stored in a data base or retrieval system, without the prior written permission of the
publisher.

1234567890 DODO 898765432

ISBN 0-07-037024-9

Library of Congress Cataloging in Publication Data

Leet, Kenneth.
    Reinforced concrete design.

    Includes bibliographical references and index.
    1. Reinforced concrete construction.   I. Title.
TA683.2.L36      624.1'8341      81-17179
ISBN 0-07-037024-9              AACR2

# CONTENTS

# PREFACE

Engineering and architectural students who complete this book will have acquired the background necessary to design the majority of reinforced concrete structures commonly encountered in professional practice.

The most significant feature of this text is its attempt to organize and clarify the discussion of the behavior of reinforced concrete, a topic that can be quite complex if not carefully presented. As an introductory text the primary goal of this book is to develop a clear understanding of basic principles, the critical groundwork that must be laid in a first-year course. In a carefully sequenced, step-by-step approach, each topic is broken down into its simplest and most essential components so that students learn how to break down and deal with any structure, no matter how complex, by looking at the basic behavior of each component part. All design equations are derived from a few fundamental principles of statics and engineering mechanics that should be familiar to the average student. The text, which has been class-tested for eight years, is geared toward the average student's needs, questions, and difficulties.

Having students always consider the basic behavior and characteristics of the material allows them to understand the rationales behind the provisions of the American Concrete Institute (ACI) Code and prepares students to deal with design situations not covered by the code.

To assist students in visualizing the behavior of reinforced concrete members, an unusually large number of figures (over 400) has been incorporated into the text. In addition, many examples have been included to illustrate basic design procedures and ACI Code requirements.

Although preference is given to U.S. customary units, which are used almost exclusively by engineering firms in the United States, basic data, important design equations, certain design aids, and most design examples are stated in both

SI and U.S. customary units so that the book will also be suitable for use in universities that use the metric system.

This text contains sufficient material for two semesters of reinforced concrete design. The material for the first course, which provides the essential background for proportioning members and detailing reinforcement, is covered in the first seven chapters. If time permits, Section 9.8 on the analysis of continuous members by ACI coefficients may be logically introduced after Section 3.14 has been covered. This will allow students who are not familiar with indeterminate analysis to calculate design moments in continuous one-way members. If time is limited, or if the instructor wishes to develop certain topics in more detail, Chapter 5 on torsion may be omitted from a first course without any loss in continuity.

In a second course, the material of Chapters 8 through 11 will give the student an understanding of the design of typical reinforced concrete structures and prestressed concrete. Chapter 8 on foundations provides the student with the opportunity to apply many of the design procedures introduced in the first seven chapters to a number of relatively simple but common structural elements. Chapters 9 and 10 on building design describe how the analysis of large highly indeterminate three-dimensional building frames for gravity load can be simplified by subdividing the structure into smaller two-dimensional subsystems. Finally, Chapter 11 contains an extensive treatment of prestressed concrete, which provides certain advantages over normal reinforced concrete because cracking can be eliminated and higher-strength materials used more effectively.

## NOTE TO READER

This text contains two types of equations. The first type, derived from principles of mechanics, is used to relate internal stresses to applied forces. The variables in these equations may be expressed in terms of any set of dimensionally consistent units. In the second type of equation, which is usually based on experimental studies, the variables are restricted to specific units. For example, except for the equation predicting crack width under service loads, all ACI equations in U.S. customary units require that the compressive strength of concrete $f_c'$ and the yield point of steel $f_y$ be expressed in pounds per inch squared (lb/in$^2$).

## ACKNOWLEDGMENTS

The author wishes to express his appreciation to friends, colleagues, reviewers of the text, and students who have contributed to the development of this book by suggesting changes or noting errors in the early editions of the manuscript. In particular, the author would like to thank Judith Leet, who edited the initial draft of the text; Saul Namyet and Robert Taylor, who critically reviewed many chapters; Harold Heins and Maen Allaham, who checked the accuracy of computations; Bruce Wile, who prepared solutions to many problems; and James

Amrheim and Boyd Ringo, who encouraged the author to expand his class notes into this text.

This book is dedicated to my children Kenneth and Annabel and to the memory of M. and R. Wilson.

*Kenneth Leet*

This view of the National Center for Atmospheric Research, Boulder, Colorado, illustrates the variety of scuptural forms and surface textures possible with reinforced concrete; I. M. Pei, architect, (*Photograph by Ezra Stoller © ESTO.*)

# INTRODUCTION

## 1.1 OVERVIEW

In almost every branch of civil engineering and architecture, extensive use is made of reinforced concrete for structures and foundations. Engineers and architects therefore require a basic knowledge of reinforced concrete design throughout their professional careers. Much of this text is directly concerned with the behavior and proportioning of the components that make up typical reinforced concrete structures—beams, columns, and slabs. Once the behavior of these individual elements is understood, the designer will have the background to analyze and design a wide range of complex structures, such as foundations, buildings, and bridges, composed of these elements.

Since reinforced concrete is a nonhomogeneous material that creeps, shrinks, and cracks, its stresses cannot be accurately predicted by the traditional equations derived in a course in strength of materials for homogeneous elastic materials. Much of reinforced concrete design is therefore *empirical*; i.e., design equations and design methods are based on experimental and time-proved results instead of being derived exclusively from theoretical formulations.

A thorough understanding of the behavior of reinforced concrete will allow the designer to convert an otherwise brittle material into tough ductile structural elements and thereby take advantage of concrete's desirable characteristics, its high compressive strength, its fire resistance, and its durability.

1

## 1.2 CONCRETE AND REINFORCED CONCRETE

Concrete, a stonelike material, is made by mixing cement, water, fine aggregate (often sand), coarse aggregate, and frequently other additives (that modify properties) into a workable mixture. In its unhardened or *plastic* state, concrete can be placed in forms to produce a large variety of structural elements. Although the hardened concrete by itself, i.e., without any reinforcement, is strong in compression, it lacks tensile strength and therefore cracks easily. Because unreinforced concrete is brittle, it cannot undergo large deformations under load and fails suddenly without warning. The addition of steel reinforcement to the concrete reduces the negative effects of its two principal inherent weaknesses, its susceptibility to cracking and its brittleness. When the reinforcement is strongly bonded to the concrete, a strong, stiff, and ductile construction material is produced. This material, called *reinforced concrete*, is used extensively to construct foundations, structural frames, storage tanks, shell roofs, highways, walls, dams, canals, and innumerable other structures and building products. Two other characteristics of concrete that are present even when concrete is reinforced are shrinkage and creep, but the negative effects of these properties can be mitigated by careful design.

## 1.3 UNITS OF MEASURE

Although the United States is committed to converting its system of measurements from U.S. customary units (foot, pound, second) to SI units (meter, newton, second), little change has taken place in the American concrete industry to date. At present engineering offices in the United States work almost exclusively in U.S. customary units (USCU). Although it appears that the SI system will not be used in professional practice in the United States in the near future, it still seems appropriate to familiarize students with SI units in preparation for future metrification. Some conversion factors are given in Appendix A.

Currently, design equations in the main body of the ACI Code are expressed exclusively in terms of U.S. customary units and SI equations and conversion factors are relegated to Appendix D of the Code. The normal units used in the equations of each system are listed in Table 1.1.

**Table 1.1  Units used in ACI Code equations**

|         | U.S. customary | SI |
|---------|----------------|----|
| Length  | in             | mm (millimeter) |
|         | ft             | m (meter) |
| Load    | lb             | N (newton) |
| Stress  | $lb/in^2$      | MPa (megapascal) |
| Density | $lb/ft^3$      | $kg/m^3$ (mass) |

In addition to working a number of problems in metric units, basic data, important equations, certain design aids, and a number of examples will be given in both systems, with the SI units typically in parentheses after the U.S. customary units. Because of roundoff error in the coefficients of the ACI metric equations, they may give values differing by several percent from those given by equivalent equations in U.S. customary units.

## 1.4 CODES

A code is a set of technical specifications and standards that control important details of design and construction. The purpose of codes is to produce sound structures so that the public will be protected from poor or inadequate design and construction.

Two types of codes exist. One type, called a *structural code*, is originated and controlled by specialists who are concerned with the proper use of a specific material or who are involved with the safe design of a particular class of structures. For the structural engineer we list several important codes:

The American Concrete Institute (ACI) Building Code 318-77, covering the design of reinforced concrete buildings

The American Institute of Steel Construction Specifications (AISC), covering the design of steel buildings

The American Association of State Highway and Transportation Officials (AASHTO), covering the design of highway bridges

The American Railroad Engineering Association (AREA), covering the design of railroad bridges

The second type of code, called a *building code*, is established to cover construction in a given region, often a city or a state. The objective of a building code is also to protect the public by accounting for the influence of local environmental conditions on construction. For example, local authorities may specify additional provisions to account for such regional conditions as earthquake, heavy snow, or tornados. National structural codes generally are incorporated into local building codes.

The ACI Code contains provisions covering all aspects of reinforced concrete manufacture, design, and construction. It includes specifications on quality of materials, details on mixing and placing concrete, design assumptions for the analysis of continuous structures, and equations for proportioning members for design forces. All design procedures used in this text are consistent with specifications of the ACI Code.

The specifications of the ACI Code, based on many years of research and field experience with reinforced concrete, represent the minimum standards required to produce safe, durable structures. As research provides additional understanding of behavior, the contents of the ACI Code are continually reviewed and updated.

Currently the ACI Code makes yearly changes by issuing supplementary provisions. Every six or seven years a comprehensive code revision is made, incorporating all revisions since the last edition.

The ACI Code, like most codes, makes provisions for departures from prescribed standards if it can be shown by test or analysis that such changes will produce a safe design.

## 1.5 SERVICE LOADS

All structures must be proportioned so they will not fail or deform excessively under any possible condition of service. Therefore it is important that an engineer use great care in anticipating all the probable loads to which a structure will be subjected during its lifetime.

ACI Code §8.2 contains provisions listing the factors that should be considered when establishing the forces in concrete stuctures. Although the design of most members is controlled typically by dead and live load acting simultaneously, consideration must also be given to the forces produced by wind, impact, shrinkage, temperature change, creep and support settlements, earthquake, and so forth.

The load associated with the weight of the structure itself and its permanent components is called the *dead load*. The dead load of concrete members, which is substantial, should never be neglected in design computations. The exact magnitude of the dead load is not known accurately until members have been sized. Since some figure for the dead load must be used in computations to size the members, its magnitude must be estimated at first. After a structure has been analyzed, the members sized, and architectural details completed, the dead load can be computed more accurately. If the computed dead load is approximately equal to the initial estimate of its value (or slightly less), the design is complete, but if a significant difference exists between the computed and estimated values of dead weight, the computations should be revised using an improved value of dead load. An accurate estimate of dead load is particularly important when spans are long, say over 75 ft (22.9 m), because dead load constitutes a major portion of the design load.

Live loads associated with building use are specified by city or state building codes. Instead of attempting to evaluate the weight of specific items of equipment and occupants in a certain area of a building, building codes specify values of uniform live load for which members are to be designed. Typical values of live load for standard buildings are listed in Table 1.2.

After the structure has been sized for vertical load, it is checked for wind in combination with dead and live load as specified in the code. Wind loads do not usually control the size of members in buildings less than 16 to 18 stories, but for tall buildings wind loads become significant and cause large forces to develop in the structures. Under these conditions economy can be achieved only by selecting a structural system that is able to transfer horizontal loads into the ground efficiently.

**Table 1.2 Typical live-load values[1]**

| Type of use | Minimum uniformly distributed live loads | |
| --- | --- | --- |
| | lb/ft$^2$ | kPa = kN/m$^2$ |
| Apartment buildings: | | |
|   Private units | 40 | 1.92 |
|   Public rooms | 100 | 4.80 |
|   Corridors | 80 | 3.84 |
| Office buildings: | | |
|   Offices | 50 | 2.40 |
|   Lobbies | 100 | 4.80 |
|   Corridors above first floor | 80 | 3.84 |
| Garages (cars only) | 50 | 2.40 |
| Stores: | | |
|   First floor | 100 | 4.8 |
|   Upper floors | 75 | 3.6 |
| Warehouse: | | |
|   Light storage | 125 | 6.0 |
|   Heavy storage | 250 | 12.0 |

In seismic zones, analysis for dynamic effects due to ground motions must be considered. Dynamic forces in the structure are a function of the building's mass and the acceleration imparted to the building by ground motions. For buildings located in zones of low to moderate seismic activity, the provisions of the main body of the ACI Code produce sufficient ductility to permit a concrete structure to withstand the shaking due to ground motions. In regions where large earthquakes have a high probability of occurrence, concrete structures must be designed and detailed in accordance with the provisions of Appendix A of the ACI Code in order to ensure high ductility and toughness. When members are subjected to large lateral forces, reversal of stresses may occur, and regions which would normally be in compression under vertical load are stressed in tension. For members subject to reversal of stress, reinforcement must be provided on both sides of members.

## 1.6 DUCTILITY VERSUS BRITTLENESS

A major objective of the ACI Code is to design concrete structures with adequate ductility since concrete is brittle without reinforcement. The term *ductility* describes the ability of a member to undergo large deformations without rupture as failure occurs. A structural-steel girder is an example of a ductile member that can be bent and twisted through large angles without rupture. Ductile stuctures may bend and deform excessively under load, but they remain, by and large, intact. This capability prevents total structural collapse and provides protection to occupants of buildings. On the other hand, the term *brittle* describes members that fail suddenly, completely, and with little warning. When a brittle member fractures, it usually disintegrates and may damage adjacent portions of the structure or overload other members, bringing on additional failures. The ability of a ductile structure to undergo large deformations before collapse produces visible evidence of impending failure and may give occupants the opportunity to relieve distress by reducing loads. In contrast, brittle failures occur suddenly, without warning, and with no time for measures to be taken to prevent damage.

The need to engineer ductility into the entire structure of a building was demonstrated dramatically in 1968 when 24 stories of the exterior corner of the prefabricated Ronan Point Tower in London collapsed after a gas explosion on the eighteenth floor had blown out a single load-bearing wall panel.[2]† This type of collapse, in which the effects of a local failure spread to the entire structure or to a significant portion of the structure, is termed a *progressive collapse*. Extensive research is now under way to establish design specifications that will prevent future failures of this type. By careful attention to the positioning and anchorage of reinforcement the ductility of most structures can be increased substantially with little increase in cost.

## 1.7 STRENGTH AND SERVICEABILITY

Members are always designed with a capacity for load that is significantly greater than that required to support anticipated service loads. This extra capacity not only provides a factor of safety against failure by accidental overload or defective construction but also limits the level of stress under service load to provide some control over deformations and cracking. Even if a member can support the design loads, it must not bend to such an extent that its function is impaired.

By providing a reserve of strength the designer recognizes that members may be subjected to loads greater than those assumed in design. For example, during construction temporary storage of building materials may create forces in certain members well above those produced by normal occupancy. Even after the building has been in use, it is still subject to overload if heavy equipment is introduced into an area that was designed for a smaller load. The extra strength also provides for the possibility that members may be constructed with a lower strength than specified because of understrength materials or poor workmanship.

Although it is imperative that structures be designed with adequate strength to reduce the probability of failure to an acceptable level, they must also function effectively under normal service loads. Deflections must be limited to ensure that floors will remain level within required tolerances and do not vibrate, to prevent plaster ceilings and masonry partitions from cracking, and to ensure that sensitive

† Numbered references appear at the end of each chapter.

equipment will not be thrown out of alignment. In addition, the width of cracks must be limited to preserve the architectural appearance of exposed surfaces and to protect reinforcement from attack by corrosion.

## Elastic Versus Strength Design

Two design approaches for sizing reinforced concrete members are available to the engineer. The first, called *elastic design*, is based on the prediction of stress in members as they support the anticipated service loads. *Service load* is the actual or maximum value of load the member is expected to carry. In elastic design, the members are sized so that service-load stresses do not exceed a prescribed, predetermined value of stress. The allowable stress is set as some fraction of the maximum stress that the material can sustain before rupture or yielding occurs.

The elastic-design approach requires an understanding of how the member behaves with the service load in place. It typically assumes that materials behave elastically and that stresses induced by loads can be accurately predicted. Elastic design does not take into consideration the inherent failure mode (ductile or brittle) of the member or account for the magnitude of the additional strength in a member between service-load capacity and the ultimate-load capacity. Thus the actual factor of safety against failure is, in fact, unknown.

The second design approach, called *ultimate-strength design* or more simply *strength design*, is based on predicting the load that will produce failure in a member rather than predicting stresses produced by service loads. Using strength design, the designer is concerned with determining the load that will bring a structure to complete collapse and with the mode of failure when this load is applied. The preferred mode of failure is to ensure a controlled local failure of the member in a ductile rather than brittle manner. Thus the engineer who uses strength design gives only indirect attention to the state of stress or the deflection that will occur in the member when service loads are in place.

Since the design of a ductile structure that will fail locally is a foremost concern in reinforced concrete design, strength design is considered the more desirable approach. By controlling the ultimate strength of each member of a structure, the designer can control the overall mode of failure of a total structural system. In this way it is possible to design structures so that in the unlikely event of unanticipated overload, failures are confined to a limited region instead of causing total collapse of the entire system.

In addition, laboratory studies confirm that the controlling failure modes of concrete members can be accurately and consistently predicted. On the other hand, predictions of actual stresses in reinforced concrete members cannot be predicted for several reasons:

1. Shrinkage stresses produced by the drying of plastic concrete induce a set of self-balancing stresses of unknown magnitude.
2. Under load, concrete cracks in erratic, unpredictable patterns so that the properties of the cross section are not known with certainty.
3. With time, reinforced concrete creeps. The creep causes stress intensities to change; typically stresses increase in the steel and decrease in the concrete.

## 1.8 THE ACI DESIGN PROCEDURE

In strength design, which is the design method recommended by the current edition of the ACI Code, members are sized for *factored loads* that are greater than the service loads. The factored load is produced by multiplying the service load by *load factors*, numbers greater than 1. The size of the load factor, which represents the part of the factor of safety applied to loads, reflects the accuracy with which the design loads can be predicted. A load whose magnitude and distribution can be established with certainty is increased by a smaller load factor than a load whose magnitude is subject to variation or whose exact intensity cannot be predicted with precision. For example, dead load, which can usually be computed very accurately, is increased by a load factor of 1.4; on the other hand, live load, which is subject to greater variation, is increased by a load factor of 1.7. Table 1.3 contains additional load factors for various types of load.

When used in the general sense, the factored load is often denoted by $U$.†
Factored loads or forces produced by factor loads are subscripted by a lowercase $u$. For example, $M_u$ represents the moment at a section produced by factored loads.

Using factored loads, the designer carries out an elastic analysis of the structure. Structures are usually analyzed both for full gravity loads and for wind in combination with reduced gravity loads. Example 1.1 illustrates the loading conditions

### Table 1.3 ACI load factors

Dead load $D$ . . . . . . . . .1.4
Live load $L$ . . . . . . . . . .1.7
Earth pressure $H$ . . . . .1.7
Fluid pressure $F$ . . . . .1.4
Earthquake $E$ . . . . . . . .Substitute $1.1E$ for $W$ in equations below

When the force in a member is due to a combination of wind load $W$ in addition to dead and live load, the factored load $U$ is produced by considering the following combination of load factors:

$$\text{Factored load } U = 0.75(1.4D + 1.7L + 1.7W)$$

where the 0.75 reduction factor accounts for the improbability of having maximum wind and live load acting simultaneously on the structure. Since the dead load is always present, the logic of applying the 0.75 factor to it is not clear. However, since most structures designed for wind by the above equation behave satisfactorily, there is little reason to modify the expression.

If the absence of live load produces a reversal of stress or increases the likelihood of overturning when the wind acts, the following load factors are used to establish the factored load

$$U = 0.9D + 1.3W$$

When the structural effects $T$ of differential settlement, creep, shrinkage, or temperature are to be considered, the factored load is established by

$$U = 0.75(1.4D + 1.4T + 1.7L) \quad \text{but not less than} \quad U = 1.4(D + T)$$

† A list of the notation used in this book is provided in Appendix C.

that must be investigated for a typical building frame. The forces in a member created by factored loads represent the *required strength* of the member.

Members are sized so that their *design strength* will be equal to the required strength. The design strength is a reduced value of the ultimate or *nominal strength* of the cross section. The nominal strength of a member is evaluated in accordance with provisions and assumptions specified by the ACI Code. Nominal strength is evaluated analytically by considering the state of stress associated with the particular mode of failure (either the steel yields or the concrete crushes) or experimentally by studies that relate the ultimate strength to the proportions of the cross section and the strength of the materials. Nominal strength is designated by the subscript *n*.

In order to account for inevitable losses in member strength due to imperfect workmanship, e.g., undersized members, bars placed out of position, or voids in the concrete, and understrength materials, the *nominal strength* of a member is reduced by multiplying by a *capacity reduction factor* $\phi$, a number less than 1, to give the *design strength*. Reduction factors are listed in Table 1.4. The magnitude of the reduction factor is also influenced by the ductility of the member, the degree of accuracy with which the member's capacity can be predicted, and the importance of the member to the overall strength of the structure. Failure of a beam produces a local failure, but failure of a column may result in the collapse of many floors. The reduction factor constitutes the second part of the factor of safety in strength design.

Load factors and reduction factors have been selected by the ACI Code so that failure of structures will initiate in beams by yielding of the tension steel. Yielding of the tension steel in properly designed beams produces sagging and heavily cracked members but does not cause total collapse of the structure. Factors of safety against other modes of failure are much higher.

**Table 1.4  Capacity reduction factor $\phi$**

| Nominal strength | Reduction factor $\phi$ |
|---|---|
| Bending with or without axial tension, and for axial tension | 0.9 |
| Shear and torsion | 0.85 |
| Bearing on concrete | 0.70 |
| Bending in plain concrete | 0.65 |
| Columns with spirals | 0.75† |
| Columns with ties | 0.70† |

† For members carrying moment and small values of axial load (less than $0.1 f_c' A_g$) the reduction factor varies linearly from 0.9 to 0.7 or to 0.75 for tied and spiral columns, respectively.

In summary, the design criteria of the strength method can be stated as

Required strength ≤ design strength

or

Required strength ≤ $\phi$(nominal strength)

The above criteria applied to a beam that is stressed only by shear and moment would state that at every section

$$V_u \leq \phi V_n \quad \text{and} \quad M_u \leq \phi M_n$$

where $V_u$ and $M_u$ represent the shear and the moment produced by factored loads and $V_n$ and $M_n$ stand for the nominal shear and the nominal flexural strengths at the same section.

To clarify the fundamental ideas of strength design, a short steel column carrying an axial load that produces uniform stress on each cross section is sized in Example 1.2.

Since the strength method of design is based on behavior at failure, it does not guarantee that behavior will also be satisfactory under service loads; therefore, the ACI Code has established additional criteria to ensure that members will also satisfy the requirements of serviceability. These requirements will be discussed in Secs. 3.4 and 3.5.

**Example 1.1: Computation of required strength**  Determine the axial forces for which the member $CD$ in Fig. 1.1$a$ should be designed given the following service loads: ($a$) dead load of 1 kip/ft on girder $BC$, ($b$) live load of 2 kips/ft on girder $BC$, ($c$) wind load of 20 kips horizontally at joint $B$. Wind may act in either direction.

SOLUTION  Since the frame is supported at $D$ by a roller, only an axial force equal to $R_d$ develops in member $CD$. $R_d$ may be computed by summing about support $A$ moments of forces due to factored loads.

Case 1: $L + D$ (see Fig. 1.1$b$):

$$w_u = 1.4D + 1.7L = 1.4(1) + 1.7(2) = 4.8 \text{ kips/ft}$$

Analysis gives 48 kips compression in member $CD$.

Case 2: $L + D + W$ (see Fig 1.1$c$):

$$w_u = 0.75(1.4D + 1.7L) = 0.75[1.4(1) + 1.7(2)] = 3.6 \text{ kips/ft}$$
$$W_u = 0.75(1.7W) = 0.75(1.7)(20 \text{ kips}) = 25.5 \text{ kips} \quad \text{acts to right}$$

Analysis gives 48.75 kips compression in member $CD$.

Case 3: $D + W$ (overturning) (see Fig. 1.1$d$):

$$w_u = 0.9D = 0.9(1) = 0.9 \text{ kips/ft}$$
$$W_u = 1.3W = 1.3(20 \text{ kips}) = 26 \text{ kips} \quad \text{acts to left}$$

Analysis gives 4 kips tension in member $CD$.

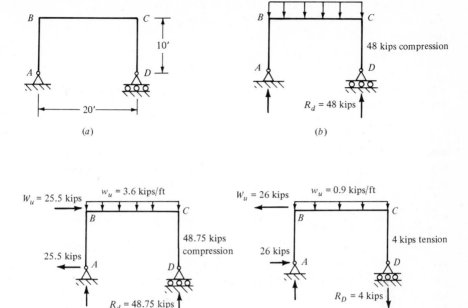

**Figure 1.1** (*a*) Frame; (*b*) dead and live load; (*c*) dead, live, and wind loads create compression in *DC*; (*d*) dead and wind; wind creates tension in *DC*.

CONCLUSION Member *CD* must be designed with a compressive strength of 48.75 kips and a tensile strength of 4 kips.

**Example 1.2: Design of a short steel column by the strength method** Determine the minimum cross-sectional area $A$ of a short steel column required to support a live load $L = 133.44$ kN (30 kips) and a dead load $D = 177.93$ kN (40 kips); see Fig. 1.2. Failure is assumed to occur when the average stress on the cross section reaches the yield-point stress $f_y$. Use a reduction factor $\phi = 0.7$ and load factors of 1.4 for dead and 1.7 for live load. $f_y = 344.7$ MPa (50 kips/in²).

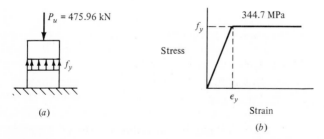

**Figure 1.2** (*a*) Failure impends; (*b*) stress-strain curve for steel.

SOLUTION Multiply the service loads by the respective load factors to produce $P_u$, the failure load, which also represents the required axial strength

$$P_u = 1.7L + 1.4D = 1.7(133.44) + 1.4(177.93) = 475.96 \text{ kN (107 kips)}$$

The design strength $\phi P_n$ of the column in terms of the area of the cross section $A$ and the yield strength of the steel $f_y$ is

$$\phi P_n = \phi A f_y = 0.7A(344.7) = 241.29A \text{ MPa (35}A \text{ kips)}$$

To establish the required area of the cross section we equate the required strength $P_u$ to the design strength $\phi P_n$, to give

$$P_u = \phi P_n$$

$$475.96 \text{ kN} = 241.29A \text{ MPa}$$

$$A = 1972.56 \text{ mm}^2 \text{ (3.06 in}^2\text{)}$$

Of course, when the full service load of 311.37 kN (70 kips) acts, the stress is well below $f_y$ and behavior is elastic. The service stresses are

$$f = \frac{P}{A} = \frac{311.37 \times 10^3 \text{ N}}{1972.56 \times 10^{-6} \text{ m}^2} = 157.85 \text{ MPa (22.9 kips/in}^2\text{)}$$

## 1.9 OVERVIEW OF THE DESIGN PROCEDURE

Before the analysis and design of individual elements (slabs, beams, and columns) of multistory structures are discussed, the overall design procedure from conception to final drawings will be outlined briefly. Although the following discussion focuses on building design, which traditionally involves close coordination between the architect and engineer, the same procedure is applicable to the design of those structures, such as dams, water-storage tanks, and bridges, which are traditionally the exclusive province of the engineer; i.e., the engineer conceives of the structural form as well as performs the structural design and analysis. The notable bridges of Maillart or shells of Nervi are well-known examples of an engineer establishing a structural form which is much admired.

### Preliminary Design

The major functional requirements of a building, whether they are a complex mechanical support system for a power plant or unbroken spaces for a basketball arena, largely determine how the architect arrives at a preliminary concept and a preliminary set of plans. The architect must also integrate the interior functional requirements with the character of the specific site chosen for the project, often making use of any natural features of the location that can contribute to the desirability and therefore the value of the building. For example, if a site permits a view across a valley or a harbor, the architect could take advantage of this feature and design a building that provides a panoramic view for as many occupants as possible. Conversely, if a building faces a blank wall, the architect will work to minimize this negative feature. The architect's preliminary plans will roughly indicate how the floor areas of the building will be divided. These layouts also establish ceiling heights and location of walls, entrances, stairs, and elevators. Once the geometry of the building has been roughly defined by its main function and by the constraints of the site, the architect will consult with a structural engineer to establish the best-conceived structural solution to the architectural problem.

During this preliminary design stage, the designer will investigate and weigh the best alternatives. Considering all possibilities of materials, structural systems, relative costs, time constraints, and availability of construction materials, the engineer searches for the structural system that will best suit the architectural scheme and will pick that system which represents a rational, efficient, and sound use of materials.

The preferred relationship between the engineer and architect is well expressed by Siegel:[3]

> The engineer ought to do more than merely make sure that what the architect designs stands up . . . . The engineer should act as the architect's critical partner, objecting strenuously wherever the design offends structural logic. . . . He should try to make the significance of the structure and its behavior clearer to the architect, so that the latter can draw upon [the engineer's] understanding for inspiration in working out the final design.

## Final Design

After the architect has incorporated the details of the structural system into the architectural drawings, and after the electrical and mechanical engineers have established the location of openings in floors and beams for pipes, conduit, ducts, and other building services, the structure and the design loads are clearly defined. In the final design phase the structural engineer will analyze and design each component of the structure for all possible loading conditions. All information—member size, reinforcement, construction joints, floor slopes, and all other details—required to build the structure will then be incorporated into the structural drawings and specifications.

## QUESTIONS

**1.1** Why is concrete design termed *empirical*?
**1.2** What is ductility and why is it important?
**1.3** What two criteria form the basis for proportioning structural members?
**1.4** Why is strength design considered more desirable than elastic design?
**1.5** What is the major deficiency of elastic design?
**1.6** What is a load factor? How is a factored load used?
**1.7** What is a reduction factor? What determines its size?
**1.8** Define *nominal strength*, *design strength*, and *required strength*.

**1.9** If strength design is used to size the members of multistory buildings, which members, columns or beams, will fail first if the structure is overloaded? Explain.

**1.10** Why is the load factor greater for soil pressure than for fluid pressure?

## PROBLEMS

**1.1** The beam in Fig. P1.1 carries uniform service loads of $w_d = 0.8$ kip/ft and $w_l = 0.6$ kip/ft. Concentrated live loads of 30 kips also act intermittently at the ends of the cantilevers. Determine the required flexural design strength at midspan and at support B. *Hint*: Different loading conditions produce maximum positive and negative moments at midspan.

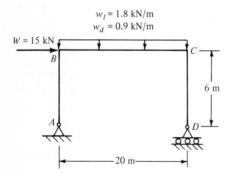

**Figure P1.1**

**1.2** The frame in Fig. Pl.2 carries service loads that consist of a wind load W as well as a uniformly distributed dead and live load on girder BC. At the section of maximum moment in column AB, determine the required flexural and axial strength when all loads act simultaneously.

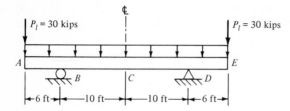

**Figure P1.2**

## REFERENCES

1. Basic Building Code, Building Officials and Code Administrators International, Chicago, 1978.
2. A. Popoff, Jr.: Design against Progressive Collapse, *Prestressed Concr. Inst.*, vol. 20, no. 2, p. 44. March-April 1975.
3. Curt Siegel: "Structure and Form in Modern Architecture," Reinhold, New York, 1962.

Concrete is being placed into forms during construction of a floor slab. (Photograph by Portland Cement Association.)

# MATERIALS

## 2.1 CONCRETE

Concrete is a carefully proportioned mixture of cement, water, fine aggregate, and coarse aggregate. To these basic components a variety of *admixtures,* i.e., chemicals that influence the reaction or modify the physical properties of the hardened concrete, are frequently added. As soon as the components of concrete have been mixed together, the cement and the water react to produce a cementing gel that bonds the fine and the coarse aggregates into a stonelike material. The chemical reaction between the cement and water, an exothermic reaction producing significant quantities of heat, is termed *hydration.*

During the initial stages of hydration, when only small amounts of gel have formed, the concrete is in a plastic state and flows easily. Throughout this stage the concrete can be deposited in forms, worked into the spaces between reinforcement, and compacted to eliminate voids, and the surfaces can be leveled and finished. As hydration continues and larger amounts of gel form, the concrete progressively stiffens, loses its workability, and gains strength.

The rate at which the gel forms is influenced by the temperature at which the reaction occurs. At high temperatures the reaction is rapid: only 10 or 15 min may be required for the concrete to stiffen. At low temperatures 10 to 12 hours may be required to produce the same degree of stiffness. The final product (Fig. 2.1) consists of the various-sized aggregates surrounded by a mortar composed of cement and the fine aggregate. Ideally, engineers would like to produce a dense homogeneous concrete that is free of voids, channels, cracks, and other defects. Such a material would have uniform strength properties, be impermeable to water, and present an attractive surface appearance. As a practical matter, however, it

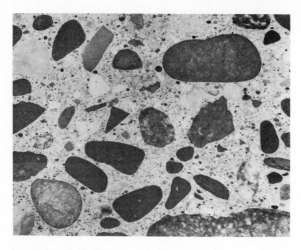

**Figure 2.1** Section through concrete showing aggregate. (*Portland Cement Association.*)

is not possible to produce concrete that is free of flaws: even the most carefully controlled concretes have a variety of small internal and surface defects. Voids are created by air bubbles produced during the mixing process and by segregation of the coarse aggregate from the mortar when concrete is deposited in forms. Since approximately twice as much water is needed for workability than is required to react with the cement, fine channels develop throughout the concrete as excess water rises to the surface when the concrete is vibrated. The rise of the water to the surface is termed *bleeding*.

Surface defects such as pits and honeycomb (the absence of mortar between aggregates) develop when the concrete contains insufficient mortar or when the concrete is not properly compacted in forms. In addition, x-ray and microscopic studies[1] show that fine *microcracks* develop as a result of the rupture of the bond between the gel and the aggregate. The break in bond is primarily attributed to drying and carbonation shrinkage (shrinkage produced by exposure to $CO_2$). The presence of these internal flaws explains in part the large difference between the tensile and the compressive strengths of concrete.

In addition to these small unavoidable defects, with time restrained concrete members may develop shrinkage and temperature cracks due to fluctuating environmental conditions. Although these cracks can never be eliminated, careful attention to design details and the use of reinforcement, called *temperature* and *shrinkage steel*, will control the size and location of cracks so that neither the function nor the appearance of the structure will be impaired.

### Concrete Materials

Most concrete produced in this country is made with portland cements. Termed *hydraulic cements* because they react with water, they have the ability to harden under water. Portland cements are available in five ASTM types (Table 2.1). They are manufactured from the same basic materials, but their properties are altered somewhat by the way the raw materials, often clays and limestones, are blended and fired before they are ground into cement powder.

**Table 2.1  Types of portland cement**

| Type | Name | Use or characteristics of cement |
|------|------|----------------------------------|
| I | Normal | General-purpose; used where no special requirements exist |
| II | Moderate | Moderate resistance to sulfate attack and moderate evolution of heat during hydration |
| III | High-early | Rapid gain in strength; permits quicker removal of forms; speeds construction |
| IV | Low-heat | Low heat generated; used in construction of massive sections, e.g., dams |
| V | Sulfate-resisting | Used where concrete exposed to high concentrations of sulfates |

Aggregates, which constitute approximately 75 percent of the concrete volume, are usually composed of well-graded gravel or crushed stone. Aggregates passing a no. 4 sieve [the wires of the screen are spaced $\frac{1}{4}$ in (6.4 mm) apart] are classified as *fine aggregates*; larger aggregates are classified as *coarse aggregates*. Since the presence of dust or chemicals can weaken the bond between the cementing gel and the aggregates, strong, durable concretes require aggregates that are clean and hard. Unreinforced concretes made with stone or gravel are normally assumed to weigh 145 lb/ft³ (2320 kg/m³). When reinforcement is added, the weight of concrete and steel is generally taken as 150 lb/ft³ (2400 kg/m³) for normal-weight aggregates.

If the designer wishes to reduce the weight of concrete members that are not subjected to heavy wear or abrasion, lightweight aggregates, made from a variety of expanded shales and slags, can be substituted for the heavier stone or gravel aggregates. Concretes that have a 28-day strength equal to or greater than 2500 lb/in² (17.24 MPa) and an air-dry weight less than 115 lb/ft³ (1840 kg/m³) are termed *structural lightweight* concrete.[3] If all aggregates (both the fine and the coarse) are lightweight, the concrete is called *all-lightweight*. If the coarse aggregate is lightweight and the fine aggregate is sand, the concrete is termed *sand-lightweight* concrete.

Unless field practice verifies that concrete with a certain size aggregate can be successfully placed and compacted. ACI Code §3.3.3 requires that the maximum size of coarse aggregate not exceed

1. One-fifth of the smallest dimension of the form
2. One-third the depth of slabs
3. Three-fourths of the minimum clear spacing between reinforcement

## Strength of Concrete

The water-cement ratio is the major variable influencing the strength and durability of concrete. When the water-cement ratio is large, a dilute, high-shrinking, weak gel is produced. Concrete containing such gels is low in strength and lacks resistance to deterioration by weathering. The variation of the compressive strength $f'_c$ with the water-cement ratio is shown in Fig. 2.2.

The rate of strength gain in concrete can be speeded by using finely ground cement, applying heat, e.g., steam curing, or by adding an accelerator such as calcium chloride. A rapid increase in strength of freshly poured concrete permits early removal of forms and speeds construction.

If freshly poured concrete is frozen, the free water present in the pores transforms into ice, producing a volume expansion which causes a significant breakup in the structure of the concrete and permanently reduces its ultimate strength. Where concrete is exposed to salt water or cycles of freezing and thawing, air-entraining agents are typically added to entrain small percentages (4 to 8) by volume of air in the form of small bubbles throughout the concrete mass. This entrapped air produces a major increase in the concrete's durability. In addition, the air bubbles serve as a lubricant to make the concrete more workable.

To enable concrete to attain maximum strength and durability, concrete should be cured, i.e., kept continuously moist for several weeks after the initial set. The presence of moisture allows any unreacted cement to hydrate further, producing additional cementing gel. Curing can be effected by continuous spraying, covering the concrete with waterproof sheets, spraying with coatings, or leaving forms in place and covering the exposed surfaces.

Figure 2.3, based on loading standard test cylinders to failure in compression, shows the gain in compressive strength of concrete with time. The cylinders on which both curves are based were made from the same concrete mix, but the cylinders of curve $A$ were continuously moist-cured until tested while those for curve $B$ were allowed to dry in air. As shown in Fig. 2.3, after 6 months, the compressive strength of the cured concrete is more than twice that of the uncured concrete.

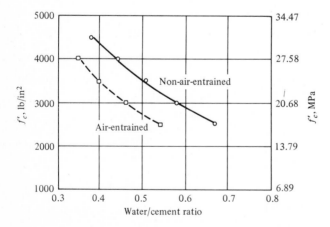

**Figure 2.2** Maximum permissible water-cement ratio, by weight, for concrete when test results from trial batches are not available. (*Adapted from ACI Code 318-71, table 4.2.4.*)

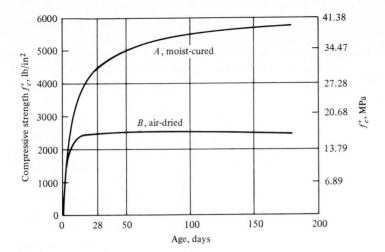

**Figure 2.3** The effect of curing on compressive strength. (*Adapted from Ref. 2.*)

## Workability

To produce high-quality concrete, sufficient water must be added to make the concrete fluid so it can be worked into all parts of the containing forms. However, if excessive amounts of water are added, a weak, high-shrinking concrete will result. The slump test is a common field test used to control the workability and quality of concrete, (see Fig. 2.4).

In the slump test a 12-in (305-mm) steel truncated cone placed on a level base is filled with fresh concrete, compacted by rodding. After the top surface has been leveled, the slump cone is slowly lifted. Lacking lateral support, the concrete slumps. The difference in height measured in inches between the top of the concrete and the top of the slump cone is called the *slump*. Concretes are typically produced with slumps ranging from 2 to 6 in (51 to 152 mm). The biggest slumps are specified for construction using narrow forms containing a large number of bars.

## Durability

If a concrete is exposed to weather, seawater, or deicing salts, durability as well as stress intensity should always be considered when the strength of the concrete is established. The resistance of a concrete to deterioration by these factors can be

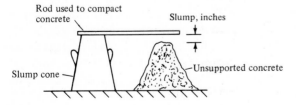

**Figure 2.4** The slump test.

significantly improved by using well-cured, carefully compacted, high-strength concretes. The following paragraphs outline the ACI Code provisions for the manufacture of concretes exposed to weather or in contact with sulfates.

When concrete structures are not only exposed to moisture but are also subject to cycles of freezing and thawing, the formation of ice in surface pores and hair cracks may create internal pressures that cause the concrete to disintegrate rapidly unless a high-strength air-entrained concrete is specified. The required limits on air content for concrete exposed to freezing temperatures and moisture is specified in Table 4.6.1 of ACI Code §4.6.1 as a function of the maximum size of the coarse aggregate. This section requires that the volume of entrained air range from 6 to 10 percent for concretes made with small aggregates [$\frac{3}{8}$ in (9.5 mm) diameter] to 2.5 to 5.5 percent for concretes made with large aggregates [2 in (51 mm) diameter]. To ensure an adequate cement content ACI Code §4.6.1.1 requires that the water-cement ratio by weight of normal-weight concretes (made from stone or gravel aggregates) not exceed 0.53. If concretes are made with lightweight aggregates, a minimum compressive strength $f'_c$ of 3 kips/in$^2$ (20.68 MPa) is required by ACI Code § 4.6.1.2. For lightweight concretes, the compressive strength rather than the water-cement ratio is considered a better measure of the quality of the concrete paste. Since the absorption of water by lightweight aggregates can be high, the amount of water available to react with the cement is uncertain.

For concrete structures exposed to water containing moderate amount of sulfates, ACI Code § 4.6.3 requires concretes to be manufactured with sulfate-resisting cement; a type II cement has adequate resistance. In addition, if the concrete is made with normal-weight aggregates, the water-cement ratio by weight should not be more than 0.5. If a lightweight aggregate is used, the compressive strength $f'_c$ must not be less than 3750 lb/in$^2$ (25.86 MPa).

When exposure to sulfates is severe, i.e., between 1500 and 10,000 ppm, a type V cement, which has the maximum sulfate resistance, should be used. To ensure a high cement content, the water-cement ratio by weight must not be more than 0.45. For concretes made with lightweight aggregates, the compressive strength $f'_c$ must not be less than 4000 lb/in$^2$ (27.58 MPa).

## 2.2 MECHANICAL PROPERTIES OF CONCRETE

### Compressive Strength and Modulus of Elasticity

The compressive strength of concrete is usually determined by loading 6-in (152-mm) diameter by 12-in (305-mm) high cylinders to failure in uniaxial compression. Cylinders are tested after they have hardened for 28 days. During the 28 days before testing the cylinders are stored under water or placed in a constant-temperature room maintained at 100 percent relative humidity. Exposure to moisture speeds the gain in strength by increasing the hydration of the cement. Additional details covering the preparation and testing of cylinders are covered by the ASTM specifications.

Typical stress-strain curves for concrete produced by a uniaxial compression test of a standard cylinder are shown in Fig. 2.5. The 28-day maximum compressive strength, which occurs at a strain of approximately 0.002, is denoted by $f'_c$. Although the ultimate strain at which total failure occurs varies from approximately 0.003 for high-strength concrete to about 0.005 for low-strength concrete, ACI Code §10.2.3 specifies that the maximum usable strain be taken as 0.003 for all strengths of concrete. While concretes with compressive strengths between 2.5 and 10 kips/in² (17.24 and 68.95 MPa) are used in construction, most concretes have a compressive strength between 3 and 4 kips/in² (20.68 and 27.58 MPa). Since the stress-strain cure for concrete is nonlinear, its modulus of elasticity (given by the slope of the stress-strain curve) varies with the intensity of stress. For concretes that weigh between 90 and 155 lb/ft³ (1440 and 2480 kg/m³), ACI Code §8.5.1 specifies an effective secant modulus of elasticity $E_c$ equal to

$$E_c = \begin{cases} w_c^{1.5}(33)\sqrt{f'_c} & \text{lb/in}^2 \\ w_c^{1.5}(0.043)\sqrt{f'_c} & \text{MPa} \end{cases} \tag{2.1}$$

where $w_c$ is the unit weight of concrete (lb/ft³ or kg/m³) and $f'_c$ is the 28-day compressive strength (lb/in² or MPa).

If concrete is made of crushed stone or gravel, $w_c = 145$ lb/ft³ (2320 kg/m³). Substituting this value into Eq. (2.1) gives

$$E_c = \begin{cases} 57{,}000\sqrt{f'_c} & \text{lb/in}^2 \\ 4730\sqrt{f'_c} & \text{MPa} \end{cases} \tag{2.2}$$

In the field the specified 28-day compressive strength $f'_c$ can be attained only if the concrete is correctly proportioned, completely mixed, carefully placed, well compacted, protected from freezing, and thoroughly cured. If any deviations from the specifications governing the manufacture and the placing of the concrete occur, the specified design strength may not develop. To improve the likelihood that the concrete will attain the required design strength, ACI Code § 4.3.1 provides that the concrete be manufactured with a greater strength than the specified $f'_c$. The increment of strength above $f'_c$ varies from 400 to 1200 lb/in² (2.76 to 8.27 MPa) and depends on the quality control of the concrete-production facility as measured by standard compression tests of concrete cylinders.

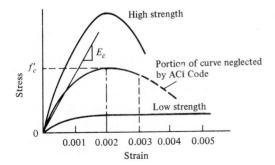

**Figure 2.5** Uniaxial stress-strain curves for concrete.

## Tensile Strength

Experimental studies show that the tensile strength of concrete is highly variable and ranges from approximately 8 to 15 percent of the compressive strength $f'_c$. The large difference between the tensile and the compressive strengths of concrete is due in part to the many fine cracks that exist throughout the concrete. At moderate levels of stress, cracks do not influence the compressive strength of concrete significantly. Since compressive stress can push the sides of the crack together and thus be transferred past a crack, both the cracked and uncracked areas are able to transmit compression stresses.

When concrete is stressed in tension, the distribution of stresses on the cross section changes. Since tension cannot be transferred across a crack, it is carried only on the uncracked area of the cross section. Because the effective area available to transmit tension is smaller than the gross area, the actual stress exceeds the average stress. As tensile stresses flow around internal cracks and voids, stress concentrations develop at the perimeter of the voids and at the tip of cracks. Even though the average tensile stress on the cross section (based on the gross area) may be small, the local stresses may be large enough to cause internal cracks to lengthen. As the uncracked area reduces in size, stresses rise rapidly and failure quickly follows. Since the size and location of cracks are unpredictable due to the nature of the material, the tensile strength is subject to large scatter (see Fig. 2.6).

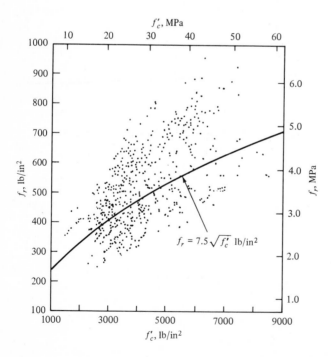

**Figure 2.6** Variation of modulus of rupture $f_r$ with $f'_c$. (*Adapted from Ref. 3.*)

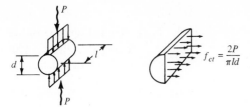

$$f_{ct} = \frac{2P}{\pi l d}$$

**Figure 2.7** The split-cylinder test.

**Tests to measure tensile strength** *The split-cylinder test* In the split-cylinder test a standard concrete cylinder, 6 in (152 mm) in diameter by 12 in (305 mm) long, is loaded to failure on its side by a line load. Except for the concrete near the outside surfaces (top and bottom), a state of pure tension exists on a vertical plane through the centerline of the section (Fig. 2.7). For stone or gravel concretes, tests show that the split-cylinder tensile strength $f_{ct}$ can be approximated by $6\sqrt{f'_c}\,\text{lb/in}^2$ or $0.50\sqrt{f'_c}\,\text{MPa}$, where $f'_c$ is the 28-day compressive strength (lb/in² or MPa).

**The standard beam test—*modulus of rupture*** In a standard beam test, an un-reinforced concrete beam, usually 4 by 4 in (102 by 102 mm) in cross section and 16 in (406 mm) long, is loaded to failure by either a concentrated load at midspan or by two loads applied at the third points. The small tensile strength of concrete means that a tensile bending failure always develops near midspan (see Fig. 2.8). Assuming that the behavior is elastic and that the bending stresses vary linearly from the neutral axis of the cross section, the maximum tensile stress at the bottom surface, termed the *modulus of rupture* $f_r$, can be computed by substituting experimental values of moment into the standard beam equation for stress at the top and bottom surfaces

$$f_r = \frac{Mc}{I}$$

Values of $f_r$ from several hundred tests of standard beams constructed of stone concrete are shown in Fig. 2.6. Based on test data, the ACI Code equation for predicting the tensile bending strength is

$$f_r = \begin{cases} 7.5\sqrt{f'_c} & \text{lb/in}^2 \\ 0.62\sqrt{f'_c} & \text{MPa} \end{cases} \tag{2.3}$$

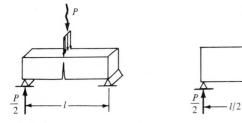

$$M = \frac{Pl}{4}$$

**Figure 2.8** Beam test for modulus of rupture.

where $f'_c$ is the 28-day compressive strength (lb/in$^2$ or MPa). A comparison of $f_r$ predicted by Eq. (2.3) with experimental values is shown in Fig. 2.6.

If the fine and coarse aggregates in the concrete are both lightweight material, $f_r$ in in Eq. (2.3) is reduced by 0.75. If the fine aggregate is sand and the coarse aggregate is lightweight, $f_r$ is reduced by 0.85.

When the split-cylinder strength of a lightweight concrete is available, the modulus of rupture can be computed from

$$f_r = 7.5 \frac{f_{ct}}{6.7} \text{ lb/in}^2 \text{ but not more than} \qquad 7.5\sqrt{f'_c} \text{ lb/in}^2$$

## Biaxial Strength

Portions of many members are in a state of *biaxial stress*, i.e., stress acting in two directions. The influence of biaxial stress on the strength of an element is shown by the curve in Fig. 2.9. This curve, based on experimental studies of plates loaded in

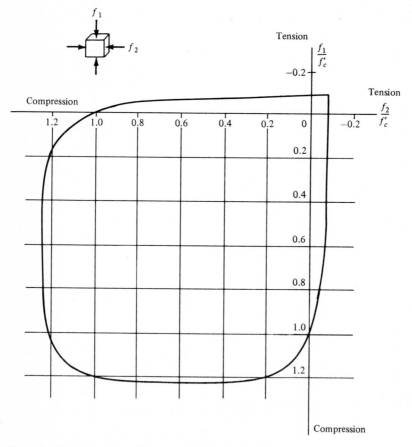

**Figure 2.9** Biaxial strength of concrete. (*Adapted from Ref. 4.*)

orthogonal directions, shows that the compressive strength increases approximately 20 percent over that of the uniaxial compressive strength when a compressive stress equal to or greater than 20 percent of the uniaxial compressive strength acts in the other principal direction. The curve also shows that tensile stresses acting perpendicular to the direction of the compression stresses can reduce the compressive strength significantly.

## Creep

Experimental studies[5] show that both axial and bending deformations of reinforced concrete members increase with time. The total deformation is often divided into two parts: (1) an initial, instantaneous deformation that occurs with the application of load, followed by (2) a time-dependent deformation, termed *creep*, that continues at a decreasing rate for a period of years. To illustrate the effect of creep, the variation of axial deformation over time for a uniformly loaded concrete cylinder is shown graphically in Fig. 2.10, which shows that after the initial deformation has taken place, the creep deformation occurs rapidly at first and then decreases steadily with time. Approximately 75 percent of the ultimate creep deformation occurs during the first year. Although measurements indicate that creep goes on for many years, it is essentially completed after 2 or 3 years. If the load is removed (see the dashed curve *BD*), immediate elastic rebound occurs, followed by a modest increment of creep recovery. However, a permanent deformation remains, denoted by the distance between points *D* and *E*.

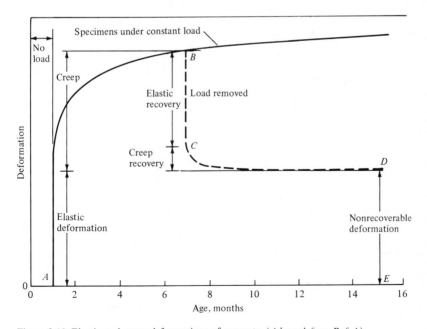

**Figure 2.10** Elastic and creep deformations of concrete. (*Adapted from Ref. 4.*)

Although creep under the normal levels of stress produced by service loads does not reduce the strength of a reinforced concrete member, it frequently influences behavior. The following cases illustrate how creep affects behavior and how its undesirable effects can be minimized

**Case 1** Because of the delayed effects of creep, the long-term deflections of a beam may be 2 or 3 times larger than the initial deflections. Thus the long-term deflections of both reinforced and prestressed concrete beams must be investigated by the designer to ensure that the structure will remain serviceable and that the anticipated creep-induced deflections will fall within accepted limits.

**Case 2** In prestressed concrete construction, members are often stressed by anchoring heavily tensioned cables, which frequently pass through ducts in the member, to the ends of members. As the compressed concrete shortens because of creep, the force in the cable reduces as the ends of the member move toward each other. Loss of prestress can lead to both increased cracking and larger deflections under service loads, which can be prevented by using high-strength steel in which final strains—even after creep losses (as well as other losses)—are still significant.

**Case 3** In a reinforced concrete column supporting a constant load, the deformations due to creep can cause the initial stress in the steel to double or triple with time, assuming that the steel remains elastic.

As the creep deformation of the concrete takes place over time, the force carried by the noncreeping steel, which is fully bonded to the concrete, gradually increases while the total force carried by the concrete reduces by an amount. Since experimental studies have shown that elastic design underestimates the stress in the steel, the design of axially loaded reinforced concrete columns has been based on the strength of the section rather than on limiting the level of stress in the

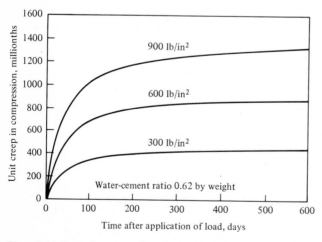

**Figure 2.11** Rate of creep as affected by the intensity of sustained load. (*Adapted from Ref. 4.*)

materials to specified allowable values. Similarly, when reinforcement is added to the compression zone of a beam to increase bending capacity, an empirical adjustment must be made to account for the additional stresses induced in the steel by creep.

Although creep deflections are often undesirable, creep is beneficial when a concrete member is heavily stressed in compression locally. As the load is applied and stresses build up, rapid creep reduces the highest stresses and increases the stresses in more lightly stressed adjacent sections, thereby relieving high local stress concentrations that might produce failure.

As long as stresses in the concrete do not exceed approximately 50 percent of the concrete strength, studies indicate that creep is nearly proportional to the magnitude of the stress (see Fig. 2.11). The ultimate value of creep is also a function of the type of aggregate, the strength of the concrete, the method of curing, the age of the concrete when initially loaded, the surface-to-volume ratio of the member, and the relative humidity. Creep can be reduced by

Deferring the application of load until the concrete gains strength
Using high-strength concrete
Keeping the volume of cement paste low relative to the volume of the aggregate
Steam curing under pressure
Adding reinforcement
Using limestone aggregates

## Shrinkage

In order to produce a workable concrete mix which will flow readily between reinforcement to fill all parts of the formwork and which can be effectively compacted to eliminate voids, nearly twice as much water as is theoretically required to hydrate the cement must be added to the concrete mix. After the concrete has been cured and begins to dry, the excessive water that has not reacted with the cement will begin to migrate from the interior of the concrete mass to the surface. As the moisture evaporates, the concrete volume shrinks. The loss of moisture from the concrete varies with distance from the surface. Drying occurs most rapidly near the surface because of the short distance the water must travel to escape and more slowly from the interior because of the increased distance from the surface. The shortening per unit length associated with the reduction in volume due to moisture loss is termed the *shrinkage strain* or *shrinkage*.

The magnitude of the ultimate shrinkage strain is primarily a function of both the initial water content of the concrete and the relative humidity of the surrounding environment. As indicated in Fig. 2.12, a nearly linear relationship exists between the magnitude of the shrinkage and the water content of the mix for a particular value of relative humidity. If the relative humidity is increased, the shrinkage of the concrete drops. When concrete is exposed to 100 percent relative humidity or submerged in water, it will actually increase in volume slightly as the gel continues to form because of the ideal conditions for hydration.

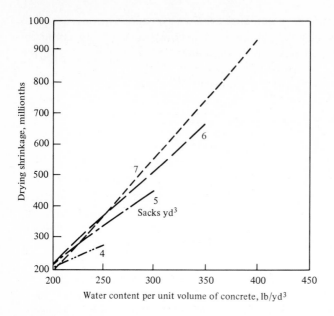

**Figure 2.12** The interrelation of shrinkage, cement content, and water content. (*Adapted from Ref. 4.*)

Both the rate at which shrinkage occurs and the magnitude of the total shrinkage (a function of the member's size and shape) increase as the ratio of surface to volume increases. The larger the surface area relative to the cross section, the more rapidly moisture can evaporate. For unreinforced concrete members exposed to average values of relative humidity (about 70 percent in the United States) ultimate values of shrinkage strain reach 0.0004 to 0.0007. If members are reinforced, the resistance to shrinkage deformation provided by the steel reduces the ultimate shrinkage strain for standard conditions to values[5] between 0.0002 and 0.0003.

Although shrinkage continues for many years, particularly for thick members, approximately 90 percent of the ultimate shrinkage occurs during the first year. Typical plots of shrinkage over time for normal and lightweight concretes are shown in Fig. 2.13, indicating that shrinkage is higher for lightweight concretes than for normal-weight concretes. The increased shrinkage is caused by the initial swelling of the more porous lightweight aggregates, which absorb somewhat more water during concrete mixing. As the concrete dries, the larger shrinkage strains reflect the combined shrinkage of the aggregates and of the cement paste.

**Stresses created by shrinkage**  Because concrete adjacent to the surface of a member dries more rapidly than concrete in the interior, shrinkage strains are initially larger near the surface than in the interior. As a result of the differential shrinkage, a set of internal self-balancing forces, i.e., compression in the interior and tension on the outside, is set up in the concrete. The approximate distribution

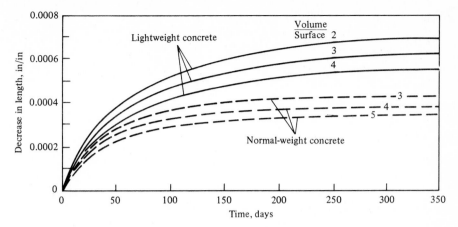

**Figure 2.13** Rate of shrinkage of lightweight and normal-weight concrete.[6]

of the longitudinal stresses associated with these factors is shown in Fig. 2.14a. The stresses induced by shrinkage can be explained by imagining that the cylindrical core of a concrete cylinder is separated from its outer shell and that the two sections are then free to shrink independently in proportion to their existing water content (see Fig. 2.14b). Since deflections must be compatible at the junction between the core and the shell, shear stresses must be created between the core and the outer shell. If free-body diagrams of the upper half of the cylinder are considered (Fig. 2.14c), it is clear that vertical equilibrium requires the shear stresses to induce compression in the core and tension in the exterior shell. With time, as the interior dries and the shrinkage strains become uniform throughout the cross section, these stresses will dissipate.

In addition to the self-balancing stresses set up by differential shrinkage, the overall shrinkage creates stresses if members are restrained in the direction in which shrinkage occurs. For example, consider a beam that is a component of a continuous frame. If the beam were free to shrink after it was poured, its length would reduce to $L'$ (see Fig. 2.15), which is less than the distance $L$ between column faces. Since the beam is connected rigidly to the joints, tensile forces will be induced in

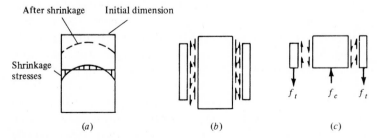

**Figure 2.14** Longitudinal stresses created by differential shrinkage: (a) stresses in a concrete cylinder; (b) shear stresses required for compatibility; (c) free-body diagrams of the upper half of the cylinder.

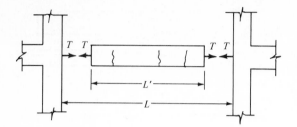

**Figure 2.15** Tension created in a restrained member by shrinkage.

the beam by the joint to resist the shortening. Because these stresses normally exceed the tensile strength of the concrete, random cracking develops. This cracking, which initiates at points of local weakness, will occur along the length of the beam.

Tensile cracking due to shrinkage will take place in any structural element (wall, a slab on grade, or a framed slab) restrained at its boundaries. It must be controlled since it permits the passage of water, is detrimental to appearance, reduces shear strength, and exposes the reinforcement to the atmosphere. To minimize cracking, the designer should act to

Minimize the water content.
Use dense nonporous aggregates.
Cure the concrete well.
Limit the area or length of concrete poured at a given time. (If walls or floor slabs are poured in small sections, a portion of the shrinkage will occur in the segments poured and will reduce the overall shrinkage.)
Use construction and expansion joints to control the location of cracks. (Figure 2.16 shows two commonly used joint details for walls and slabs. If construction joints are used to reduce shrinkage in beams or framed slabs, they should be located in regions of low shear near midspan.[7]
Add reinforcement to limit the width of cracks. A well-distributed grid of bars or welded-wire mesh, called *shrinkage steel*, reduces crack width and results in the development of many small hairline cracks that are often invisible to the naked eye.

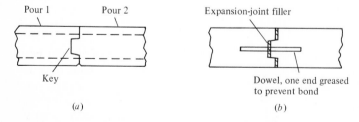

**Figure 2.16** Details of (*a*) construction and (*b*) expansion joints.[7]

## Temperature Change

Concrete expands as the temperature increases and contracts as the temperature decreases. If a member is restrained, a drop in temperature produces tensile stresses and an increase in temperature produces compression stresses. The effects of temperature change are handled very much like those of shrinkage. The designer must either make provision for the deformations to occur so that stresses are not induced in the structure, or the members must be designed for the induced forces. The coefficient of temperature expansion for concrete is equal approximately to $5.5 \times 10^{-6}$ in/in/°F, a value only slightly less than that of steel.

## Summary

The discussion of concrete properties in this chapter should provide the designer with the background required to prepare the concrete specifications that are a part of each project. For a broader treatment of the properties of concrete see Ref. 10. Although care is obviously required at each step in the manufacture and placing of concrete, attention to *curing* is probably the single most important requirement for the production of strong, durable, watertight concrete.

## 2.3 REINFORCEMENT

The low tensile strength of unreinforced concrete, a brittle material, results in limited structural applications since most structural elements carry forces, e.g., shear and moment, that create tensile stresses of significant magnitude. The addition of high-strength ductile reinforcement that bonds strongly to concrete produces a tough ductile material capable of transmitting tension and suitable for constructing many types of structural elements, e.g., slabs, beams, and columns.

   Although a variety of materials such as glass fibers and plastic filaments have been used as reinforcement, most concrete members are reinforced with steel, in the form of bars, wire mesh, or strand. Thanks to its high strength, ductility, and stiffness, steel reinforcement imparts great strength and toughness to concrete. Reinforcement also reduces creep and minimizes the width of cracks. Slender reinforcing bars are easily fabricated into a variety of shapes that can be used to construct members of any configuration.

## Characteristics of Reinforcing Bars

The largest volume of steel reinforcement is supplied in the form of circular bars whose surface has been imprinted with protruding ribs to improve the bond between the steel and concrete. Several examples of deformation patterns are shown in Fig. 2.17. Although manufacturers are not restricted to a specific pattern of ribs, the height, length, shape, and spacing of the ribs are controlled by ASTM specifications.

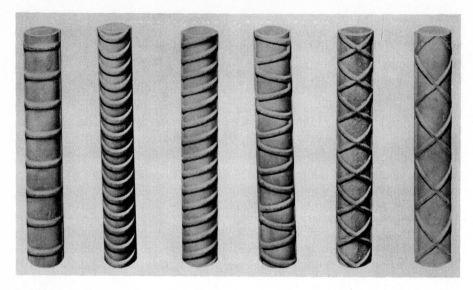

**Figure 2.17** Examples of deformed bars. (*Concrete Reinforcing Steel Institute.*)

Deformed bars are manufactured in diameters from $\frac{3}{8}$ to $2\frac{1}{4}$ in (9.5 to 57.33 mm). Bar size is designated by the number of eighths inches contained in the bar diameter. For example, a $\frac{3}{4}$-in-diameter bar, which has six $\frac{1}{8}$-in in its diameter, is a no. 6 bar. The standard bars and their properties are shown in Table 2.2.

Bars are supplied in lengths up to 60 ft (18.3 m). Bars of greater length would bend excessively when picked up because of their flexibility. Although most bars are rolled from new billet steel, bars are also fabricated from worn rails of train

**Table 2.2 Properties of standard reinforcing bars**

| Bar no.† | Nominal diameter | | Cross-sectional area | | Perimeter | | Nominal weight | |
|---|---|---|---|---|---|---|---|---|
| | in | mm | in² | mm² | in | mm | lb/ft | kg/m |
| 3 | 0.375 | 9.52 | 0.11 | 71 | 1.178 | 29.9 | 0.376 | 0.560 |
| 4 | 0.500 | 12.70 | 0.20 | 129 | 1.571 | 39.9 | 0.668 | 0.994 |
| 5 | 0.625 | 15.88 | 0.31 | 200 | 1.963 | 49.9 | 1.043 | 1.552 |
| 6 | 0.750 | 19.05 | 0.44 | 284 | 2.356 | 59.8 | 1.502 | 2.235 |
| 7 | 0.875 | 22.22 | 0.60 | 387 | 2.749 | 69.8 | 2.044 | 3.042 |
| 8 | 1.000 | 25.40 | 0.79 | 510 | 3.142 | 79.8 | 2.670 | 3.973 |
| 9 | 1.128 | 28.65 | 1.00 | 645 | 3.544 | 90.0 | 3.400 | 5.060 |
| 10 | 1.270 | 32.26 | 1.27 | 819 | 3.990 | 101.4 | 4.303 | 6.404 |
| 11 | 1.410 | 35.81 | 1.56 | 1006 | 4.430 | 112.5 | 5.313 | 7.907 |
| 14 | 1.693 | 43.00 | 2.25 | 1452 | 5.32 | 135.1 | 7.650 | 11.38 |
| 18 | 2.257 | 57.33 | 4.00 | 2581 | 7.09 | 180.1 | 13.600 | 20.24 |

† Based on the number of eighths of an inch included in the nominal diameter of the bars.

tracks or from axles of old locomotives, termed *rail* and *axle steel* respectively. The properties of bars made from various ASTM grades of steel are listed in Table 2.3. As might be expected, rail and axle steel, heavily cold-worked by the repeated impact of train wheels, is the least ductile. If reinforcement must be welded, low-alloy steels, which have a limit on carbon content and other embrittling elements, are specified.

Since the cost of steel with a yield point of 60 kips/in$^2$ is about the same as that of steel with a yield point of 40 kips/in$^2$, the 60-yield-point steel is most commonly used in the manufacture of reinforcing bars. For certain structures in which stresses are to be limited to reduce crack width or to limit deflections, the lower-yield-point steels may be specified because of their increased ductility.

To identify the characteristics of reinforcing bars, all bars are stamped with a code that identifies the type of steel, the bar size, the yield point, and the manufacturer (Fig. 2.18).

## Physical Properties

The strength and stiffness of reinforcing bars are determined from a stress-strain curve of a reinforcing bar specimen loaded to failure in uniaxial tension. For steels whose yield point does not exceed 60 kips/in$^2$ (413.7 MPa), the stress-strain curve is assumed to be composed of two straight lines (see Fig. 2.19). This assumption neglects the curvature of the stress-strain curve near the yield point and the slope of the strain-hardening region. (see the dashed lines in Fig. 2.19).

**Table 2.3  Physical requirements of standard ASTM deformed reinforcing bars**

| Type of steel and ASTM specification no. | Bar size no. | Grade | Minimum tensile strength | | Minimum yield point or yield strength | |
|---|---|---|---|---|---|---|
| | | | kips/in$^2$ | MPa | kips/in$^2$ | MPa |
| Billet, A615-81† | 3–11 | 40 | 70 | 482.7 | 40 | 275.8 |
| | 3–11 | 60 | 90 | 620.6 | 60 | 413.7 |
| | 14, 18 | | | | | |
| Rail, A616-79 | 3–11 | 50 | 80 | 551.6 | 50 | 344.7 |
| | 3–11 | 60 | 90 | 620.6 | 60 | 413.7 |
| Axle, A617-79 | 3–11 | 40 | 70 | 482.7 | 40 | 275.8 |
| | 3–11 | 60 | 90 | 620.6 | 60 | 413.7 |
| Low Alloy, A706-80 | 3–11 | 60 | 80 | 551.6 | 60 | 413.7 |
| | 14, 18 | | | | | |

† Bars marked with the symbol S satisfy more restrictive supplementary requirements of ASTM A615-79; other bars marked with an N.
*Source*: Adapted from Ref. 8.

N   for New Billet
S   for Supplemental Requirements A615
A   for Axle
I   for Rail
W   for Low Alloy
*(a)*

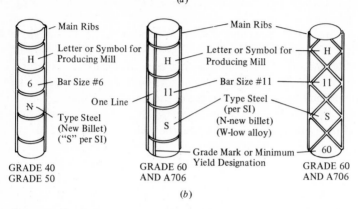

GRADE 40
GRADE 50

GRADE 60
AND A706

GRADE 60
AND A706

*(b)*

**Figure 2.18** Identification marks rolled into the surface of standard bars to denote the producer's mill, bar size, type of steel, and yield strength: (*a*) types of steel; (*b*) grade marks by line or number system. (*Adapted from Ref. 8.*)

The modulus of elasticity of the steel $E_s$ (the slope of the stress-strain curve in the elastic region) measures 29,000 kips/in$^2$ (200 GPa) for all grades of steel.

For higher-yield-point steels, the stress-strain curve may not develop a horizontal yield region, continuing instead to slope upward at a reduced slope when the steel is strained beyond the elastic region. For stress-strain curves of this shape the yield point is defined as the stress associated with a strain of 0.35 percent (ACI Code §3.5.3.3) (see Fig. 2.20).

ACI Code §9.4 permits the use of reinforcing bars with yield points as high as 80 kips/in$^2$ (551.6 MPa). When the yield point of a steel exceeds 60 kips/in$^2$ (413.7 MPa), its properties must satisfy the provisions of ACI Code § 3.5.3.3.

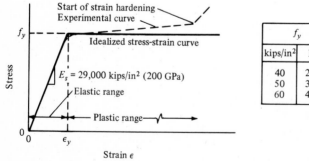

| | $f_y$ | | $\epsilon_y$ |
|---|---|---|---|
| | kips/in$^2$ | MPa | |
| | 40 | 275.8 | .00138 |
| | 50 | 344.7 | .00172 |
| | 60 | 413.7 | .00207 |

**Figure 2.19** Idealized stress-strain curve for reinforcing bars.

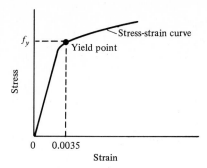

**Figure 2.20** Stress-strain curve of a high-strength steel that lacks a horizontal yield range.

## Fabrication of Bars

Once detailed plans and specifications for a reinforced concrete structure have been prepared, the builder will contract with a steel fabricator to supply the reinforcing bars, which must be cut and bent to fit inside the forms. Fabricators are normally able to deliver bars to the site within several weeks. As a first step in the fabrication of the reinforcing steel, detailed drawings (called shop drawings) are prepared of every reinforcing bar that will go into the structure. Examples of common bar configurations produced by fabricators are shown in Fig. 2.21. After the design engineer has reviewed and approved the fabricator's shop drawings, the bars are cut and bent to the specified dimensions. The fabricated bars, tagged to identify their location in the structure, are then shipped to the construction site in bundles.

To ensure that bars will not be knocked out of position when concrete is poured into forms and vibrated, ironworkers tie the reinforcement together securely with steel wire to form rigid mats or cages, which are positioned and secured in the forms with a variety of steel accessories. Figure 2.22 shows several types of supports for reinforcement in slabs and beams.

If bars are exposed to rain, dew, or snow, rust develops rapidly on the surface of the bar. Normally the rusting is light and need not be removed (in fact, light rust, which pits and roughens the reinforcement surface, improves bond). However, if steel is exposed to moisture for long periods, heavy surface rust that produces flaking may develop. To ensure a proper bond the excess rust must be removed.

**Figure 2.21** Typical bar bends. (*Adapted from Ref. 8.*)

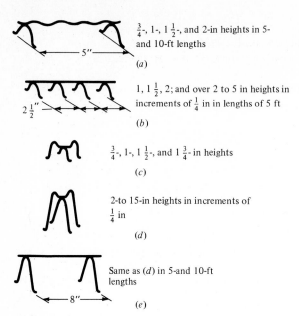

$\frac{3}{4}$-, 1-, 1 $\frac{1}{2}$-, and 2-in heights in 5- and 10-ft lengths

*(a)*

1, 1 $\frac{1}{2}$, 2; and over 2 to 5 in heights in increments of $\frac{1}{4}$ in in lengths of 5 ft

*(b)*

$\frac{3}{4}$-, 1-, 1 $\frac{1}{2}$-, and 1 $\frac{3}{4}$- in heights

*(c)*

2-to 15-in heights in increments of $\frac{1}{4}$ in

*(d)*

Same as *(d)* in 5-and 10-ft lengths

*(e)*

**Figure 2.22** Examples of standard bar supports: *(a)* slab bolster, *(b)* beam bolster, *(c)* individual bar chair, *(d)* individual high chair, and *(e)* continuous high chair. (*Adapted from Ref. 8.*)

Before concrete is poured, the reinforcement is given a final inspection by the design engineer or the field inspector to make certain that the steel is positioned as required by the plans and specifications. This final check is most important. Many structural failures can be traced to the omission or incorrect positioning of reinforcement.

## QUESTIONS

**2.1** Define *hydration*. What is *bleeding* of concrete?

**2.2** What accounts for the large difference between the tensile and compressive strength of concrete?

**2.3** What size aggregates divides coarse from fine aggregates?

**2.4** What characteristics define structural lightweight concrete?

**2.5** What is the major parameter that determines the strength of concrete?

**2.6** What benefits are produced by using air entraining in concrete?

**2.7** Define *curing*. How is it done?

**2.8** What field test is used to measure workability?

**2.9** What type of concrete is required for high durability?

**2.10** What type of cement should be used to manufacture concrete exposed to sulfates?

**2.11** Define $f'_c$. How is it established?

**2.12** What is the modulus of elasticity of a lightweight concrete that weighs 110 lb/ft³ and has an $f'_c = 4000 \, \text{lb/in}^2$.

**2.13** What is the modulus of rupture? How is it measured?

**2.14** Define creep. List two negative effects of creep.

**2.15** What steps can be taken to reduce creep?

**2.16** What is the major factor that influences the magnitude of shrinkage?

**2.17** What are the major problems produced by shrinkage? How can they be mitigated?

**2.18** What is a control joint?

**2.19** What is the purpose of using reinforcing bars with surface deformations?

**2.20** If a steel has a yield point of 70 kips/in$^2$, at what strain will it yield?

# REFERENCES

1. T. C. Hsu, F. O. Slate, G. M. Sturman, and G. Winter: Microcracking of Plain Concrete and the Shape of the Stress-Strain Curve, *J. ACI*, vol. 60, p. 209, February 1963.
2. "Concrete Manual," 7th ed., U.S. Bureau of Reclamation, Denver, Colorado, 1963.
3. S. A. Mirza, M. Hatzinkolas, and J. MacGregor: Statistical Description of Strength of Concrete, *Proc. ASCE*, vol. 105, no., ST6, p. 1021, June 1979.
4. H. Kupfer, H. K. Hilsdorf, and H. Rusch: Behavior of Concrete under Biaxial Stresses, *J. ACI*, vol. 66, p. 656, August 1969.
5. "Design and Control of Concrete Mixtures," 11th ed., Portland Cement Association, Skokie, Ill., July 1968.
6. "PCI Design Handbook," Prestressed Concrete Institute, Chicago, 1971.
7. ACI Committee 504: Guide to Joint Sealants for Concrete Structures, *J. ACI*, vol. 67, p. 489, July 1970.
8. "Manual of Standard Practice," 23rd ed., Concrete Reinforcing Steel Institute, Chicago, 1981.
9. A. M. Neville: Hardened Concrete: Physical and Mechanical Aspects, *ACI Monogr.* 6, Detroit, 1971.
10. G. E. Troxell, H. E. Davis, and J. W. Kelly, "Composition and Properties of Concrete," 2nd ed., McGraw-Hill, New York, 1968.

Cascade Orchards Bridge over the Wenatchee River, Leavenworth, Washington. Continuous prestressed concrete bridge with slinder beams designed by Avid Grant Associates. (*Photograph by Arvid Grant.*)

# THREE

## DESIGN OF BEAMS FOR FLEXURE

### 3.1 INTRODUCTION

In this chapter we examine in detail the behavior of shallow beams and their response to moment. Shallow beams are one of the most common elements in reinforced concrete design. The same principles that apply to shallow beams can be readily extended and applied to the design of more complicated elements, such as slabs, footings, and the girders of continuous frames. If the designer is thoroughly familiar with the bending behavior of shallow beams, these more complex structural elements should present no unfamiliar problems.

Since the flexural strength normally controls the dimensions of beams, the beam is initially designed for moment and only later checked for shear and reinforced additionally if required. The check for shear occurs at a later stage and is independent of the design for moment. The effects of shear will be covered in detail in Chap. 4. Only in the most extreme cases (short spans and heavy loads) are the proportions of the beam controlled by shear.

Before the final selection of reinforcing steel, the designer must also check the bond between the reinforcement and the concrete to ensure that no slippage will take place. In this chapter on shallow beams the bond will be assumed to be adequate; the subject of bond will be covered fully in Chap. 6.

Usually the designer wants beams to be as shallow as possible to maximize headroom, even though deeper beams are more rigid and structurally efficient. The ACI Code defines a shallow beam as one whose depth-to-span ratio is less than $\frac{2}{5}$ for continuous spans and less than $\frac{4}{5}$ for simple spans. With few exceptions beams used in buildings and bridges fall into this category.

Although equations for proportioning beams will be derived for cross sections of any shape, considerable emphasis will be placed on beams with rectangular compression zones since most beams actually constructed have rectangular proportions. Special attention will be paid to the area of steel reinforcement required to produce ductile flexural members so that structures will have a high degree of toughness. Since proportioning by strength design accounts only for the ultimate moment capacity of the cross section and does not consider the state of stress in the materials or the behavior under service loads, the ACI Code has also established criteria to limit crack width and to control deflections. Beams that are excessively flexible (even if their strength is adequate) tend to vibrate, present an unsatisfactory sagging appearance, and may damage attached or supported non-structural elements.

Although concrete beams are typically constructed as part of a continuous floor system (Fig. 3.1), the discussion in this chapter will focus for the most part on the design of statically determinate beams to simplify the analysis. Once the design of a determinate beam is understood, continuous beams can be designed with no difficulty. The analysis and design of continuous beams that are an integral part of a floor system of a building frame will be covered in Chaps. 9 and 10.

Since most beams receive continuous lateral support from the floor system, they never fail in lateral torsional buckling. Should a beam not be rigidly connected to a slab but exist as a separate unit, lateral torsional buckling will not occur as long as the distance between lateral supports does not exceed 50 times the width of the compression flange. Because the designer will probably never encounter a situation in which lateral torsional buckling controls the moment capacity of a beam, only failure of a cross section by overstress in flexure will be considered.

## Organization of Chapter

Understanding the flexural behavior of reinforced concrete beams at all significant stages of loading is essential to provide the engineer with the background needed to design beams which have an adequate factor of safety and behave well under service loads. The sections that follow examine in detail the behavior of a concrete beam at progressively higher levels of load.

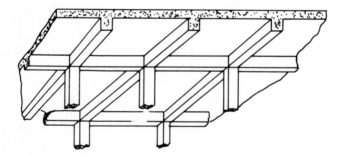

**Figure 3.1** Beams as supporting elements of a beam-and-slab floor system.

At the first level of loading to be considered, the reinforced concrete beam is loaded just to the verge of cracking but remains uncracked. Information gathered from this stage of loading will have application in the derivation of minimum steel requirements and in deflection computations. The second level of loading will be at the stage where service loads are applied. With service loads in place, serviceability requirements such as magnitude of deflections and width of cracks rather than strength are of primary concern. At the third level the beam is loaded to the verge of failure, i.e., to the point where its capacity to carry moment is fully mobilized. The proportioning of beams for moment is based on ensuring that the flexural design strength of the beam will exceed the moment produced by factored service loads at all sections. Once reinforced concrete behavior has been examined at each of these significant levels of moment, the basis of the ACI design procedures, which are developed in later sections of this chapter, will be clarified.

## 3.2 THE UNCRACKED CROSS SECTION; COMPUTATION OF THE CRACKING MOMENT

The cross-sectional area of a reinforced concrete beam that is effective in carrying flexural stresses varies with the magnitude of the moment. If the moment is small, the tensile bending stresses in the concrete are low (less than the tensile bending strength of the concrete) and no cracking of the concrete occurs (Fig. 3.2a). For this condition the entire cross section carries flexural stresses. On the other hand, if the moment is large, the maximum tensile stresses in the concrete exceed the tensile strength of the concrete and cracking occurs. Once the beam cracks, only the longitudinal steel in the tension zone carries tensile stresses (see Fig. 3.2b).

Since the area of flexural reinforcement is small, on the order of 1 percent of the gross cross-sectional area, the influence of the steel area on the flexural stresses in the uncracked concrete is small and may be neglected. In other words, the analysis for bending stresses in the *uncracked* beam can be based on the properties of the gross cross-sectional area using elastic equations. Stresses in the concrete can then be predicted by the standard beam equation

$$f = \frac{My}{I_g} \tag{3.1}$$

where $M$ = bending moment on cross section
$y$ = distance from centroid of the gross cross-sectional area to point at which stress $f$ is to be evaluated
$I_g$ = moment of inertia of gross section about its centroidal axis
$f$ = stress at distance $y$ from the centroidal axis

Equation (3.1) holds as long as the maximum tensile stress in the concrete does not exceed the modulus of rupture $f_r$. If a moment is applied that causes the maximum tensile stress in the concrete just to reach the modulus of rupture, the cross section will be on the verge of cracking. The moment that produces a tensile stress just

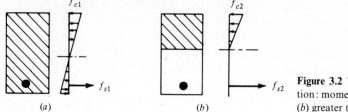

**Figure 3.2** The effective cross section: moment (a) less than $M_{cr}$ and (b) greater than $M_{cr}$.

equal to the modulus of rupture is termed the *cracking moment* $M_{cr}$. $M_{cr}$ can be evaluated by setting $f$ equal to $f_r$ in Eq. (3.1). In ACI notation

$$f_r = \frac{M_{cr} y_t}{I_g}$$

and

$$M_{cr} = \frac{I_g f_r}{y_t} \tag{3.2}$$

where $y_t$ is the distance from the centroid of the uncracked cross section to the outside tension surface. Note that cracking of a reinforced concrete cross section does not imply failure as long as the area of steel in the tension zone has the capacity to carry the tension force created by the applied moment. Example 3.1 illustrates the use of Eq. (3.2) to compute the cracking moment of a reinforced cross section.

**Example 3.1:** Compute the cracking moment $M_{cr}$ for the cross section in Fig. 3.3 if $f_c' = 3,600$ lb/in² (24.82 MPa), $f_y = 60$ kips/in² (413.7 MPa), and stone concrete is used.

SOLUTION Locate the neutral axis of the gross section. Neglect $A_s$; break the cross section into three rectangular areas (Fig. 3.3); and sum the moment of areas about a horizontal axis through the base of the cross section

$$A_{tot} \bar{Y} = \Sigma A_n y_n$$

$$[7(20)(2) + 6(6)]\bar{Y} = 140(10)(2) + 36(3)$$

$$\bar{Y} = 9.2 \text{ in (233.68 mm)}$$

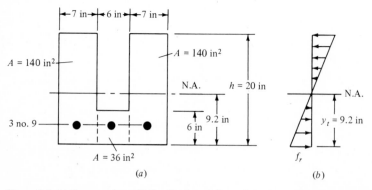

**Figure 3.3** (a) Cross section; (b) distribution of flexural stresses as cracking impends.

Compute

$$I_g = \sum \frac{bh^3}{12} + Ad^2$$

$$= \left[\frac{7(20^3)}{12} + 7(20)(10 - 9.2)^2\right](2) + \left[\frac{6(6^3)}{12} + 36(6.2^2)\right]$$

$$= 11{,}004.4 \text{ in}^4$$

Then

$$f_r = 7.5\sqrt{f_c'} = 7.5\sqrt{3600} = 450 \text{ lb/in}^2 = 0.45 \text{ kips/in}^2 \text{ (3.1 MPa)}$$

From Eq. (3.2)

$$M_{cr} = \frac{f_r I_g}{y_t} = \frac{0.45(11{,}004.4)}{9.2} = 538.3 \text{ in} \cdot \text{kips} = 44.85 \text{ ft} \cdot \text{kips (60.8 kN} \cdot \text{m)}$$

When the total service loads are applied to a concrete beam, moments develop that are considerably larger than the cracking moment; therefore Eq. (3.2) is of no use in sizing reinforced concrete members since the section will be cracked throughout most of the beam's length. In deflection computations, however, use is made of the cracking moment in calculating the effective moment of inertia for use in elastic-deflection equations. A concrete beam of constant depth will, in fact, behave as if it were composed of segments with two different moments of inertia (Fig. 3.4). In regions where the moment is less than $M_{cr}$, the cross section is uncracked, and the moment of inertia is based on the gross area. On the other hand, in regions where the moment exceeds $M_{cr}$, the beam is cracked, and properties of the cross section are a function of the properties of the cracked section.

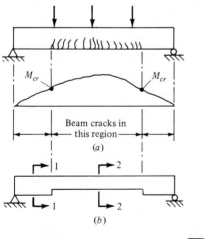

(a)

(b)

Section 1-1 · Section 2-2

**Figure 3.4** Influence of cracking on cross-sectional properties: (a) moment curve, (b) effective beam.

**Example 3.2:** The concrete beam in Fig. 3.5 is made of an all-lightweight concrete with a unit weight $\gamma = 110$ lb/ft$^3$ (1762 kg/m$^3$). (a) Compute the value of the cracking moment and (b) estimate the value of the concentrated load $P$ at midspan that causes the beam to crack (include the dead weight of the beam) $f_c' = 3000$ lb/in$^2$ (20.68 MPa) and $f_{ct} = 320$ lb/in$^2$ (2.2 MPa).

SOLUTION (a) The weight of the beam per foot is

$$\gamma \times \text{volume} = 110(\tfrac{176}{144}) = 134.4 \text{ lb/ft} \ (1.96 \text{ kN/m})$$

Compute $f_r$ with Eq. (2.3)

$$f_r = 7.5\frac{f_{ct}}{6.7}$$

but not more than

$$7.5\sqrt{f_c'} = 411 \text{ lb/in}^2 \ (2.83 \text{ MPa})$$

$$f_r = 7.5\frac{320}{6.7} = 358 \text{ lb/in}^2 < 411 \text{ lb/in}^2 \qquad \text{OK}$$

From Eq. (3.2)

$$M_{\text{cr}} = \frac{f_r I_g}{y_t} = 0.358 \text{ kip/in}^2 \frac{5190.1}{12.18} = 152.5 \text{ in} \cdot \text{kips} = 12.7 \text{ ft} \cdot \text{kips} \ (17.22 \text{ kN} \cdot \text{m})$$

(b) Compute the value of $P$ by equating the midspan moment due to external loads (Fig. 3.5) to $M_{\text{cr}}$

$$3.3 + 3.5P = 12.7 \text{ ft} \cdot \text{kips}$$

$$P = 2.69 \text{ kips} \ (11.97 \text{ kN})$$

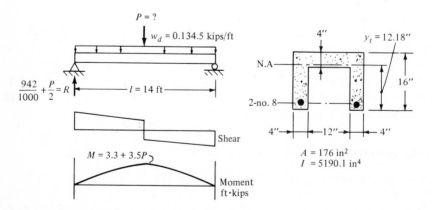

**Figure 3.5**

# 3.3 WORKING-STRESS DESIGN; SERVICE-LOAD STRESSES

## Introduction

Before strength-design provisions were introduced into the ACI Code in 1963, beams were proportioned by *working stress* or elastic design. This method is based on the designer's holding computed stresses produced by service loads below specified values of allowable stress. The derivation of design equations for the working-stress method is based on the assumption that reinforced concrete behaves like a linear-elastic material and cracks uniformly across the length of the beam whenever service moments exceed cracking moments. However, the intensity of the stresses predicted by working-stress design does not account for initial stresses produced by shrinkage or the redistribution of stresses that occurs with time as a result of concrete creep.

As strength design replaced working-stress design as the preferred method, elastic design was steadily deemphasized in the ACI Code. Although the 1963 ACI Code gave equal coverage to both design methods, the 1971 Code was devoted almost exclusively to strength design. In the 1971 Code less than one page was devoted to working-stress design, now termed the "alternate design method." In the 1977 edition of the Code, working-stress design has been moved out of the main body of the Code and relegated to an appendix; this move by the Code committee reflects the present obsolescence of the method. Now only the design of sanitary structures holding fluids is based on working-stress design, since keeping stresses low is a logical way to limit cracking and prevent leakage. Even for this application many concrete designers feel that ultimate-strength design with the proper controls on cracking would be superior to working-stress design.

Although limited use is made of working-stress design today, the method is introduced here to develop an understanding of behavior when service loads are applied. Service loads cause the development of extensive flexural cracking in sections where the service moments exceed the cracking moment of the cross section. Once the cross section cracks, the steel alone must carry the tensile stress produced by moment.

The derivation of expressions that relate stresses in the steel and the concrete to the service-load moments and to the properties of the cracked cross section will introduce the concept of the *transformed cracked section*. Once the beam develops flexural cracks, its bending stiffness reduces significantly. For problems where the bending stiffness influences behavior, such as deflection computations, the influence of cracking must be considered.

## Equations Relating Flexural Stresses and Service-Load Moments

To derive equations that relate the bending stresses at a particular point in a reinforced concrete beam to the dimensions of the cross section and the service-load moment, we shall consider the flexural deformations of a short segment of the simply supported beam shown in Fig. 3.6a. The segment, cut from the unstressed

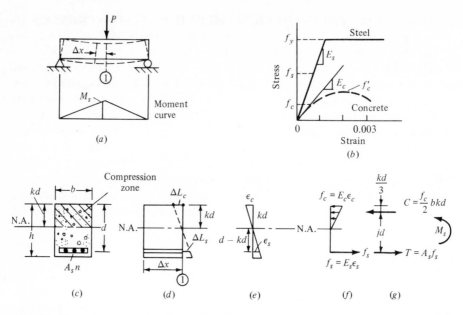

**Figure 3.6** Flexural stresses and strains produced by service-load moments: (*a*) simply supported beam with service load, (*b*) stress-strain curves of steel and concrete, (*c*) cross section of beam, (*d*) flexural deformations of a small beam segment, (*e*) strain distribution, (*f*) stress distribution, and (*g*) internal couple.

beam by two vertical planes a short distance $\Delta x$ apart, is shown by dashed lines in Fig. 3.6a. As the service load is applied, the beam bends and deflects and the faces on each side of the segment rotate about a horizontal axis in the plane of the cross section. This axis, located a distance $kd$ from the top surface (Fig. 3.6c), is called the *neutral axis* (N.A.). Since the longitudinal deformations at the level of the neutral axis are zero, no longitudinal stresses develop at that depth. Above the neutral axis, longitudinal elements of the segment shorten, an indication of compression. Below the neutral axis longitudinal elements elongate, an indication of tension. These deformations create the longitudinal stresses and strains shown in Fig. 3.6e and f. The variation of the stresses and strains is based on the following assumptions given in appendix B of the ACI Code:

1. Plane sections before bending remain plane after bending; therefore, strains vary linearly from the neutral axis.
2. Hooke's law, that stress is proportional to strain, applies.
3. The concrete in the tension zone is fully cracked, and only the steel carries tension.
4. No slip occurs between the reinforcing steel and the concrete.

To relate $f_c$, the maximum stress in the concrete at the top of the compression surface (Fig. 3.6f), and $f_s$, the stress in the steel, to the internal service-load moment

$M_s$, we first locate the position of the neutral axis. Then the position of the resultant of the compressive stresses (Fig. 3.6$g$) can be established ($C$ of course is equal to $T$ since the two forces constitute the equal and oppositely directed components of the internal couple). Once the arm $jd$ of the internal couple is known, the components of the internal couple can be evaluated by equating the internal moment $Cjd$ or $Tjd$ to the known value of the external service-load moment at the section.

The position of the neutral axis can be located by equating the components of the internal couple, to give

$$C = T \tag{3.3}$$

Expressing the forces in Eq. (3.3) in terms of the stresses (Fig. 3.6$g$) gives

$$kdb\frac{f_c}{2} = A_s f_s \tag{3.4}$$

where $b$ = width of cross section
$d$ = distance from compression surface to centroid of tension steel
$f_c$ = maximum stress in concrete
$A_s$ = total area of tension steel
$f_s$ = average stress in steel

Equation (3.4) can be simplified by using Hooke's law to eliminate the stresses. Substituting $f_s = E_s \epsilon_s$ and $f_c = E_c \epsilon_c$ into Eq. (3.4) gives

$$kdbE_c\frac{\epsilon_c}{2} = A_s E_s \epsilon_s \tag{3.5}$$

where $E_c$ = modulus of elasticity of concrete
$E_s$ = modulus of elasticity of steel
$\epsilon_c$ = strain in concrete in top fiber of compression zone
$\epsilon_s$ = strain in steel

Since elastic behavior is assumed, the stresses in the steel and concrete are assumed to be in the elastic region of the stress-strain curves of the two materials (Fig. 3.6$b$).

From the geometry of the strain distribution (Fig. 3.6$e$), the strains in the steel and concrete can be related by

$$\frac{\epsilon_c}{kd} = \frac{\epsilon_s}{d - kd}$$

and
$$\epsilon_c = \frac{kd}{d - kd} \epsilon_s \tag{3.6}$$

Substituting Eq. (3.6) into the left side of Eq. (3.5), dividing both sides by $E_c$, and simplifying gives

$$kdb\frac{kd}{2}\epsilon_s = A_s \epsilon_s \frac{E_s}{E_c}(d - kd) \tag{3.7}$$

Then set

$$\frac{E_s}{E_c} = n \tag{3.8}$$

where $n$, called the *modular ratio*, may be rounded off to the nearest whole number but must not be less than 6. Substituting Eq. (3.8) into (3.7) and cancelling $\epsilon_s$ produces

$$kdb\frac{kd}{2} = nA_s(d - kd) \tag{3.9}$$

which can be further simplified to yield an expression for $k$ by dividing each side by $bd$ to give

$$\frac{kdbkd}{bd(2)} = \frac{A_s nd(1 - k)}{bd} \tag{3.10}$$

Now set

$$\frac{A_s}{bd} = \rho \tag{3.11}$$

where $\rho$, called the *reinforcement ratio*, represents the area of steel per unit area of concrete (a large value of $\rho$, for example, indicates a heavily reinforced cross section). Substituting Eq. (3.11) into (3.10) and solving for $k$ gives

$$k = -\rho n + \sqrt{(\rho n)^2 + 2\rho n} \tag{3.12}$$

where $\rho = A_s/bd$ and $n = E_s/E_c$. It should be emphasized that Eq. (3.12) can be used to establish the position of the neutral axis only for a beam with a *rectangular compression zone*.

Equation (3.9) shows that the area of steel $A_s$ influences the position of the neutral axis: on the left side of Eq. (3.9) the quantity $kdb$ equals the area of the compression zone, and $kd/2$ equals the distance from the centroid of the compression zone to the neutral axis. Thus, the left side of Eq. (3.9) represents the moment of the area of the compression zone about the neutral axis. On the right side of Eq. (3.9) the term $nA_s$, called the *transformed area of steel*, is multiplied by $d-kd$, which represents the distance of the centroid of the steel from the neutral axis. Thus, the right-hand side of Eq. (3.9) represents the moment of the transformed steel area about the neutral axis. As the area of steel used to reinforce a particular cross section becomes less, the neutral axis must move away from the centroid of the steel area, reducing the size of the compression zone in order to maintain the equality of the moments of the respective areas about the neutral axis. Since the magnitudes of the strains in the steel and the concrete vary linearly from the neutral axis (Fig. 3.6e), the strain in the steel increases and the strain in the concrete decreases as the neutral axis moves away from the steel. If the neutral axis is positioned so that the strains in the steel created by flexure are large relative to those in the concrete, the beam will behave in a ductile manner as the beam bends under increasing load because the concrete, which fails at a low value of strain (0.003 is usually assumed), will not crush until the steel has undergone large deformations.

Although Eq. (3.9) was derived for a rectangular cross section, the basic concept of locating the position of the neutral axis by summing moments of the appropriate areas about the neutral axis is applicable to cross sections of any shape. To locate the neutral axis the designer defines the distance $kd$ of the neutral axis from the compression surface by a single variable, say $Y$, and then expresses the moments of both the transformed steel and the compression zone about the neutral axis in terms of this variable. This procedure results in a quadratic equation in terms of the single variable $Y$, which can be solved by any convenient method. The application of this principle to a nonrectangular section is illustrated in Example 3.3, where the quadratic equation is solved by *completing the squares*.

To derive an expression for $f_s$, the average stress in the steel produced by the service load moment $M_s$, we can sum moments of the internal forces on the cross section about $C$ (Fig. 3.6g). The internal moment is then equated to $M_s$

$$M_s = T \times \text{arm} = A_s f_s jd$$

and solving for $f_s$ leads to

$$f_s = \frac{M_s}{A_s jd} \tag{3.13}$$

Since the compressive stresses are assumed to have a triangular distribution, $C$ is located a distance $kd/3$ from the top of the compression zone at the centroid of the compressive stresses; therefore, the arm of the internal couple $jd$ is

$$jd = d - \frac{kd}{3} = d\left(1 - \frac{k}{3}\right) \tag{3.14}$$

Similarly, an expression for the maximum compressive stress at the top of the compression zone $f_c$ can be established by summing moments of the internal forces about $T$ to give

$$M_s = C \times \text{arm} = \left(\frac{f_c}{2} kdb\right) jd$$

Solving for $f_c$, we have

$$f_c = \frac{2M_s}{jkbd^2} \tag{3.15}$$

where all terms in Eqs. (3.13) and (3.15) are illustrated in Fig. 3.6 and $M_s$ represents the value of service moment at the section where stresses are being computed. These equations apply only to a beam with a rectangular compression zone. The use of Eqs. (3.13) and (3.15) will be illustrated in Example 3.5.

**Example 3.3:** For the T-shaped section in Fig. 3.7 (*a*) locate the position of the neutral axis using the fact that, with respect to the neutral axis, the moment of the area of the compression zone equals the moment of the area of the transformed steel. (*b*) Next

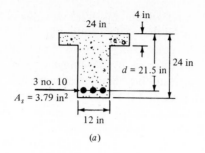

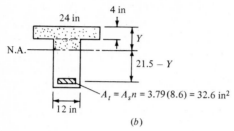

$A_t = A_s n = 3.79(8.6) = 32.6 \text{ in}^2$

**Figure 3.7** (*a*) Cross section, (*b*) transformed section.

compute the moment of inertia of the *cracked transformed cross section* $I_{cr}$, which equals the sum of the moment of inertia of the compression zone and the moment of inertia of the transformed steel both with respect to the neutral axis. The moment of inertia of the transformed steel about its own axis is insignificant and may be neglected; $f'_c = 3.5$ kips/in$^2$ and $n = 8.6$.

SOLUTION (*a*) Assume that the position of the neutral axis is in the web. The distance of the neutral axis from the top of the compression zone is denoted by $Y$ (Fig. 3.7*b*). Sum moments of areas about the neutral axis

$$(24 \times 4)(Y - 2) + (Y - 4)12 \frac{Y - 4}{2} = 32.6(21.5 - Y)$$

Simplifying gives

$$Y^2 + 13.43Y = 132.82$$

Solve by completing the square, to get

$$(Y + 6.72)^2 = 132.82 + 6.72^2$$

$$Y = 6.62 \text{ in}$$

The neutral axis lies in the web, as assumed.

(*b*)

$$I_{cr} = nA_s(d - Y)^2 + I_{web} + I_{flange}$$

$$= 32.6(14.88^2) + \frac{12(2.62^3)}{3} + \left[ \frac{24(4^3)}{12} + 4(24)(4.62^2) \right]$$

$$= 9{,}467.11 \text{ in}^4$$

**Properties of a doubly reinforced section**  The compression zone of a reinforced concrete beam is occasionally reinforced with steel to raise the member's bending strength or to reduce long-term deflections produced by creep. Steel located in the compression zone, or *compression steel*, is denoted by $A'_s$. Beams reinforced with both tension and compression steel are *doubly reinforced*.

To compute the properties of the cracked transformed section of a doubly reinforced beam the tension steel is multiplied by the modular ratio $n = E_s/E_c$ but the compression steel is multiplied by $2n$ (see ACI Code §B.5.5). Doubling the modular ratio of the compression steel accounts for the increase in stress that occurs with time as the concrete in the compression zone creeps. The creep deformation of the concrete produces additional strain in the compression steel and gradually raises the level of stress to approximately twice that of the initial value.

**Example 3.4:** (*a*) Locate the neutral axis and (*b*) compute the moment of inertia of the cracked transformed section of the doubly reinforced beam in Fig. 3.8, $f'_c = 3600\ \text{lb/in}^2$, and stone concrete is used.

SOLUTION (*a*) Compute the modular ratio $n$

$$n = \frac{E_s}{E_c} = \frac{29,000,000\ \text{lb/in}^2}{57,000\sqrt{3,600}} = 8.5 \quad \text{use 8}$$

Locate the neutral axis. Equate moments of compression area and transformed compression steel to the moment of the transformed tension steel. Refer moments to the neutral axis

$$Y(10)\frac{Y}{2} + 2nA'_s(Y - 3) = nA_s(17 - Y)$$

$$5Y^2 + 16(1.2\ \text{in}^2)(Y - 3) = 8(3\ \text{in}^2)(17 - Y)$$

$$Y = 6.25\ \text{in}$$

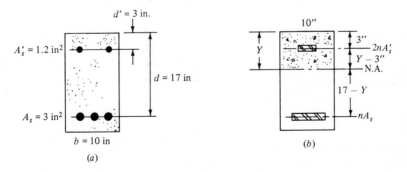

**Figure 3.8** (*a*) Double reinforced cross section, (*b*) cracked transformed section.

(b) Compute

$$I_{cr} = \frac{bY^3}{3} + 2nA_s'(Y - 3)^2 + nA_s(17 - Y)^2$$

$$= \frac{10(6.25^3)}{3} + 16(1.2)(6.25 - 3)^2 + 8(3)(17 - 6.25)^2$$

$$= 3790.1 \text{ in}^4$$

**Example 3.5:** The rectangular cross section in Fig. 3.15 is 12 by 24 in (305 by 610 mm) deep and carries a service-load moment $M_s$ of 150 ft·kips (203.4 kN·m). Determine (a) the maximum compressive stress in the concrete and (b) the stress in the steel if $f_c' = 3$ kips/in$^2$ (20.68 MPa) and $f_y = 60$ kips/in$^2$ (413.7 MPa).

SOLUTION (a) Establish properties of the cross section:

$$n = \frac{E_s}{E_c} = \frac{29,000,000}{57,000\sqrt{3000}} = 9.25 \quad \text{use 9}$$

$$\rho = \frac{A_s}{bd} = \frac{5.06}{12(21.6)} = 0.0195$$

Use Eq. (3.12) to compute $k$

$$k = -\rho n + \sqrt{(\rho n)^2 + 2\rho n}$$

$$= -0.0195(9) + \sqrt{[0.0195(9)]^2 + 2(0.0195)(9)} = 0.442 \quad \text{use 0.44}$$

By Eq. (3.14)

$$jd = d\left(1 - \frac{k}{3}\right) = 21.6\left(1 - \frac{0.44}{3}\right) = 18.43 \text{ in (467.36 mm)}$$

and $j = 18.43/21.6 = 0.85$.

(b) Use Eq. (3.13) to compute $f_s$, the stress in the steel

$$f_s = \frac{M_s}{A_s jd} = \frac{(150)12}{5.06(18.43)} = 19.3 \text{ kips/in}^2 \text{ (133.1 MPa)}$$

Use Eq. (3.15) to compute $f_c$, the stress at the top of the compression zone

$$f_c = \frac{2M_s}{jkbd^2} = \frac{2(150)(12)}{0.85(0.44)(12)(21.6^2)} = 1.72 \text{ kips/in}^2 \text{ (11.65 MPa)}$$

## 3.4 DEFLECTIONS UNDER SERVICE LOADS

### Introduction

To be designed properly, reinforced concrete beams must have adequate stiffness as well as strength. Under service loads, deflections must be limite so that attached nonstructural elements, e.g., partitions, pipes, plaster ceilings, and windows, will

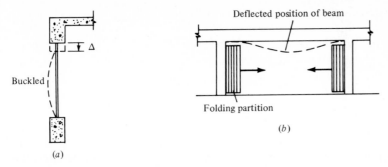

**Figure 3.9** Examples of damage or loss of function to nonstructural elements attached to beams that undergo large deflections (*a*) deflecting beam buckles window; (*b*) excessive deflections of beam cause attached plaster ceiling to sag and crack and the folding partition cannot be closed.

not be damaged or rendered inoperative by large deflections (Fig. 3.9). Obviously floor beams that sag excessively or vibrate as live loads are applied are not satisfactory.

A study of deflections of reinforced concrete beams must account for the instantaneous elastic deflections as loads are first applied as well as for the long-term deflections which develop due to creep and shrinkage and which continue to increase over a period of several years. Under a constant value of load, by the time long-term deflections reach their maximum size they are generally of the order of twice the magnitude of the initial elastic deflections.

Although the research engineer in the laboratory is able to carry out carefully controlled loading tests in which measured instantaneous elastic deflections are within 20 or 30 percent of those predicted by empirical equations for deflection,[1] the practicing engineer must expect deviations greater than 30 percent between predicted and measured deflections of beams constructed under actual field conditions. Deflections are minimized when beams are carefully constructed out of high-strength, low-slump concretes that are well compacted and effectively cured. In the field the engineer has a certain limited control over construction methods and procedures by means of the plans and specifications covering the design of the concrete mix and details of placing steel and concrete; however, what the designer specifies and what the construction crews produce can differ widely. Water content may be increased at the job site, incomplete compaction may leave voids and honeycombing, and reinforcing bars may be improperly positioned. By reducing the quality of the concrete this and other construction procedures can produce deflections larger than expected.

For example, if forms are removed from a beam before the concrete has developed its full design strength, the dead weight of the beam and superimposed construction loads can increase the amount of initial creep deflection by several hundred percent over that of a beam whose concrete has fully hardened before the removal of forms. In addition, reinforced concrete beams loaded at an early curing age, while the modulus of rupture and the modulus of elasticity remain low, may crack and deflect extensively in regions that would normally not be cracked when a properly cured beam is put into service and carries its designated design loads.

The common practice of storing construction materials on unshored floors may subject beams and slabs to short-term loads far in excess of anticipated service loads. Although the beams may not fail, the temporary loads can produce large moments that cause more extensive cracking than the normal service loads would have caused. This premature cracking can reduce the bending stiffness of a cross section permanently by 50 percent or more.

In spite of the many expedient construction practices that can reduce the stiffness of reinforced concrete beams and increase the creep deflections, few serious problems develop that are attributable to excessive deflections if the well-considered recommendations of the ACI Code are followed. Concrete beams that are part of continuous frames are naturally stiff because of negative end moments, which provide substantial resistance to deflections.

When shallow beams are required whose deflections are difficult to control, the use of prestressed concrete should be considered. The prestressing of concrete beams eliminates cracking and makes the moment of inertia of the entire cross section effective in resisting bending deflections. Moreover, by proper positioning of prestressed steel, deflections—opposite in direction to, and of the same magnitude as, those produced by the applied loads—can be induced to produce deflection-free beams for one specific loading condition.

In the next sections the ACI Code procedures for controlling and predicting deflections are examined, and a number of examples are worked.

### Computation of Immediate Deflections

Elastic equations (Fig. 3.10) are used to compute the immediate deflections of a reinforced concrete beam. Other useful deflections equations are tabulated in engineering textbooks and design manuals, e.g., the AISC Steel Construction Manual.

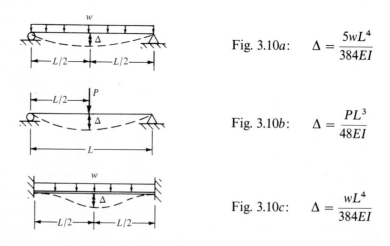

Fig. 3.10a: $\Delta = \dfrac{5wL^4}{384EI}$

Fig. 3.10b: $\Delta = \dfrac{PL^3}{48EI}$

Fig. 3.10c: $\Delta = \dfrac{wL^4}{384EI}$

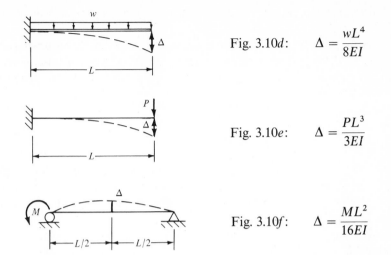

Fig. 3.10$d$:     $\Delta = \dfrac{wL^4}{8EI}$

Fig. 3.10$e$:     $\Delta = \dfrac{PL^3}{3EI}$

Fig. 3.10$f$:     $\Delta = \dfrac{ML^2}{16EI}$

**Figure 3.10** Deflection equations.

**Effective moment of inertia**   In a reinforced concrete beam, the effective cross section varies along the length of the member. In regions of low moment, where no cracks exist, the effective moment of inertia should logically be based on the gross transformed area of the cross section. At sections of high moment, where cracking is extensive, the effective moment of inertia is more properly based on the properties of the cracked transformed cross section. To account for the variation of the moment of inertia along the beam axis, ACI §9.5.2.3 requires the use in elastic-deflection equations of an effective moment of inertia $I_e$, which is computed by Eq. (3.16). This empirical equation is based on a statistical analysis of deflection measurements from extensive beam tests. In Eq. (3.16) the moment of inertia of the gross cross-sectional area is used to approximate the moment of inertia of the uncracked transformed cross section

$$I_e = \left(\frac{M_{cr}}{M_a}\right)^3 I_g + \left[1 - \left(\frac{M_{cr}}{M_a}\right)^3\right] I_{cr} \leq I_g \qquad (3.16)$$

where $f_r$ = modulus of rupture = $7.5\sqrt{f'_c}$ lb/in$^2$ or $0.62\sqrt{f'_c}$ MPa [Eq. (2.3)]

   $y_t$ = distance from centroid of gross section to extreme fiber in tension

   $M_{cr}$ = cracking moment = $f_r I_g / y_t$ [Eq. (3.2)]

   $M_a$ = maximum moment in member at stage for which deflection is being computed

   $I_g$ = moment of inertia of gross section neglecting area of tension steel

   $I_{cr}$ = moment of inertia of transformed cracked cross section

Equation (3.16) should be used when $1 \leq M_a/M_{cr} \leq 3$. If $M_a/M_{cr} > 3$, the cracking is extensive and $I_e$ can be taken equal to $I_{cr}$ with no significant error. If $M_a/M_{cr} < 1$, no cracking is likely and $I_e$ can be taken as equal to $I_g$ (Fig. 3.11).

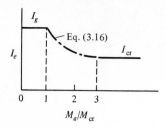

**Figure 3.11** Variation of effective moment of inertia with maximum moment.

To compute deflections in a span of a continuous beam, the effective moment of inertia may be taken as the average of the values computed for the sections of maximum negative and maximum positive moment (see ACI Code §9.5.2.4).

**Example 3.6:** Compute the deflection of the beam at midspan under service loads shown in Fig. 3.12;

$$f'_c = 3600 \text{ lb/in}^2 \quad \text{and} \quad f_y = 60 \text{ kips/in}^2.$$

SOLUTION  Compute $M_{cr}$

$$I_g = \frac{bh^3}{12} = \frac{12(24^3)}{12} = 13,824 \text{ in}^4$$

$$M_{cr} = \frac{f_r I_g}{y_t} = \frac{7.5\sqrt{3600}(13,824)}{(12 \text{ in})(12,000)} = 43.2 \text{ ft} \cdot \text{kips}$$

$$\frac{M_a}{M_{cr}} = \frac{84}{43.2} = 1.94 > 1$$

Use Eq. (3.16) to compute $I_e$.

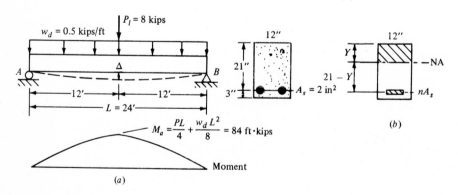

$$M_a = \frac{PL}{4} + \frac{w_d L^2}{8} = 84 \text{ ft} \cdot \text{kips}$$

(a)

(b)

**Figure 3.12**

Compute $I_{cr}$ (Fig. 3.12b). Locate the neutral axis using Eq. (3.12)

$$n = \frac{E_s}{E_c} = \frac{29,000,000}{57,000\sqrt{3600}} = 8.5 \qquad \rho = \frac{A_s}{bd} = \frac{2}{12(21)} = 0.0079$$

$$k = -\rho n + \sqrt{(\rho n)^2 + 2\rho n} = -0.0672 + \sqrt{(0.0672)^2 + 2(0.0672)} = 0.306$$

$$Y = kd = 0.306(21) = 6.43 \text{ in}$$

$$I_{cr} = \frac{bY^3}{3} + nA_s(d - Y)^2 = \frac{12(6.43)^3}{3} + 8.5(2)(21 - 6.43)^2 = 4672 \text{ in}^4$$

$$I_e = \left(\frac{M_{cr}}{M_a}\right)^3 I_g + \left[1 - \left(\frac{M_{cr}}{M_a}\right)^3\right]I_{cr}$$

$$= \left(\frac{1}{1.94}\right)^3 13{,}824 + \left[1 - \left(\frac{1}{1.94}\right)^3\right]4672 = 5925 \text{ in}^4$$

The deflection at midspan is

$$\Delta = \frac{5wL^4}{384EI_e} + \frac{PL^3}{48EI_e} = \frac{5(0.5)(24^4)(1728)}{384(3.42 \times 10^3)(5925)} + \frac{8(24^3)(1728)}{48(3.42 \times 10^3)(5,925)}$$

$$= 0.184 \text{ in} + 0.196 \text{ in}$$

$$= 0.38 \text{ in}$$

**Example 3.7:** A 10-ft-long cantilever beam carries service loads consisting of a uniform dead load of 0.3 kips/ft and a 4.5-kip concentrated live load at the tip. Compute the immediate deflection at the tip for (a) uniform load and (b) total load if

$$I_g = 3600 \text{ in}^4, \ I_{cr} = 2000 \text{ in}^4, \ M_{cr} = 25 \text{ ft} \cdot \text{kips, and } E_c = 3000 \text{ kips/in}^2$$

SOLUTION (a) For deflection at point B due to uniform dead load (Fig. 3.13a)

$$\frac{M_a}{M_{cr}} = \frac{15 \text{ ft} \cdot \text{kips}}{25 \text{ ft} \cdot \text{kips}} = 0.6 < 1$$

Since the beam is not cracked, use $I_e = I_g$.

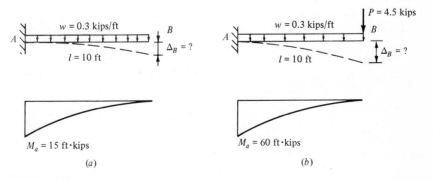

Figure 3.13 (a) Dead load only; (b) dead and live load.

Compute deflection using Fig. 3.10d

$$\Delta_B = \frac{wl^4}{8E_c I_e} = \frac{(0.3)(10^4)(1728)}{8(3000)(3600)} = 0.06 \text{ in}$$

(b) For deflection point $B$ due to dead and live load (Fig. 3.13b)

$$\frac{M_a}{M_{cr}} = \frac{60 \text{ ft} \cdot \text{kips}}{25 \text{ ft} \cdot \text{kips}} = 2.4 > 1;$$

Since the beam is cracked, use Eq. (3.16) to compute $I_e$

$$I_e = \left(\frac{M_{cr}}{M_a}\right)^3 I_g + \left[1 - \left(\frac{M_{cr}}{M_a}\right)^3\right] I_{cr}$$

$$= \left(\frac{1}{2.4}\right)^3 3600 \text{ in}^4 + \left[1 - \left(\frac{1}{2.4}\right)^3\right] 2000 \text{ in}^4$$

$$= 2116 \text{ in}^4$$

Use Fig. 3.10 $d$ and $e$ to compute deflection at the tip

$$\Delta = \frac{wl^4}{8E_c I_e} + \frac{Pl^3}{3E_c I_e}$$

$$\Delta_B = \frac{(0.3)(10^4)(1728)}{8(3000)(2116)} + \frac{(4.5)(10^3)(1728)}{3(3000)(2116)} = 0.51 \text{ in}$$

Once a beam has been cracked by a large moment, it can never return to its original uncracked state; therefore, the effective moment of inertia $I_e$ that should be used in deflection computations must always be equal to the effective moment of inertia associated with the maximum past moment to which the beam has been subjected. Often this moment is impossible to determine for most beams.

**Example 3.8:** Consider the beam in Fig. 3.13 again. Computations indicated that the effective moment of inertia for the beam is $I_e = 2116$ in$^4$ when the beam is subject to a maximum moment $M_a = 60$ ft · kips. If the live load is now removed, so that only the dead load acts, $M_a$ decreases to 15 ft · kips. What value of $I_e$ should be used to compute the deflection at the tip of the cantilever?

SOLUTION Use $I_e = 2116$ in$^4$ for all computations in which $M_a$ is equal to or less than 60 ft · kips.

## Long-Term Deflections

Deflections of concrete beams consist of two components, an *initial deflection* that occurs simultaneously with the application of load and a *long-term* or *additional* increment of deflection (produced by creep and shrinkage) that takes place over time. This latter component of deflection, which occurs rapidly at first and then slows, is largely complete after 1 or 2 years. The increase with time of the long-term deflection is shown in Fig. 3.14.

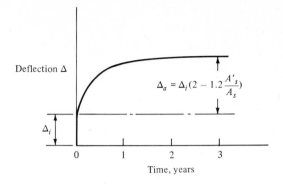

**Figure 3.14** Variation of deflection with time. $\Delta_i$ = initial elastic deflection, $\Delta_a$ = long-term deflection.

To estimate the magnitude of the long-term deflection, ACI Code §9.5.2.5 specifies that the initial deflection, computed by elastic equations, be multiplied by the empirical factor

$$2 - 1.2A_s'/A_s \qquad \text{but not less than } 0.6 \qquad (3.17)$$

where $A_s'$ is the area of the compression steel and $A_s$ is the area of the tension steel. If no compression steel is used ($A_s' = 0$), Eq. (3.17) indicates that the long-term component of the deflection will be twice that of the initial deflection. To estimate the long-term deflection in computations, only the deflection produced by the sustained portion of the load should be multiplied by Eq. (3.17).

If the long-term deflection exceeds the value permitted by Table 3.2, the designer may either increase the depth of members or add additional compression steel. If the sag produced by the long-term deflections is objectionable from an architectural or functional point of view, forms may be raised (*cambered*) a distance equal to that of the anticipated deflection. Example 3.9 illustrates the use of Eq. (3.17) in deflection computations.

**Example 3.9: Investigation of deflections** The beam in Fig. 3.15 carries uniform service loads of 1 kip/ft dead load and 1.5 kips/ft (21.89 kN/m) live load. The dead load, which includes the weight of the beam, is always in place, but only an average of 60 percent of the live load acts continuously. (*a*) Compute the initial deflection due to total load and (*b*) estimate additional long-term deflection; $f_c' = 3000$ lb/in² (20.68 MPa), $f_y = 50$ kips/in² (334.7 MPa), $I_g = 13,824$ in⁴ (5752 × 10⁶ mm⁴), $I_{cr} = 10,300$ in⁴ (4287 × 10⁶ mm⁴), and $f_r = 7.5\sqrt{f_c'} = 411$ lb/in² (2.83 MPa).

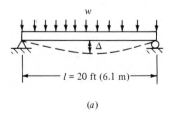

(*a*)

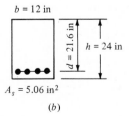

(*b*)

**Figure 3.15**

SOLUTION (a)

$$M_a = \frac{wl^2}{8} = \frac{2.5(20^2)}{8} = 125 \text{ ft} \cdot \text{kips} = 1500 \text{ in} \cdot \text{kips (169.5 kN} \cdot \text{m)}$$

$$M_{cr} = \frac{f_r I_g}{y_t} = \frac{0.411(13,824)}{12} = 473 \text{ in} \cdot \text{kips (53.45 kN} \cdot \text{m)}$$

$$\frac{M_a}{M_{cr}} = \frac{1500}{473} = 3.17 > 3$$

Therefore use

$$I_e = I_{cr} = 10,300 \text{ in}^4$$

$$E_c = \frac{57,000\sqrt{3000}}{1000} = 3120 \text{ k/in}^2 \text{ (21,500 MPa)}$$

$$\Delta = \frac{5wl^4}{384E_c I_e} = \frac{5(2.5)(20^4)(1728)}{384(3120)(10,300)} = 0.28 \text{ in (6.6 mm)}$$

(b) Base computations on the sustained portion of the load: dead load plus 60 percent of the live load. The sustained load is

$$1 \text{ kip/ft} + 0.6(1.5 \text{ kip/ft}) = 1.9 \text{ kips/ft (27.7 kN/m)}$$

Since except for the value of load, all terms in the deflection equation are identical to those in the computation of total load deflection, deflections are proportional to load and the initial deflection is

$$\Delta_i = \frac{1.9(0.28 \text{ in})}{2.5} = 0.21 \text{ in (5.3 mm)}$$

The additional long-term deflection is then

$$\Delta_a = \left(2 - \frac{1.2A_s'}{A_s}\right)\Delta_i = 2(0.21) = 0.42 \text{ in. (10.7 mm)}$$

where $A_s' = 0$.

## Control of Deflections

Since deflection computations can take some time to carry out, the ACI Code has established a procedure to limit deflections by placing restrictions on the minimum depth of a beam.[2] This method eliminates the need to compute deflections and controls deflections by requiring that beam depths not be less than a specified fraction of the span length. Minimum depths for a variety of common support conditions are given in Table 3.1. The theoretical basis for this method of controlling deflections can be established by considering the calculations for the mid-

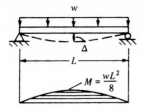

**Figure 3.16**

span deflection of a simply supported, uniformly loaded beam that behaves elastically (Fig. 3.16)

$$\Delta = \frac{K_1 w L^4}{EI} \qquad \text{where } K_1 = \frac{5}{384} \tag{1}$$

where $w$ = uniform load
$L$ = span length
$E$ = modulus of elasticity
$I$ = effective moment of inertia

Equation (1) can be expressed in terms of the maximum moment $M$ at midspan by substituting $wL^2 = 8M$ into Eq. (1)

$$\Delta = \frac{K_1 8 M L^2}{EI} = \frac{K_2 M L^2}{EI} \qquad \text{where } 8K_1 = K_2 \tag{2}$$

When a linear variation of compressive stress is assumed, the maximum concrete stress can be expressed as

$$f = \frac{Mc}{I} \tag{3}$$

where $c$ is the distance from the neutral axis to the outside compression surface and $f$ is the maximum service stress at a distance $c$ from the neutral axis.

Solving Eq. (3) for $M = fI/c$ and substituting into Eq. (2) gives

$$\Delta = \frac{K_2 fI}{c} \frac{L^2}{EI} \tag{4}$$

$f$ and $E$ in Eq. (4) can be expressed as functions of $f'_c$ (in earlier editions of the ACI Code $E$ was expressed as $1000f'_c$)

$$f = K_3 f'_c \qquad \text{and} \qquad E = K_4 f'_c$$

and

$$\frac{f}{E} = \frac{K_3}{K_4} = K_5 \tag{5}$$

Using the relationship in (5) to eliminate $f$ and $E$ in (4) gives

$$\Delta = \frac{K_2 K_5 L^2}{c} \tag{6}$$

When $c$ is expressed as a fraction of the overall depth $h$ by $c = K_6 h$, Eq. (6) can be written as

$$\Delta = \frac{K_2 K_5 L^2}{K_6 h} = \frac{K L^2}{h} \qquad \text{where } K = \frac{K_2 K_5}{K_6} \tag{7}$$

Based on a study of the maximum deflections reinforced concrete beams can safely undergo without producing damage to attached construction, maximum values of deflection have been established as a fraction of the span length $L$ for a variety of design situations. For example, if plaster ceilings are not to crack, experience indicates that the maximum live-load deflection of beams must not exceed $L/360$. In general terms

$$\Delta \leq \frac{L}{C_1} \tag{8}$$

where $\Delta$ is the maximum permitted deflection and $C_1$ is a constant that depends on the maximum deflection the beam can tolerate if it is to serve its intended purpose satisfactorily.

Using Eq. (7) to eliminate $\Delta$ in Eq. (8) gives

$$\frac{K L^2}{h} \leq \frac{L}{C_1}$$

or
$$h \geq K C_1 L \tag{9}$$

To eliminate a decimal, the product of the two constants $K$ and $C_1$ is set equal to $1/N$, where $N$ is a number greater than 1. Then Eq. (9) can be written

$$h \geq \frac{L}{N} \tag{10}$$

Values of $N$ that ensure the serviceability of beams and one-way slabs are tabulated in Table 3.1. By the proper choice of $N$ values provision can be made for the effects of creep and shrinkage on long-term deflections. Since the method is approximate, the ACI Code limits its use to beams that support or are attached to nonstructural elements capable of tolerating relatively large deflections without incurring damage. If (for architectural or other functional reasons) the designer must use a beam whose depth is less than the values required by Table 3.1, the resulting deflections must be computed and compared with the maximum permitted values of deflection tabulated in Table 3.2.

**Example 3.10:** Determine the minimum depth required by the ACI Code if deflection computations are to be avoided for the lightweight-concrete beam in Fig. 3.17; $f'_c = 3$ kips/in² (20.68 MPa), $f_y = 50$ kips/in² (344.7 MPa), and $w = 110$ lb/ft³ (1762 kg/m³).

SOLUTION From Table 3.1 under "Simply supported" we read

$$h_{\text{min}} = \frac{l}{16} (1.65 - 0.005w) \left( 0.4 + \frac{f_y}{100,000} \right)$$

**Table 3.1 Minimum thickness of beams or one-way slabs unless deflections are computed (ACI Code Table 9.5a)†**

Members not supporting or attached to partitions or other construction likely to be damaged by large deflections

| | Minimum thickness $h$ | | | |
|---|---|---|---|---|
| | Simply supported | One end continuous | Both ends continuous | Cantilever |
| Solid one-way slabs | $l/20$ | $l/24$ | $l/28$ | $l/10$ |
| Beams or ribbed one-way slabs | $l/16$ | $l/18.5$ | $l/21$ | $l/8$ |

† Span length $l$ in inches. Values in the table apply to normal-weight concrete reinforced with steel of $f_y = 60{,}000$ lb/in² (413.7 MPa). For structural lightweight concrete with a unit weight between 90 and 120 lb/ft³ (1440 and 1920 kg/m³) multiply the table values by $1.65 - 0.005w$ or $1.65 - 0.0003w$, respectively, but not less than 1.09; the unit weight $w$ is in lb/ft³ (kg/m³). For reinforcement having a yield point other than 60,000 lb/in² (413.7 MPa), multiply the table values by $0.4 + f_y/100{,}000$ $(0.4 + f_y/690)$ with $f_y$ in lb/in² (MPa).

and
$$1.65 - 0.005w \geq 1.09$$

$$= \frac{20(12)}{16} [1.65 - 0.005(110)] \left(0.4 + \frac{50{,}000}{100{,}000}\right)$$

$$= 14.85 \text{ in } (377.2 \text{ mm})$$

**Example 3.11:** Determine the minimum depth of the two-span continuous beam in Fig. 3.18 if deflection computations are to be avoided; $f_c' = 4$ kips/in² (27.58 MPa), $f_y = 60$ kips/in² (413.7 MPa), and $w = 145$ lb/ft³ (2320 kg/m³).

SOLUTION  Since the concrete is normal weight and $f_y = 60$ kips/in², no modifiers are required. From Table 3.1 under "One end continuous" we read

$$h_{\min} = \frac{l}{18.5} = \frac{20(12)}{18.5} = 13 \text{ in } (330.2 \text{ mm})$$

**Example 3.12:**  What is the minimum depth required to control deflections of the three-span continuous beam in Fig. 3.19 stone concrete and $f_y = 60$ kips/in²?

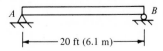

Figure 3.17

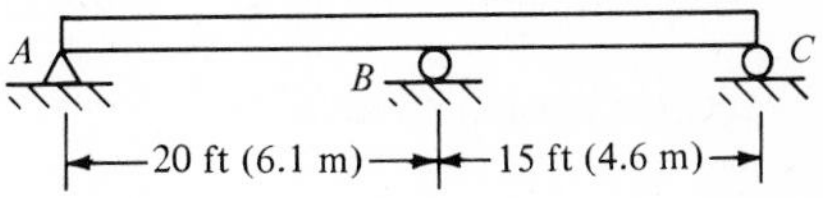

Figure 3.18

**Table 3.2 Maximum allowable computed deflections (ACI Code Table 9.5b)**

| Type of member | Deflection to be considered | Deflection limitation |
|---|---|---|
| Flat roofs not supporting or attached to nonstructural elements likely to be damaged by large deflections | Immediate deflection due to the live load $L$ | $\dfrac{l\dagger}{180}$ |
| Floors not supporting or attached to nonstructural elements likely to be damaged by large deflections | Immediate deflection due to the live load $L$ | $\dfrac{l}{360}$ |
| Roof or floor construction supporting or attached to nonstructural elements likely to be damaged by large deflections | That part of the total deflection which occurs after attachment of the nonstructural elements, the sum of the long-time deflection due to all sustained loads, and the immediate deflection due to any additional live load§ | $\dfrac{l\ddagger}{480}$ |
| Roof or floor construction supporting or attached to nonstructural elements not likely to be damaged by large deflections | | $\dfrac{l\P}{240}$ |

† This limit is not intended to safeguard against ponding. Ponding should be checked by suitable calculations of deflection, including the added deflections due to ponded water, and considering long-time effects of all sustained loads, camber, construction tolerances, and reliability of provisions for drainage.
‡ The long-time deflection is determined in accordance with ACI Code §9.5.2.3 or §9.5.4.2 but may be reduced by the amount of deflection which occurs before attachment of the nonstructural elements. This amount is determined on the basis of accepted engineering data relating to the time-deflection characteristics of members similar to those being considered.
§ This limit may be exceeded if adequate measures are taken to prevent damage to supported or attached elements.
¶ But not greater than the tolerance provided for the nonstructural elements. This limit may be exceeded if camber is provided so that the total deflection minus the camber does not exceed the limitation.

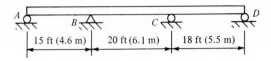

Figure 3.19

SOLUTION Span *BC* for two ends continuous

$$h_{min} = \frac{1}{21} = \frac{20(12)}{21} = 11.43 \text{ in } (290.3 \text{ mm})$$

Span *CD* for one end continuous

$$h_{min} = \frac{l}{18.5} = \frac{18(12)}{18.5} = 11.68 \text{ in } (296.7 \text{ mm}) \qquad \text{controls.}$$

## 3.5 CONTROL OF CRACKING UNDER SERVICE LOADS

As a reinforced concrete beam deflects, the tension side of the beam cracks wherever the low tensile strength of the concrete is exceeded. The more the beam deflects, the greater the length and width of cracks. Although cracking cannot be prevented, it is possible by careful detailing of the steel to produce beams that develop narrow, closely spaced cracks in preference to a few wide cracks.

Excessive cracking of the concrete that covers the reinforcement is of considerable concern: the protection of the reinforcement from the environment depends on the integrity of the concrete cover. If members have been constructed of high-quality, well-compacted concrete, studies indicate that the presence of narrow cracks in the cover will not significantly reduce the effectiveness of the concrete to cover and to protect the reinforcement from deterioration by weather or by other corrosive elements, such as seawater. On the other hand, if wide cracks develop in the cover, the reinforcement will be exposed to the atmosphere and be vulnerable to continuing deterioration by corrosion. The maximum crack width the designer should permit depends on exposure conditions.[3] If concrete is exposed to seawater or cycles of wetting and drying, the maximum width of any crack should not exceed 0.006 in (0.15 mm) or at the far limit 0.008 in (0.2 mm). For members protected against weather, crack widths up to 0.016 in (0.41 mm) are permitted by the ACI Code. Where concrete surfaces are exposed to view, excessive cracking looks unsightly, the integrity and solidity of the member is compromised, and the architect's intention of a smooth monolithic concrete surface is not achieved.

In past years, when concrete design was based on working-stress theory, low-yield-point steels were used, and the stresses in the reinforcement under service loads were typically low. When this design technique was used, the low strains in the steel and the adjacent concrete minimized cracking. Under these circumstances, limited attention was paid to control of crack width. Since strength design has superseded elastic design and the use of higher-yield-point steels has increased, stress and strain levels in the reinforcement and in the adjacent concrete have also increased. These higher allowable levels have resulted in greater cracking, the control of which has required provisions to be introduced into the ACI Code limiting the width of cracks when steels with yield points over 40 kips/in² (276 MPa) are used.

Although volumetric changes, such as result from shrinkage or temperature change, can contribute to the formation of cracks, the ACI Code does not consider their effects to be significant in the design process; only cracks produced by flexural stresses that result from bending need be taken into account.

At any section of a beam where the moment just exceeds the cracking moment, flexural microcracks develop. They are hairline thin and visible only by microscopic study. As the moment continues to increase to its maximum value under service load, the width and length of the cracks increase. Under sustained loads, creep causes cracks to widen an additional 30 to 40 percent over the initial width produced by service loads. Experimental studies show that the width of cracks varies directly with the magnitude of the steel stress and demonstrate that a large number of small bars well distributed through the tension zone of the beam is more effective in reducing the width of cracks than a small number of larger-diameter bars used to supply the same area of steel. These tests also show that reducing the distance between the first row of bars and the surface reduces the crack width. Using a statistical analysis of six series of beam tests, Gergely and Lutz[8] established an expression for predicting the most probable maximum width of crack at the surface of a reinforced concrete beam

$$w = 0.076\beta f_s \sqrt[3]{d_c A} \tag{3.18}$$

where $w$ = maximum width of crack, thousandths of an inch
  $\beta$ = distance from neutral axis of cracked section to outside surface divided by distance from neutral axis to centroid of the steel; can be taken as 1.2 for beams and 1.35 for slabs
  $f_s$ = stress in steel due to service loads, kips/in²
  $d_c$ = distance from tension surface to center of reinforcing bar closest to outside surface
  $A$ = effective tension area of concrete divided by number of reinforcing bars

The effective tension area of concrete (Fig. 3.20) is the product of the web width and a height of web equal to twice the distance between the centroid of the steel and tension surface. When the reinforcement consists of more than one size of bar, the number of bars is expressed in terms of the size of the largest bar

$$\text{Number of bars} = \frac{\text{total area of steel}}{\text{area of largest bar}}$$

Using a value of $\beta = 1.2$, the Gergely–Lutz equation (3.18) can be written for beams as

$$w = 0.091 f_s \sqrt[3]{d_c A} \tag{3.19}$$

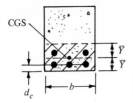

**Figure 3.20** Effective tension area (crosshatched); $\bar{Y}$ = distance from tension surface to CGS (center of gravity of steel).

Since the width of cracks shows considerable scatter, values of $w$ produced by Eq. (3.19) indicate the order of magnitude of the maximum crack width rather than actual crack width.

ACI Code §10.6.4 uses Eq. (3.19) as the basis of its provision to control the width of cracks at sections of maximum positive and negative moment when the yield point of the reinforcement exceeds 40 kips/in² (276 MPa).

When we let $w/0.091 = z$, Eq. (3.19) becomes

$$z = f_s \sqrt[3]{d_c A} \tag{3.20}$$

where $f_s$, the stress in the steel in kips per square inch, may be taken as $0.6f_y$.

ACI Code §10.6.4 specifies that $z$ is not to exceed 145 kips/in (25.4 MN/m) for exterior exposure or 175 kips/in (30.6 MN/m) for interior exposure. These values corresponding to maximum crack widths of 0.013 in (0.33 mm) and 0.016 in (0.41 mm), respectively.

## Additional Reinforcement for Deep Beams

If the depth of the web exceeds 3 ft (0.91 m) ACI Code §10.6.7 requires that an area of steel equal to 10 percent of the main flexural steel area be added to the tension side of the web adjacent to the vertical faces. The spacing of the additional steel is not to exceed 12 in (305 mm) or the width of the web. Failure to add this steel will permit the width of cracks in the upper part of the tension zone to exceed the crack width at the level of the main steel (see Fig. 3.21).

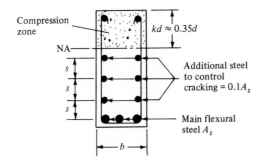

**Figure 3.21** Additional steel to control cracking of web; $s$ not to exceed 12 in or $b$.

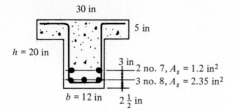

Figure 3.22

**Example 3.13:** Determine whether the reinforcement pattern in Fig. 3.22 satisfies ACI Code requirements for crack control; exterior exposure, $f_y = 50$ kips/in².

SOLUTION  Locate the center of gravity of steel by summing moments of bar areas about an axis through the base of the cross section

$$A_{st}\,\overline{Y} = \Sigma A_n \bar{y}_n$$

$$(3.55 \text{ in}^2)(\overline{Y}) = (2.35 \text{ in}^2)(2.5 \text{ in}) + (1.2 \text{ in}^2)(5.5 \text{ in})$$

$$\overline{Y} = 3.51 \text{ in}$$

Use Eq. (3.20) to compute $z$

$$A = \frac{\text{area of concrete}}{\text{number of bars}} = \frac{12(3.51)(2)}{3.55/0.79} = 18.75$$

$$f_s = 0.6f_y = 30 \text{ kips/in}^2 \qquad d_c = 2.5 \text{ in}$$

$$z = f_s\sqrt[3]{d_c A} = (30 \text{ kips/in}^2)\sqrt[3]{2.5(18.75)} = 108 < 145 \qquad \text{OK}$$

**Example 3.14:** If the cross section in Fig. 3.23 is reinforced with 2 two-bar bundles, does the reinforcement pattern satisfy the ACI Code requirements for crack control? Exterior exposure, $f_y = 60$ kips/in².

SOLUTION  Treat each bundle as a single bar

$$A = \frac{\text{area of concrete}}{\text{number of bars}} = \frac{3(2)(12)}{2} = 36$$

$$f_s = 0.6f_y = 36 \text{ kips/in}^2 \qquad d_c = 3 \text{ in}$$

$$z = f_s\sqrt[3]{d_c A} = 36\sqrt[3]{3(36)} = 171 > 145 \qquad \text{no good}$$

If the bars are separated, $A = 18$ and

$$z = 136 < 145 \qquad \text{OK}$$

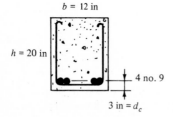

4 no. 9

3 in = $d_c$        Figure 3.23

# 3.6 FAILURE MODES AND FLEXURAL STRENGTH OF REINFORCED CONCRETE BEAMS

## Introduction

When a reinforced concrete beam is loaded to failure, three modes of bending failure are possible. The particular mode of failure is determined by the percentage of steel located in the tension zone. Two of these modes are brittle and one is ductile. Since the designer's prime concern is to produce ductile beams with a high capacity for energy absorption, beams must be proportioned to ensure that only the ductile failure mode is possible.

**Case 1**    The beam is *overreinforced* and the failure mode is a sudden, brittle failure, which the engineer must carefully guard against in design. When the overreinforced beam is loaded to failure, the failure is initiated by the crushing of the concrete followed by the sudden disintegration of the compression zone while the stress in the relatively large area of steel has not reached its yield point. To prevent a brittle failure, the reinforcement must yield while the strain in the concrete is less than the failure strain of 0.003.

**Case 2**    The beam has a moderate percentage of steel, and the failure mode is initiated by a yielding of the steel while the strains in the concrete are relatively low. Such beams can continue to carry load and are able to undergo large deflections before final collapse occurs; this ductile mode of failure is the only acceptable mode.

**Case 3**    The beam is lightly reinforced with a very small percentage of steel, and the failure mode is also brittle. When the tensile stress in the concrete exceeds the modulus of rupture (the tensile strength), the concrete cracks and immediately releases the tensile force it carries; the lightly stressed steel must then absorb this increment of load. If the area of steel provided is too small to carry this added force, the steel will snap and total rupture of the section will occur suddenly.

To ensure ductile failures, upper and lower limits on the permitted area of reinforcing steel are established by the ACI Code. The lower limit ensures that enough steel will be used to prevent the steel from snapping suddenly and causing the beam to split. The upper limit on steel area prevents the design of overreinforced beams.

Since the presence of shear force has little influence on the moment capacity of a cross section, shear is not considered in the design of members for bending.

## Failure of an Overreinforced Concrete Beam

To study the bending failure of a simply supported overreinforced concrete beam, the behavior of the beam in Fig. 3.24 is observed as it is loaded to failure by progressively increasing load increments at midspan. In a beam of constant cross

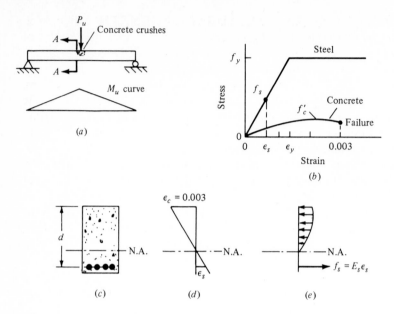

**Figure 3.24** Failure of an overreinforced beam: (*a*) beam with failure load, (*b*) stress-strain curves showing stress in the reinforcing steel and maximum stress in the concrete at failure, (*c*) section *A-A*, (*d*) strains, (*e*) stresses.

section, failure will occur at the section of maximum moment where stresses are greatest. Deformations, strains, and stresses discussed below are all associated with the section of maximum moment. As discussed in Section 3.3, the position of the neutral axis in a reinforced concrete beam is a function of the area of the reinforcing steel. In an overreinforced beam, the large area of steel positions the neutral axis close to the tension steel (Fig. 3.24). As load is applied, the beam bends and cross sections rotate about the neutral axis. The deformations and, correspondingly, the bending strains are zero at the neutral axis and vary linearly with distance from the neutral axis. The maximum compressive strain in the concrete occurs at the top surface of the compression zone. The maximum compressive strain is much greater than the tensile strain in the steel since the upper surface of the concrete is farther from the neutral axis than the steel. When the strain in the concrete at the top surface reaches 0.003, failure occurs. It is characterized by a large depth of the concrete compression zone crushing rapidly while the steel, strained below the yield point, is still elastic, as shown in Fig. 3.24*b*. When steel is not designed to yield at failure, designs are uneconomical because the potential strength of the steel is not used.

To sum up, failure in an overreinforced beam is initiated by crushing of the concrete while deflections are relatively small and before the appearance of extensive cracks in the tension zone. Before failure, deflections and cracking are small because of the low strains in both the concrete and the steel. Since concrete is brittle, failure will be sudden and will occur with limited visual warning. When the concrete crushes and falls out, the beam's capacity to carry load is destroyed. Such local failure can be dangerous if the falling concrete overloads other portions of the structure and precipitates failure of other members.

To guard against the design of brittle overreinforced beams, the ACI Code establishes a limit on the maximum amount of steel that can be inserted in a beam. Limiting the percentage of reinforcing steel moves the neutral axis toward the compression zone. As a result, the strain in the steel will increase more rapidly than the strain in the concrete, and the beam will fail in a ductile manner by yielding of the steel before the concrete crushes. Such beams are termed *underreinforced*. The difference in behavior between an underreinforced and an overreinforced beam is shown by the load-deflection curves in Fig. 3.25. Except for the area of reinforcing steel, both simply supported beams in Fig. 3.25 have the same dimensions. Notice that at low values of load, before the concrete has cracked, both beams deflect approximately the same amount since behavior is governed primarily by the properties of the gross area of the concrete.

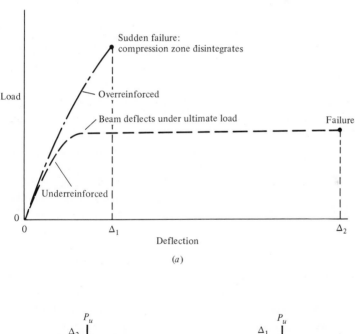

**Figure 3.25** (*a*) Comparison of behavior between an underreinforced (*b*) and overreinforced (*c*) beam with identical dimensions.

## Flexural Strength of an Underreinforced Beam

To understand the flexural behavior of an underreinforced beam, we shall study the variation of the longitudinal stresses on a vertical section at the center of a simply supported beam with a concentrated load at midspan (Fig. 3.26a). This study will cover the changes in stress at the section of maximum moment as the internal bending moment increases from the service-load level to the ultimate flexural strength of the section. The stress distribution associated with failure is of particular importance since this distribution is needed in the derivation of design equations

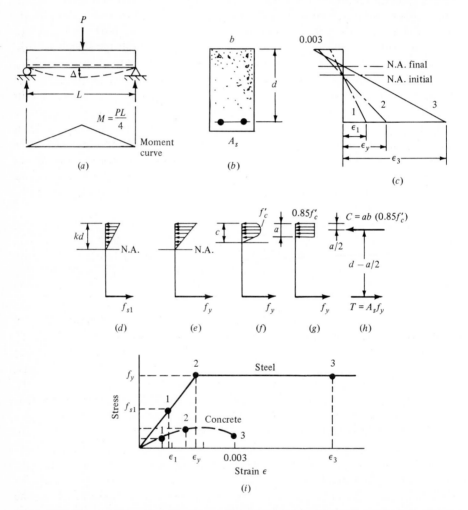

**Figure 3.26** The variation of bending stress with moment in an underreinforced beam loaded progressively to failure: (a) beam and moment curve; (b) cross section; (c) strain; (d) service-load stresses; (e) steel yields, initial failure; (f) secondary compression failure; (g) equivalent stress block, $a = \beta_1 c$; (h) internal couple; (i) stress-strain curves for steel and concrete.

relating the ultimate flexural strength, which the ACI Code calls the *nominal moment capacity*, to the cross-sectional dimensions, the area of reinforcement, and the strength of the materials (as measured by $f'_c$ and $f_y$). Although the beam is assumed to have a rectangular cross section, the most common shape in practice, the procedure for evaluating the nominal moment capacity can easily be extended to cross sections of any shape.

Experimental studies verify that flexural strains in a reinforced concrete beam vary linearly from the neutral axis even when the beam is heavily cracked and the materials strained into the inelastic region. Because of the relatively small area of steel in an underreinforced beam, the neutral axis is positioned closer to the outside compression surface than to the tension steel. As a result, when the beam bends, the tensile strain in the steel is greater than the maximum compressive strain in the concrete at the top surface. If the concentrated load at midspan is initially assumed to be equal to the service load, the stress distribution shown in Fig. 3.26$d$ develops on the cross section at midspan. As indicated in this figure, the concrete stress, proportional to the strain, varies linearly from zero at the neutral axis to its maximum value at the outside compression face. In the tension zone, the concrete is assumed to be cracked, so that all the tension force is carried by the steel, which is stressed to $f_{s_1}$, typically about half the yield-point stress. The magnitude of the stresses induced in the steel and in the top surface of the concrete by the service load is noted (Fig. 3.26$i$) on the stress-strain curves of the materials by points labeled 1. At this level of load, deflections are small and tension cracks in the concrete are typically invisible to the naked eye.

If the concentrated load at midspan is now increased, the beam will bend and deflect farther, due to the increase in moment, which is proportional to the load. Strain and stresses on vertical cross sections will increase. The new distribution of strain is denoted by line 2 in Fig. 3.26$c$, and the corresponding stress state by the distribution in Fig. 3.26$e$. This figure indicates that the total load has the particular value needed to raise the steel stress just to the yield point. Points labeled 2 in Fig. 3.26$i$ show the level of stress and strain in the steel and in the concrete at the top surface. Since the strains vary linearly from the neutral axis, the strain in the steel is greater than the strain at the compression surface of the concrete; therefore when the steel is strained just to its yield point, the greatest concrete strain is still below the value of 0.003, the strain at failure. Since the forces in the steel and in the concrete constitute the components of the internal moment, they must be equal in magnitude under all stress conditions. Yielding of the tension steel does not produce failure of the beam since a small additional amount of moment capacity can still develop: this additional moment capacity is created by the ability of the arm between the steel force $T$ and the resultant compression force $C$ in the concrete to increase in length a small amount as the beam is further deflected and strained by additional load. The increased strain causes the concrete located away from the neutral axis to be stressed into the nonlinear range as shown in Fig. 3.26$f$. The additional strain cannot increase the steel stress, which is in the horizontal portion of the stress-strain curve, where stress remains constant with increasing strain; the tension force, equal to $A_s f_y$, therefore remains constant. For this reason the compression force $C$, the other component of the internal couple, must also remain constant in magnitude. With the stresses in the concrete increasing, the equality of internal forces can be maintained only by a decrease in the area of the compression zone. For the area to reduce in size, the neutral axis must shift upward. The effect of this move is to increase the distance between the neutral axis and the steel. As a result the strain in the steel increases at an accelerating rate. In the same way the reduced distance between the neutral axis and the compression surface causes the concrete strain to increase at a decreasing rate. Associated with the shift of the neutral axis are an increase in the length and width of tension cracks and a substantial increase in the magnitude of beam deflection. Total failure of the beam is assumed to occur when the moment is large enough to produce a strain of 0.003 in the concrete at the top surface of the beam. At this level of strain the concrete breaks up and the beam fails. Figure 3.26*f* shows the stress distribution on the beam cross section just before failure occurs; points in Fig. 3.26*i* marked with a 3 denote the final state of stress.

The variation of concrete stress in the compression zone between the neutral axis and the outside surface of the concrete is identical to that of the stress-strain curve of the concrete in Fig. 3.26*i*. Since a nonlinear stress distribution complicates the derivation of design equations, the state of stress is approximated by a uniform stress of $0.85 f'_c$ acting over the upper portion of the compression zone (Fig. 3.26*g*). The area on which the uniform stress acts is called the *stress block*. Whitney[4] established the magnitude of the uniform stress and the depth of the stress block required to produce a resultant compression force equal in magnitude and located at the same depth as that produced by the actual nonlinear stress distribution in Fig. 3.26*f*. From analyses of test results, the depth of the stress block $a$ was related to $c$, the distance between the outside compression surface and the neutral axis, by

$$a = \beta_1 c \tag{3.21}$$

where

$$\beta_1 = \begin{cases} 0.85 & \text{if } f'_c \leq 4 \text{ kips/in}^2 \ (27.58 \text{ MPa}) \quad (3.22a) \\ 0.85 - 0.05[f'_c \, (\text{kips/in}^2) - 4] & \text{if } f'_c > 4 \text{ kips/in}^2 \ (27.58 \text{ MPa}) \quad (3.22b) \end{cases}$$

but

$$\beta_1 \geq 0.65$$

These expressions apply to compression zones of any shape.

The resultant compression force in the concrete $C$ acts at the centroid of the stress block. For a rectangular compression zone, $C$ is located a distance of $a/2$ from the top surface. With the internal forces established, the nominal bending strength of the section at failure $M_n$ can be evaluated by summing moments about $T$ to give

$$M_n = C\left(d - \frac{a}{2}\right) = 0.85 f'_c ab\left(d - \frac{a}{2}\right) \tag{3.23a}$$

or summing moments about $C$ to give

$$M_n = T\left(d - \frac{a}{2}\right) = A_s f_y\left(d - \frac{a}{2}\right) \tag{3.23b}$$

If the components of the internal couple are equated, that is, $C = T$, the depth of the equivalent compression zone $a$ can be expressed in terms of the dimensions of the cross section and the strength of the materials as

$$a = \frac{A_s f_y}{0.85 f'_c b} \qquad (3.24)$$

Multiplying the top and the bottom of the right-hand side of Eq. (3.24) by $d$ and setting $A_s/bd = \rho$ gives

$$a = \frac{A_s f_y d}{bd(0.85 f'_c)} = \frac{\rho f_y d}{0.85 f'_c} = \frac{qd}{0.85} \qquad (3.25)$$

where $q = \rho f_y / f'_c$. The *tension reinforcement index* $q$ is a measure of the strength of the reinforcement. Substituting Eq. (3.25) into Eq. (3.23a) and simplifying gives

$$M_n = f'_c bd^2 q(1 - 0.59q) \qquad (3.26)$$

Experimental studies verify that the equations derived in this section predict the ultimate flexural strength of underreinforced beams accurately. In Fig. 3.27 the nominal flexural strength represented by Eq. (3.26) and plotted as a solid line is compared with the flexural strength at failure measured in 364 beam tests. To permit test results of beam constructed of different strength concretes and with a variety of steel yield points to be compared with values of moment predicted by Eq. (3.26), nondimensional variables are established by dividing the flexural strength by $f'_c bd^2$. Figure 3.27 indicates that the actual flexural strength, which rarely deviates more than 10 percent from the calculated value, will usually be within 4 or 5 percent of the predicted value.

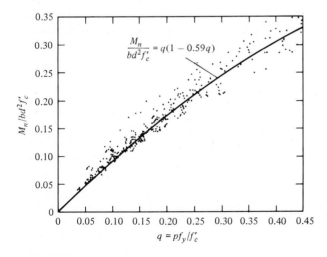

**Figure 3.27** Variation of flexural strength with reinforcement index.[4]

**Example 3.15:** Determine the theoretical value of the concentrated load $P_n$ that will cause the underreinforced beam in Fig. 3.28 to fail in flexure. Base computations on the direct evaluation of the internal couple from the assumed stress condition at failure; $f'_c = 3$ kips/in² (20.68 MPa) $f_y = 60$ kips/in² (413.7 MPa).

SOLUTION The midspan moment due to applied load at midspan is

$$\frac{P_n l}{4} = \frac{P_n(16)}{4} = 4P_n \quad \text{ft} \cdot \text{kips}$$

$$T = A_s f_y = (2 \text{ in}^2)(60 \text{ kips/in}^2) = 120 \text{ kips (533.76 kN)}$$

Equate $T = C$

$$120 \text{ kips} = ab(0.85f'_c)$$

$$a = 4.7 \text{ in (119.4 mm)}$$

Nominal moment capacity summing moments about $C$ equals

$$M_n = T\left(d - \frac{a}{2}\right) = 120\left(18 - \frac{4.7}{2}\right)\frac{1}{12} = 156.5 \text{ ft} \cdot \text{kips (212.2 kN} \cdot \text{m)}$$

Equate moment due to applied loads to $M_n$

$$4P_n = 156.5 \text{ ft} \cdot \text{kips}$$

$$P_n = 39.1 \text{ kips (173.9 kN)}$$

**The plastic hinge** A beam on the verge of failure from the strain of its ultimate load is said to have developed a *plastic hinge* when the beam can be displaced up or down by the application of an infinitesimal force. Such behavior implies that the application of ultimate load has reduced the bending stiffness of a beam to zero with respect to its response to any *additional* load. The stress distribution in Fig. 3.26g constitutes the state of stress associated with the formation of a plastic hinge. A method of analysis for indeterminate structures, called *plastic design*, makes use of the plastic-hinge concept. Although plastic design is not currently permitted for the analysis of reinforced concrete structures, the ACI Code §8.4 in its pro-

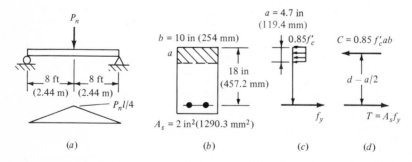

**Figure 3.28** (a) Beam and moment diagram; (b) cross section; (c) stresses at failure; (d) internal couple.

visions on moment redistribution recognizes that indeterminate, ductile concrete structures which are overstressed at one section can use the reserve bending capacity of adjacent sections to carry additional load.

To illustrate the concept of a plastic hinge, a load-deflection curve of a simply supported, underreinforced beam with a concentrated load at its center is shown in Fig. 3.29a. Deflection at midspan is plotted against applied load $P$ or against the internal moment $M$ at midspan in Fig. 3.29b. Of course, $M$, a direct function of $P$, is equal to $PL/4$. As the load increases in size from zero to its ultimate value $P_u$, the load-deflection curve initially rises at a steep slope. When the load is large enough to create bending moments greater than the cracking moment of the cross section, the slope of the load-deflection curve reduces slightly, due to the reduced stiffness of the cracked beam. Point 1 in Fig. 3.29b is the point at which the slope changes. Above point 1 the load-deflection curve continues to rise almost linearly to point $B$, where the load $P_y$ has created sufficient deformation of the beam to strain the steel to its yield point. Between points $A$ and $B$ an increment of deflection, such as $\Delta_2$, requires a substantial increment of load $\Delta P_2$. In this region of the curve the beam can be considered to have a stiffness $K_2 = \Delta P_2/\Delta_2$. As shown in Fig. 3.29b, $K_2$ is equal to the slope of the load-deflection curve.

At point $B$ on the curve, the load $P_y$ has just raised the stress in the steel to the yield point. Additional load above $P_y$ causes the neutral axis to shift toward the compression surface of the beam. Once the steel yields, little additional moment capacity is available and the slope of the load-deflection curve (still a measure of bending stiffness) reduces rapidly and approaches zero. When the beam is loaded into the region between points $C$ and $D$ where the stiffness reduces to zero, a plastic hinge is said to have formed. In the horizontal region between $C$ and $D$ in Fig. 3.29 an increment of deflection $\Delta_3$ is produced by an infinitesimally small change in

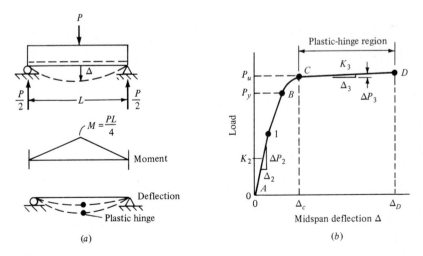

(a)                           (b)

**Figure 3.29** Behavior of an underreinforced beam: (a) moment curve and deflected shape, (b) load-deflection curve.

load $\Delta P_3$ which is essentially zero in magnitude. If we again define stiffness of the member by the slope of the load-deflection curve, we can see that the stiffness is approximately zero. In other words, with the failure load in place, the beam loses all resistance to deflection by *additional* load. Such a beam, which may be supporting many tons of load, can be moved up or down by the lightest pressure. The term *plastic hinge* is applied to this physical state since it is similar to the behavior of a hinged joint, which also is characterized by zero resistance to rotation.

### Minimum Reinforcement to Prevent a Brittle Failure

For most beams the cracking moment, whose magnitude is normally much smaller than that of the maximum moment produced by service loads, is a small fraction (in the 10 to 15 percent range) of the nominal flexural strength. As long as the maximum tensile stress in the concrete is less than the tensile strength of the concrete, given by the modulus of rupture, the beam remains uncracked and the stress in the tension steel is low, of the order of 2 or 3 kips/in² (13.8 to 20.68 MPa). Although the stress in the reinforcement will increase after a crack forms because the steel must also carry the tension formerly carried by the concrete in addition to its initial stress, the steel stress is typically well below the yield-point stress; therefore, the initial cracking of the concrete and the subsequent redistribution of tensile stress to the steel normally occur without overstressing the reinforcement.

On occasion, architectural or functional considerations may require beam dimensions to be set much larger than those required for flexural strength or for deflection control. Because of the large arm between the components of the internal couple, a beam of this type may require a very small area of reinforcement. As a result its nominal flexural strength may be less than the cracking moment of the cross section. If the cracking moment in a beam of this type is ever exceeded, e.g., by accidental overload, the beam will fail immediately by rupture of the steel. To prevent this type of brittle failure, ACI Code §10.5.1 requires that all beams be reinforced with an area of steel at least equal to $200b_w d/f_y$ (USCU) or $1.4b_w d/f_y$ (SI) to ensure that the nominal flexural strength will exceed the cracking moment by a safe margin.

Although the ACI equation for minimum steel is applied to cross sections of all shapes, the form of the variables and the approximate magnitude of the factor of safety can be established by equating the cracking moment to the nominal flexural strength for a rectangular cross section (see Fig. 3.30b). Using the internal-couple method, we can express the cracking moment as

$$M_{cr} = T_c(\tfrac{2}{3}h) \tag{3.27}$$

where $T_c$ is the tension force carried by the concrete.

Expressing $T_c$ in terms of the maximum tensile stress $f_r$ in the concrete gives

$$M_{cr} = \frac{f_r}{2} \frac{hb_w}{2} \left(\frac{2}{3}h\right) \tag{3.28}$$

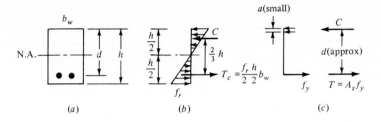

**Figure 3.30** State of stress in a rectangular beam: (a) cross section, (b) cracking impends, (c) at failure.

If the area of the steel is small relative to the gross area of the section, the depth of the stress block at failure will also be small; therefore, to evaluate the nominal flexural capacity of the cracked cross section, the arm of the internal couple can be taken equal to $d$ approximately (see Fig. 3.30c), and

$$M_n = A_s f_y d \qquad (3.29)$$

Equating the cracking moment given by Eq. (3.28) to the nominal flexural strength given by Eq. (3.29), expressing $f_r$ as $7.5\sqrt{f'_c}$, and noting that $h \approx d$ gives

$$\frac{7.5\sqrt{f'_c}\,db_w}{4}\frac{2}{3}d = A_s f_y d$$

Solving for $A_s$ produces

$$A_s = \frac{1.25\sqrt{f'_c}\,b_w d}{f_y} \qquad (3.30)$$

To establish a minimum area of steel for a concrete of average strength, assume that $f'_c = 4000 \text{ lb/in}^2$ in Eq. (3.30); then

$$A_s = \frac{79 b_w d}{f_y} \qquad (3.31)$$

If the right-hand side of Eq. (3.31) is multiplied by a factor of safety equal to approximately 2.5, $A_s$ (now called $A_{s,\text{min}}$) equals

$$A_{s,\text{min}} = \frac{200 b_w d}{f_y \,(\text{lb/in}^2)} \qquad (3.32)$$

Dividing both sides of Eq. (3.32) by $b_w d$ gives the ACI equation 10.3 for the minimum reinforcement ratio as

$$\rho_{\text{min}} = \begin{cases} \dfrac{200}{f_y} & \text{USCU} \\[2ex] \dfrac{1.4}{f_y} & \text{SI} \end{cases} \qquad (3.33)$$

Since the square root of $f'_c$ rather than $f'_c$ is used in the computation for minimum steel, the factor of safety is not as sensitive to concrete strength $f'_c$ as it would be if it were a direct function of $f'_c$. Had a smaller value of $f'_c$ been used instead of 4 kips/in$^2$ (27.58 MPa), a factor of safety slightly larger than 2.5 would be required to produce the coefficient 200 in the numerator of Eq. (3.33). Keeping the coefficient in Eq. (3.33) constant means that the factor of safety will in effect increase as the quality of the concrete decreases.

The provision for minimum steel does not have to be followed if the area of steel provided at every section, both positive and negative, is at least one-third greater than that required by analysis. It is felt this extra area of steel provides sufficient extra capacity to ensure a safety design. For beams with large cross sections even the minimum steel requirement would produce an extremely large amount of steel.

## 3.7 BALANCED FAILURE OF A BEAM WITH A RECTANGULAR CROSS SECTION

### Introduction

To ensure ductile behavior and controlled failures, the ACI Code permits only the design of underreinforced beams. The maximum area of steel permitted in a cross section can be developed from a consideration of the level of strain in the concrete at the instant the steel initially yields.

Ductility depends primarily on the magnitude of the maximum strain in the concrete as the tension steel is stressed just to its yield point. If the strain in the concrete is small when the steel begins to yield, the beam can undergo considerable additional bending deformation before the concrete is strained to failure. As shown in Fig. 3.26c, after the tension steel yields, additional moment applied to an underreinforced beam causes the neutral axis to shift toward the compression surface. This movement of the neutral axis toward the compression surfaces causes the concrete strains to increase less rapidly while the strains in the steel increase more rapidly. Under these conditions, the beam behaves in a ductile manner and is able to undergo considerable deflection before failure of the concrete in the compression zone.

As the percentage of steel used for reinforcement increases, the neutral axis moves toward the centroid of the tension steel and away from the surface of the compression zone. Since strains vary directly with distance from the neutral axis, the greater the distance between the neutral axis and the compression surface, the greater the strains in the concrete as the beam bends and the cross section rotates about the neutral axis (Fig. 3.31).

Ductility disappears when the percentage of steel, called *balanced steel* $A_{sb}$, is large enough to position the neutral axis so that the strain in the concrete reaches 0.003 just as the steel is strained to its yield point (Fig. 3.31c). With the concrete reaching its ultimate strain and crushing just as the steel yields, the moment capacity of the section is destroyed, and a brittle failure occurs.

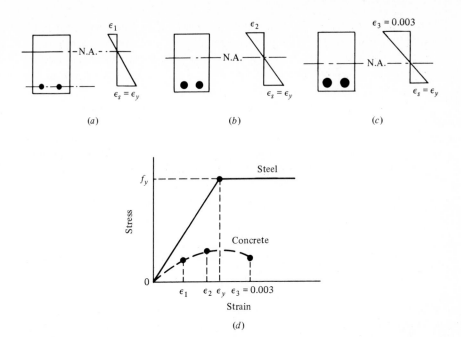

Figure 3.31 Influence of steel area on the magnitude of the strain in the top concrete fiber at the initial yielding of steel: (a) small $A_s$; (b) moderate $A_s$; (c) balanced $A_s$; (d) stress-strain curves.

In the next section the area of balanced steel will be established for a rectangular cross section. Computation of balanced steel for a rectangular cross section is important, since most beams have this shape.

## Balanced Steel for a Rectangular Cross Section

With balanced steel as reinforcement, the neutral axis is positioned so that the strain distribution in Fig. 3.32b results; the strain at the surface of the compression zone is 0.003, and the steel strain is $\epsilon_y$. The position of the neutral axis can be established from the geometry of the strain diagram.

By similar triangles in Fig. 3.32b

$$\frac{0.003}{c_b} = \frac{\epsilon_y}{d - c_b} \tag{1}$$

Solve for $c_b$

$$c_b = \frac{0.003}{0.003 + \epsilon_y} d \tag{2}$$

Substitute $\epsilon_y = f_y/E_s$

$$c_b = \frac{0.003}{0.003 + f_y/E_s} d \tag{3}$$

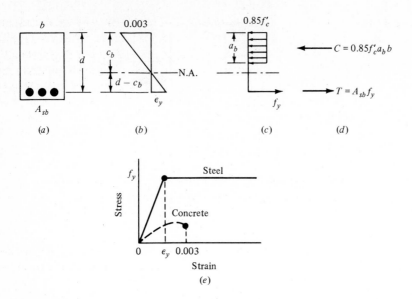

**Figure 3.32** Balanced failure of a rectangular beam: (*a*) cross section, (*b*) strains, (*c*) stresses, (*d*) internal couple, (*e*) stress-strain curves; dots indicate the level of stress at failure in steel and in the top concrete fiber.

Multiply the top and bottom of Eq. (3) by $E_s$

$$c_b = \frac{0.003 E_s}{0.003 E_s + f_y} d \tag{4}$$

Letting $E_s = 29{,}000{,}000 \ \text{lb/in}^2$ in Eq. (4) gives

$$c_b = \frac{87{,}000}{87{,}000 + f_y \ (\text{lb/in}^2)} d \tag{3.34}$$

Equation (3.34), which applies only to the case of a balanced failure, must not be used to establish the position of the neutral axis when less than balanced steel is used.

The stress distribution shown in Fig. 3.32*c* follows from the strain in Fig. 3.32*b*. Multiplying the areas of steel and concrete by the stresses acting on them gives the equal and opposite components of internal moment, $C$ and $T$ (Fig. 3.32*d*). By equating $C$ and $T$, $A_{sb}$ can be established

$$T = C$$

$$A_{sb} f_y = 0.85 f'_c a_b b$$

$$A_{sb} = \frac{0.85 f'_c b a_b}{f_y} \tag{3.35}$$

To eliminate $a_b$, the depth of the stress block from Eq. (3.35), use $a_b = \beta_1 c_b$ and the value of $c_b$ from Eq. (3.34) to give

$$A_{sb} = \frac{0.85\beta_1 f'_c bd}{f_y} \frac{87,000}{87,000 + f_y} \tag{3.36}$$

Equation (3.36) can be put into a more useful form by dividing each side by $bd$

$$\rho_b = \frac{A_{sb}}{bd} = \begin{cases} \dfrac{0.85\beta_1 f'_c}{f_y} \dfrac{87,000}{87,000 + f_y} & \text{USCU} \\[2ex] 0.85\beta_1 \dfrac{f'_c}{f_y} \dfrac{600}{600 + f_y} & \text{SI} \end{cases} \tag{3.37}$$

The term $\rho_b$, the balanced reinforcement ratio, represents the area of steel per unit area of the concrete cross section. By using $\rho_b$ instead of $A_{sb}$ the dimensions of the cross section are eliminated.

To ensure that concrete beams fail in a ductile manner, ACI Code §10.3.3 requires that beams be reinforced with an area of steel $A_s$ which is not to exceed $\frac{3}{4}A_{sb}$

$$A_s \le \tfrac{3}{4}A_{sb} \tag{3.38}$$

or, dividing both sides by $bd$.

$$\frac{A_s}{bd} \le \frac{\frac{3}{4}A_{sb}}{bd}$$

$$\rho \le \tfrac{3}{4}\rho_b \tag{3.39}$$

where $\rho_b$ is given by Eq. (3.37).

## 3.8 STRENGTH DESIGN OF RECTANGULAR BEAMS FOR MOMENT

From the basic principles and equations established in the preceding sections we now develop a procedure for designing a rectangular beam. Since most reinforced concrete beams used in construction are rectangular, this procedure will be used repeatedly by the designer. All steps are consistent with the requirements of the current ACI Code §318-77.

All beams are designed to ensure that the moment produced by factored loads does not exceed the available flexural design strength of the cross section at any point along the length of the beam. If the flexural design strength $\phi M_n$ just equals the required flexural strength $M_u$, the criterion for design can be stated as

$$M_u = \phi M_n \tag{3.40}$$

where $\phi = 0.9$ and $M_n$ is the nominal moment capacity of the cross section. This criterion can be developed into a design equation if we express $M_n$ in terms of the

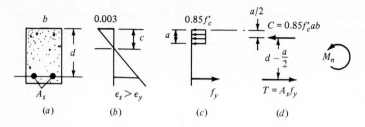

**Figure 3.33** (a) State of stress in an underreinforced beam at failure: (b) strains, (c) stresses, (d) internal couple.

material and the geometric properties of a rectangular cross section (Fig. 3.33). If we then sum moments about the centroid of the tension steel, $M_n$ can be expressed as

$$M_n = C\left(d - \frac{a}{2}\right) \tag{3.41}$$

where $C$ is the resultant of the compressive stresses and $a$ is the depth of the rectangular stress block. As indicated in Fig. 3.33d, $C = 0.85f'_c ab$. Substituting this value of $C$ into Eq. (3.41) gives

$$M_n = 0.85f'_c ab\left(d - \frac{a}{2}\right) \tag{3.42}$$

To express $a$ in Eq. (3.42) in terms of the dimensions of the cross section and the properties of the material, $f'_c$ and $f_y$, we set $T = C$ and solve for $a$, to give

$$a = \frac{A_s f_y}{b(0.85f'_c)} \tag{3.43}$$

Multiplying both top and bottom of Eq. (3.43) by $d$ and setting $A_s/bd = \rho$ leads to

$$a = \frac{A_s f_y d}{bd(0.85f'_c)} = \frac{\rho f_y d}{0.85f'_c} \tag{3.44}$$

Substituting Eq. (3.44) into (3.42) and simplifying gives

$$M_n = \rho f_y bd^2\left(1 - \frac{\rho f_y}{1.7f'_c}\right) \tag{3.45}$$

Finally Equation (3.45) is substituted into Equation (3.40) to give the basic beam design equation

$$M_u = \phi\rho f_y bd^2\left(1 - \frac{\rho f_y}{1.7f'_c}\right) \tag{3.46}$$

where $\rho$ must not be greater than $\frac{3}{4}\rho_b$ or less than $\rho_{min} = 200/f_y$ for U.S. customary units or $(1.4/f_y)$ for SI units. The first requirement ensures that the beam will be underreinforced and will fail in a ductile manner; the second requirement prevents a brittle failure, i.e., the snapping of the steel when the beam cracks initially.

Equation (3.46) can be used either to investigate the capacity of a cross section if the dimensions and material properties are known or to design a cross section (i.e., to establish the width $b$, the depth $d$, and the area of steel $A_s$) if the value of the factored moment $M_u$ is specified. Although Eq. (3.46) can be used to establish the flexural design strength of a cross section since all terms on the right side of the equation are known, the designer may prefer to work directly with the internal forces on the rectangular cross section to evaluate $M_n$ because of the simplicity of the calculations. In the latter procedure, $T = A_s f_y$ is first evaluated, then the depth of the stress block is computed by equating $T = C$, and finally the internal couple is evaluated by multiplying $T$ by the arm $d - a/2$ between $T$ and $C$. Both procedures are illustrated in Example 3.16.

**Example 3.16: Determining the flexural strength of a reinforced cross section** Determine the maximum value of factored moment $M_u$ the cross section in Fig. 3.34 can support; $f'_c = 20.68$ MPa, and $f_y = 344.7$ MPa.

SOLUTION Compute

$$\rho = \frac{A_s}{bd} = \frac{2280}{300(400)} = 0.019$$

From Eq. (3.46)

$$M_u = \phi \rho f_y (bd^2) \left(1 - \frac{\rho f_y}{1.7 f'_c}\right)$$

$$= 0.9(0.019)(334.7 \times 10^6 \text{ N/m}^2)(0.3 \text{ m})(0.4 \text{ m})^2 \left[1 - \frac{0.019(344.7)}{1.7(20.68)}\right]$$

$$= 0.230 \times 10^6 \text{ N} \cdot \text{m} = 230 \text{ kN} \cdot \text{m}$$

Compute $M_u$ based on the statics of the internal forces (see Fig. 3.34):

$$T = A_s f_y = (2280 \times 10^{-6} \text{ m}^2)(344.7 \times 10^6 \text{ N/m}^2) = 785.9 \text{ kN}$$

Since $T = C$,

$$785.9 \times 10^3 \text{ N} = ab(0.85 f'_c) = a(0.3 \text{ m})[0.85(20.68 \times 10^6 \text{ N/m}^2)]$$

$$a = 0.149 \text{ m}$$

$$M_u = \phi T\left(d - \frac{a}{2}\right) = 0.9(785.9 \text{ kN})\left(0.4 - \frac{0.149}{2}\right) = 230 \text{ kN} \cdot \text{m}$$

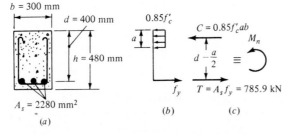

Figure 3.34 (*a*) Cross section; (*b*) stresses at failure; (*c*) internal couple.

The procedure for establishing the proportions of a cross section for a particular value of $M_u$ requires the designer to make a number of sizing decisions in certain variables of Eq. (3.46) at various stages of the design. (Additional discussion of the factors that influence the choice of materials and the dimensions follows this brief discussion of the design procedure.) First, the designer selects $f_c'$ and $f_y$. On the right side of Eq. (3.46) the properties $b$, $d$, and $\rho$ of the cross section are unknown. With one equation involving three unknowns, the designer must specify two of the variables to achieve a solution. First, the designer selects a value of $\rho$—any value of $\rho$ between $\rho_{min}$ and $\frac{3}{4}\rho_b$ is satisfactory. The larger the value of $\rho$ selected, the smaller the cross section will be. After $\rho$ has been selected, Eq. (3.46) is solved for $bd^2$. To produce specific values of $b$ and $d$ the designer then specifies $b$ and solves for $d$. Typically, an even value of $b$ is selected, such as 10 or 12 in (254 or 305 mm). The width must provide enough clearance for the insertion of the specified reinforcing bars. For many floor systems of moderate span—say 15 to 25 ft (4.6 to 7.6 m)—an economical beam results when the depth equals $1\frac{1}{2}$ to 2 times the width. For long spans, deep narrow beams are often most efficient. If no limit exists on the depth, the depth may be made 3 or 4 times the width of the beam.

After $d$ has been selected, the overall depth of the cross section is established by adding about $2\frac{1}{2}$ in (64 mm) to the effective depth $d$. This additional depth provides approximately 2 in (51 mm) of cover to protect the steel from fire and corrosion. The final dimension for the depth should be rounded off to the nearest whole inch (25 mm) to simplify fabrication of the forms. If the resulting values of width and depth seem reasonable, the design is completed by computing $A_s = \rho bd$, in which $\rho$ represents the value of the reinforcement ratio previously selected by the designer. If the proportions of the cross section are not satisfactory (the beam is too shallow to control deflections, or the width of the web is too narrow to fit in all the steel), the designer must repeat the computations after making appropriate modifications.

After the first cycle of the design, the designer will know the approximate dimensions of the final cross section. In the second cycle, the designer may either specify the values of $b$ and $d$ and solve for $\rho$ or select a new value of $\rho$. In either case, the designer should be able to arrive at acceptable proportions by the second cycle of design. The basic procedure for proportioning a cross section for a given moment is illustrated in Example 3.17. Example 3.18 shows the complete design procedure of a beam with the required checks for serviceability. A checklist for the steps required to design a rectangular beam is given in the summary following Example 3.17. The designer should also keep in mind that bond stresses must be checked to verify that a particular size of bar is properly anchored. Bond is covered fully in Chap. 6 and will be neglected at this stage, but it must be checked before the design is complete.

**Example 3.17: Proportioning a rectangular cross section for moment** Use Eq. (3.46) to establish the width, depth, and area of steel for a rectangular cross section that must carry a moment $M_u = 150$ ft·kips (203.4 kN·m); $f_c' = 3$ kips/in² (20.68 MPa), $f_y = 60$ kips/in² (413.7 MPa), $\frac{3}{4}\rho_b = 0.016$, and $\rho_{min} = 200/f_y = 0.0033$.

SOLUTION  The designer is free to select any value of $\rho$ between $\rho_{min}$ and $\frac{3}{4}\rho_b$; try $\rho = 0.012$. From Eq. (3.46)

$$150(12) \text{ in} \cdot \text{kips} = 0.9(0.012)(60bd^2)\left[1 - \frac{0.012(60)}{1.7(3)}\right]$$

$$bd^2 = 3234.4 \text{ in}^3$$

Try $b = 10$ in; then $d^2 = 3234.4/10$, and $d = 18.0$ in

$$A_s = \rho bd = 0.012(10)(18.0) = 2.16 \text{ in}^2$$

The total depth is

$$h = d + 2\tfrac{1}{2} \text{ in} = 18.0 + 2\tfrac{1}{2} = 20.5 \text{ in} \qquad \text{use 21 in (533.4 mm)}$$

As another possible design, try $b = 12$ in; then $d^2 = 3234.4/12$ and $d = 16.4$ in.

$$A_s = \rho bd = 0.012(12)(16.4) = 2.36 \text{ in}^2(1522.7 \text{ mm}^2)$$

The total depth is

$$h = d + 2\tfrac{1}{2} \text{ in} = 16.4 + 2\tfrac{1}{2} = 18.9 \text{ in} \qquad \text{use 19 in (482.6 mm)}$$

If, after reviewing the dimensions of the previous cross sections, the designer decides to use a beam with $b = 12$ in (305 mm) and $h = 18$ in (457 mm), what area of steel is required? Estimate $d$ by deducting $2\tfrac{1}{2}$ in from $h$; $d = 18 - 2\tfrac{1}{2} = 15.5$ in; then use Eq. (3.46) to solve for $\rho$

$$150(12) = 0.9\rho(60)(12)(15.5^2)\left[1 - \frac{\rho(60)}{1.7(3)}\right]$$

$$\rho = 0.0138$$

$$A_s = \rho bd = 0.0138(12)(15.5) = 2.57 \text{ in}^2 \ (1658.2 \text{ mm}^2)$$

## Summary of the Design Procedure for Proportioning Beams

**Step 1**  The designer selects $f'_c$ and $f_y$. Concrete with $f'_c = 3000$ lb/in$^2$ (20.68 MPa) is commonly used for the design of beams protected from weather and spanning moderate distances. Concrete of this quality not only has sufficient strength to carry moment and shear but also contains enough cement to be workable, i.e., the concrete is easily placed and easily compacted. If spans are more than moderate length or loads unusually heavy, a higher-strength concrete may be justified. Higher-strength materials produce smaller, lighter members and will reduce the dead load the structure must support. If concrete is to be exposed to weather, seawater, or other aggressive environments, durability considerations may require the use of a high-strength air-entrained concrete with an $f'_c$ of 3.5 to 4 kips/in$^2$ (24.13 to 27.58 MPa).

Steel with a yield point of 60 kips/in$^2$ (413.7 MPa) is normally used in design. When a beam must be designed for large bending moments, higher-yield-point steels may be desirable to reduce the area of reinforcement required. Smaller areas of a high-strength steel often simplify reinforcing details and eliminate the congestion that results when a larger number of lower-strength bars must be used to supply the required tension force. Fewer bars increase clearances between reinforcement and facilitate the flow of concrete into all sections of the forms. If cracking is to be minimized, stresses must be kept low in the steel. For such designs, low-yield-point steel should be used because there is no advantage in using a high-yield-point steel that cannot be stressed to its full capacity.

**Step 2** Establish service loads and multiply by load factors to produce ultimate design loads.

**Step 3** Check minimum depth of beam required to prevent excessive deflections (see Table 3.1 or 3.2).

**Step 4** Analyze the structure assuming elastic behavior. Although concrete is an inelastic material, ACI Code §8.3.1 permits analysis by elastic theory because experience indicates that satisfactory designs are produced.

**Step 5** Establish width $b$ and depth $h$. Once the moment curve has been established, Eq. (3.46) can be used to select $b$ and $d$. The overall depth is then established by adding an allowance of several inches of concrete cover to protect the reinforcement.

$$M_u = \phi \rho f_y bd^2 \left(1 - \frac{\rho f_y}{1.7 f'_c}\right) \qquad (3.46)$$

Substituting $f'_c$, $f_y$, and the maximum value of $M_u$ into the above equation gives an expression with three unknowns, $\rho$, $b$, and $d$. To arrive at values for $b$ and $d$, the designer selects a value of $\rho \le \frac{3}{4}\rho_b$, and solves for $bd^2$. Combinations of $b$ and $d$ that satisfy the resulting expression are generated.

Proportions are then selected which are satisfactory from an architectural point of view and which limit deflections. For short spans the ratio of $d/b$ will usually run between 1.5 and 2. For longer spans the $d/b$ ratio will increase substantially. The overall depth and width should be expressed to the nearest inch. Larger beams are usually dimensioned to the nearest even inch.

**Step 6** Select $A_s$. The required area of reinforcement $A_s$ is computed by

$$A_s = \rho bd$$

but $\rho$ must not be less than

$$\rho = \begin{cases} \dfrac{200}{f_y} & \text{USCU} \\[2ex] \dfrac{1.4}{f_y} & \text{SI} \end{cases}$$

Reinforcing bars are then selected to supply the required area of steel. If they can be placed in the beam with proper clearances and cover, the design is complete.

When the area of steel is large, several rows of steel may be required or the bars may be bundled. If these measures still do not permit the reinforcement to be placed, a smaller value of $\rho$ should be selected and Eq. (3.46) solved again to arrive at a cross section with larger dimensions. Table B.1 in the appendix may be used to facilitate the selection of the reinforcing bars.

**Step 7**  Check to ensure that the width of cracks is not excessive.

$$z = f_s \sqrt[3]{d_c A}$$

where $z$ must not exceed 145 kips/in (25.4 MN/m) for exterior exposure or 175 kips/in (30.6 MN/m) for interior exposure.

**Step 8**  Required clearances for beam reinforcement. ACI Code §7.7 establishes minimum values of concrete cover required to protect reinforcing bars from corrosion by weather or loss of strength from exposure to fire. Some minimum values of cover for cast-in-place beams are given below:

|  | Minimum cover | |
| --- | --- | --- |
| Bar number | in | mm |
| 6–18 | 2 | 51 |
| 5 and under | $1\frac{1}{2}$ | 38 |
| All: | | |
| Not exposed to weather | $1\frac{1}{2}$ | 38 |
| Cast against, and permanently exposed to, earth | 3 | 76 |

If bars are bundled, the minimum cover required is to be taken equal to the equivalent diameter of the bundle but need not be more than 2 in (51 mm) or the minimum values given above, whichever is greater. If beams are exposed to very corrosive environments, the concrete may require protective coatings of plastic, asphalt, or other suitable compounds.

ACI Code §7.6.1 requires that the clear distance between bars be not less than the nominal diameter of the bars or 1 in (25.4 mm). When bars are placed in several rows (layers), the upper bars should be positioned directly over the lower bars. A clear distance of at least 1 in (25.4 mm) should be provided between rows of steel.

**Example 3.18:** A simply supported beam with a 24-ft (7.32-m) span carries a uniform factored load $w_u = 2.25$ kips/ft. (32.84 kN/m). Consider flexure only and design a rectangular cross section that satisfies the provisions of the ACI Code. Exterior exposure, $f'_c = 3$ kips/in$^2$ (20.68 MPa), $f_y = 50$ kips/in$^2$ (344.7 MPa), and $\frac{3}{4}\rho_b = 0.0204$.

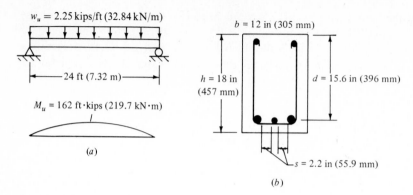

**Figure 3.35**

SOLUTION  Design for the maximum moment at midspan (see Fig. 3.35a)

$$M_u = \frac{w_u l^2}{8} = \frac{2.25(24^2)}{8} = 162 \text{ ft} \cdot \text{kips (219.67 kN} \cdot \text{m)}$$

The minimum depth to limit deflections is

$$h_{\min} = \frac{l}{16}\left(0.4 + \frac{f_y}{100,000}\right) = \frac{24(12)}{16}(0.4 + 0.5) = 16.2 \text{ in (411.5 mm)}$$

Use Eq. (3.46) to establish $b$ and $d$. Try $\rho = 0.018$:

$$M_u = \phi\rho f_y bd^2\left(1 - \frac{\rho f_y}{1.7f'_c}\right)$$

$$162(12)\text{in} \cdot \text{kips} = 0.9(0.018)(50bd^2)\left[1 - \frac{0.018(50)}{1.7(3)}\right]$$

$$bd^2 = 2914.3 \text{ in}^3$$

Try several combinations of $b$ and $d$. If $b = 12$ in, $d = 15.6$ in; add 2.4 in for cover and $h = 18$ in (457 mm). If $b = 10$ in, $d = 17.1$ in; add 2.4 for cover and $h = 19.5$ in, say 20 in (508 mm). The depth of both cross sections exceeds $h_{\min}$. Although the deeper beam is a little more economical, use $b = 12$ in and $h = 18$ in to maximize headroom.

$$A_s = \rho bd = 0.018(12)(15.6) = 3.37 \text{ in}^2 \text{ (2174 mm}^2)$$

Use two no. 10 and one no. 9; $A_s$ supplied $= 3.53$ in². Although a depth of 18 in will result in a slightly smaller $d$ if 2 in of cover is maintained, the moment capacity will still be adequate since the $A_s$ supplied is greater than the $A_s$ required. Normally the designer can neglect small differences of this magnitude.

Check $z$ for control of crack width

$$z = f_s\sqrt[3]{d_c A} = 30\sqrt[3]{2.4(20.7)} = 110 < 145 \qquad \text{OK}$$

where

$$f_s = 0.6f_y = 30 \text{ kips/in}^2$$

$$d_c = 2.4 \text{ in}$$

$$A = \frac{\text{area of tension concrete}}{\text{number of bars}} = \frac{2.4(2)(12)}{3.53/1.27} = 20.7$$

Check that reinforcing bars fit into a 12-in width of beam stem. Spacing between bars must not be less than 1 in or one bar diameter. Allowing 2 in of side cover, the available spacing between bars equals

$$\frac{12 - [2 \text{ in side cover})(2) + 2(\tfrac{10}{8}) + \tfrac{9}{8}]}{2 \text{ spaces}} = 2.2 \text{ in} > \tfrac{10}{8} = d_b \qquad \text{OK}$$

The final design is detailed in Fig. 3.35b. In addition to the flexural steel, U-shaped reinforcement, called *stirrups*, is provided for shear. To ensure that the reinforcement will remain in position when the concrete is placed, a rigid cage is formed by wiring the stirrups to the main flexural steel at the bottom and to small-diameter bars, used to anchor the stirrups, at the top. The top steel is not designed but arbitrarily set by the designer.

## Design Aids

To reduce the time required to generate values of $b$ and $d$ when Eq. (3.46) is used to proportion a cross section for a specific value of moment, the ACI has prepared several design aids in the form of tables and graphs that give properties of the cross section as a function of $\rho$, the reinforcement ratio. Several of these design aids for proportioning beams will be discussed briefly. Because design aids do not contribute to a basic understanding of concrete behavior, they are not used extensively in this text, but the student should be aware of their existence since they are used extensively in actual practice.

Insight into the terms in Table 3.3, which permits the designer to solve Eq. (3.46) rapidly for $b$ and $d$, can be established by regrouping the variables in Eq. (3.46) to give

$$M_u = bd^2 \left[ \phi \rho f_y \left( 1 - \frac{\rho f_y}{1.7f'_c} \right) \right]$$

Then setting $K_u = [\phi \rho f_y (1 - \rho f_y/1.7f'_c)]$, we can express Eq. (3.46) as

$$M_u = bd^2 K_u \tag{3.47}$$

In the factor $K_u$, terms $\rho$, $f_y$, and $f'_c$ are specified by the designer and $\phi$, which is equal to 0.9, is given by the ACI Code. In Table 3.3 $K_u$ is tabulated as a function of $\rho$ for values of $\rho$ between $\rho_{min}$ and $\tfrac{3}{4}\rho_b$. After the designer has specified a value of $\rho$, $K_u$, given in the first column of Table 3.3, can be determined by locating $\rho$ in the column under the specified value of $f_y$ and then moving horizontally to the left to read $K_u$ in the first column. Table 3.3 is limited to the design of members constructed with $f'_c = 3$ kips/in$^2$ (20.68 MPa). Other tables are available for other

## Table 3.3 Coefficients for rectangular sections without compression reinforcement,[6] $f_c' = 3000 \text{ lb/in}^2$†

$$K_u = \phi \rho f_y \left(1 - \frac{\rho f_y}{1.7 f_c'}\right) \qquad \rho = \frac{A_s}{bd} \qquad A_s = \frac{M_u}{a_u d}$$

$$a_u = \frac{\phi f_y (1 - 0.59 \rho f_y / f_c')}{12,000} \qquad F = \frac{M_u}{K_u}$$

| | $f_y = 40,000$ | | $f_y = 50,000$ | | $f_y = 60,000$ | |
|---|---|---|---|---|---|---|
| $K_u$ | $\rho$ | $a_u$ | $\rho$ | $a_u$ | $\rho$ | $a_u$ |
| 53 | 0.0015 | 2.96 | 0.0012 | 3.71 | 0.0010 | 4.45 |
| 80 | 0.0023 | 2.95 | 0.0018 | 3.68 | 0.0015 | 4.42 |
| 105 | 0.0030 | 2.93 | 0.0024 | 3.66 | 0.0020 | 4.39 |
| 131 | 0.0038 | 2.91 | 0.0030 | 3.64 | 0.0025 | 4.37 |
| 156 | 0.0045 | 2.89 | 0.0036 | 3.62 | 0.0030 | 4.34 |
| 181 | 0.0053 | 2.88 | 0.0042 | 3.60 | 0.0035 | 4.31 |
| 206 | 0.0060 | 2.86 | 0.0048 | 3.57 | 0.0040 | 4.29 |
| 230 | 0.0068 | 2.84 | 0.0054 | 3.55 | 0.0045 | 4.26 |
| 254 | 0.0075 | 2.82 | 0.0060 | 3.53 | 0.0050 | 4.23 |
| 278 | 0.0083 | 2.81 | 0.0066 | 3.51 | 0.0055 | 4.21 |
| 301 | 0.0090 | 2.79 | 0.0072 | 3.48 | 0.0060 | 4.18 |
| 324 | 0.0098 | 2.77 | 0.0078 | 3.46 | 0.0065 | 4.15 |
| 347 | 0.0105 | 2.75 | 0.0084 | 3.44 | 0.0070 | 4.13 |
| 369 | 0.0113 | 2.73 | 0.0090 | 3.42 | 0.0075 | 4.10 |
| 391 | 0.0120 | 2.72 | 0.0096 | 3.40 | 0.0080 | 4.08 |
| 413 | 0.0128 | 2.70 | 0.0102 | 3.37 | 0.0085 | 4.05 |
| 434 | 0.0135 | 2.68 | 0.0108 | 3.35 | 0.0090 | 4.02 |
| 455 | 0.0143 | 2.66 | 0.0114 | 3.33 | 0.0095 | 4.00 |
| 476 | 0.0150 | 2.65 | 0.0120 | 3.31 | 0.0100 | 3.97 |
| 497 | 0.0158 | 2.63 | 0.0126 | 3.29 | 0.0105 | 3.94 |
| 517 | 0.0165 | 2.61 | 0.0132 | 3.26 | 0.0110 | 3.92 |
| 537 | 0.0173 | 2.59 | 0.0138 | 3.24 | 0.0115 | 3.89 |
| 556 | 0.0180 | 2.58 | 0.0144 | 3.22 | 0.0120 | 3.86 |
| 575 | 0.0188 | 2.56 | 0.0150 | 3.20 | 0.0125 | 3.84 |
| 594 | 0.0195 | 2.54 | 0.0156 | 3.17 | 0.0130 | 3.81 |
| 613 | 0.0203 | 2.52 | 0.0162 | 3.15 | 0.0135 | 3.78 |
| 631 | 0.0210 | 2.50 | 0.0168 | 3.13 | 0.0140 | 3.77 |
| 469 | 0.0218 | 2.49 | 0.0174 | 3.11 | 0.0145 | 3.73 |
| 667 | 0.0225 | 2.47 | 0.0180 | 3.09 | 0.0150 | 3.70 |
| 684 | 0.0233 | 2.45 | 0.0186 | 3.06 | 0.0155 | 3.68 |
| 701 | 0.0240 | 2.43 | 0.0192 | 3.04 | 0.0160‡ | 3.65 |
| 718 | 0.0248 | 2.42 | 0.0198 | 3.02 | | |
| 734 | 0.0255 | 2.40 | 0.0204‡ | 3.00 | | |
| 750 | 0.0263 | 2.38 | | | | |
| 766 | 0.0270 | 2.36 | | | | |
| 781 | 0.0278‡ | 2.35 | | | | |

† $A_s$ is in square inches and $M_u$ is in foot/kips.

‡ The last value of $\rho$ in each column $= \frac{3}{4}\rho_b$.

## Table 3.4 Coefficient $F$ for resisting moments of rectangular and T sections[6]

$$\text{Values of } F = \frac{bd^2}{12,000}$$

a. Enter table with known values of $F = M_u/K_u$; select $b$ and $d$ in.
b. Enter table with known value of $b$ and $d$; compute resisting moment in concrete: $K_u F$ ft · kips

| d | Width of compressive area b | | | | | | | | | | | | | | | | | | | |
|---|---|---|---|---|---|---|---|---|---|---|---|---|---|---|---|---|---|---|---|---|
| | 4.0 | 5.0 | 6.0 | 7.0 | 7.5 | 8.0 | 9.0 | 9.5 | 10.0 | 11.0 | 12.0 | 13.0 | 15.0 | 17.0 | 19.0 | 21.0 | 23.0 | 25.0 | 30.0 | 36.0 |
| 5.0 | .008 | .010 | .013 | .015 | .016 | .017 | .019 | .020 | .021 | .023 | .025 | .027 | .031 | .035 | .040 | .044 | .048 | .052 | 0.63 | .075 |
| 5.5 | .010 | .013 | .015 | .018 | .019 | .020 | .023 | .024 | .025 | .028 | .030 | .033 | .038 | .043 | .048 | .053 | .058 | .063 | .076 | .091 |
| 6.0 | .012 | .015 | .018 | .021 | .023 | .024 | .027 | .029 | .030 | .033 | .036 | .039 | .045 | .051 | .057 | .063 | .069 | .075 | .090 | .108 |
| 6.5 | .014 | .018 | .021 | .025 | .026 | .028 | .032 | .033 | .035 | .039 | .042 | .046 | .053 | .060 | .067 | .074 | .081 | .088 | .016 | .127 |
| 7.0 | .016 | .020 | .025 | .029 | .031 | .033 | .037 | .039 | .041 | .045 | .049 | .053 | .061 | .069 | .078 | .086 | .094 | .102 | .123 | .147 |
| 7.5 | .019 | .023 | .028 | .033 | .035 | .038 | .042 | .045 | .047 | .052 | .056 | .061 | .070 | .080 | .089 | .098 | .108 | .117 | .141 | .169 |
| 8.0 | .021 | .027 | .032 | .037 | .040 | .043 | .048 | .051 | .053 | .059 | .064 | .069 | .080 | .091 | .101 | .112 | .123 | .133 | .160 | .192 |
| 8.5 | .024 | .030 | .036 | .042 | .045 | .048 | .054 | .057 | .060 | .066 | .072 | .078 | .090 | .102 | .114 | .126 | .138 | .151 | .181 | .217 |
| 9.0 | .027 | .034 | .041 | .047 | .051 | .054 | .061 | .064 | .068 | .074 | .081 | .088 | .101 | .115 | .128 | .142 | .155 | .169 | .203 | .243 |
| 9.5 | .030 | .038 | .045 | .053 | .056 | .060 | .068 | .071 | .075 | .083 | .090 | .098 | .113 | .128 | .143 | .158 | .173 | .188 | .226 | .271 |
| 10.0 | .033 | .042 | .050 | .058 | .063 | .067 | .075 | .079 | .083 | .092 | .100 | .108 | .125 | .142 | .158 | .175 | .192 | .208 | .250 | .300 |
| 10.5 | .037 | .046 | .055 | .064 | .069 | .074 | .083 | .087 | .092 | .101 | .110 | .119 | .138 | .156 | .175 | .193 | .211 | .230 | .276 | .331 |
| 11.0 | .040 | .050 | .061 | .071 | .076 | .081 | .091 | .096 | .101 | .111 | .121 | .131 | .151 | .171 | .192 | .212 | .232 | .252 | .303 | .373 |
| 11.5 | .044 | .055 | .066 | .077 | .083 | .088 | .099 | .105 | .110 | .121 | .132 | .143 | .165 | .187 | .209 | .231 | .253 | .276 | .331 | .397 |
| 12.0 | .048 | .060 | .072 | .084 | .090 | .096 | .108 | .114 | .120 | .132 | .144 | .156 | .180 | .204 | .228 | .252 | .276 | .300 | .360 | .432 |
| 12.5 | .052 | .065 | .078 | .091 | .098 | .104 | .107 | .124 | .130 | .143 | .156 | .169 | .195 | .221 | .247 | .273 | .299 | .326 | .391 | .469 |
| 13.0 | .056 | .070 | .085 | .099 | .106 | .113 | .127 | .134 | .141 | .155 | .169 | .183 | .211 | .239 | .268 | .296 | .324 | .352 | .423 | .507 |
| 13.5 | .061 | .076 | .091 | .106 | .114 | .122 | .137 | .144 | .152 | .167 | .182 | .197 | .228 | .258 | .289 | .319 | .349 | .380 | .456 | .547 |
| 14.0 | .065 | .082 | .098 | .114 | .123 | .131 | .147 | .155 | .163 | .180 | .196 | .212 | .245 | .278 | .310 | .343 | .376 | .408 | .490 | .588 |
| 14.5 | .070 | .088 | .105 | .123 | .131 | .140 | .158 | .166 | .175 | .193 | .210 | .228 | .263 | .298 | .333 | .368 | .403 | .438 | .526 | .631 |
| 15.0 | .075 | .094 | .113 | .131 | .141 | .150 | .169 | .178 | .188 | .206 | .225 | .244 | .281 | .319 | .356 | .394 | .431 | .469 | .563 | .675 |
| 15.5 | .080 | .100 | .120 | .140 | .150 | .160 | .180 | .190 | .200 | .220 | .240 | .260 | .300 | .340 | .380 | .420 | .460 | .501 | .601 | .721 |
| 16.0 | .085 | .107 | .128 | .149 | .160 | .171 | .192 | .203 | .213 | .235 | .256 | .277 | .320 | .363 | .405 | .448 | .491 | .533 | .640 | .768 |
| 16.5 | .091 | .113 | .136 | .159 | .170 | .182 | .204 | .216 | .227 | .250 | .272 | .295 | .340 | .386 | .431 | .476 | .522 | .567 | .681 | .817 |
| 17.0 | | .120 | .145 | .169 | .181 | .193 | .217 | .229 | .241 | .265 | .289 | .313 | .361 | .409 | .458 | .506 | .554 | .602 | .723 | .867 |
| 17.5 | | .128 | .153 | .179 | .191 | .204 | .230 | .242 | .255 | .281 | .306 | .332 | .383 | .434 | .485 | .536 | .587 | .638 | .766 | .919 |
| 18.0 | | .135 | .162 | .189 | .203 | .216 | .243 | .257 | .270 | .297 | .324 | .351 | .405 | .459 | .513 | .567 | .621 | .675 | .810 | .972 |
| 18.5 | | .143 | .171 | .200 | .214 | .228 | .257 | .271 | .285 | .314 | .342 | .371 | .428 | .485 | .542 | .599 | .656 | .713 | .856 | 1.03 |
| 19.0 | | | .181 | .211 | .226 | .241 | .271 | .286 | .301 | .331 | .361 | .391 | .451 | .511 | .572 | .632 | .692 | .752 | .903 | 1.08 |
| 20.0 | | | .200 | .233 | .250 | .267 | .300 | .317 | .333 | .367 | .400 | .433 | .500 | .567 | .633 | .700 | .767 | .833 | 1.00 | 1.20 |
| 21.0 | | | .221 | .257 | .276 | .294 | .331 | .349 | .368 | .404 | .441 | .478 | .551 | .625 | .698 | .772 | .845 | .919 | 1.10 | 1.32 |
| 22.0 | | | .242 | .282 | .303 | .323 | .363 | .383 | .403 | .444 | .484 | .524 | .605 | .686 | .766 | .847 | .928 | 1.01 | 1.21 | 1.45 |
| 23.0 | | | | .309 | .331 | .353 | .397 | .419 | .441 | .485 | .529 | .573 | .661 | .749 | .838 | .926 | 1.01 | 1.10 | 1.32 | 1.59 |
| 24.0 | | | | .336 | .360 | .384 | .432 | .456 | .480 | .528 | .576 | .624 | .720 | .816 | .912 | 1.01 | 1.10 | 1.20 | 1.44 | 1.73 |
| 25.0 | | | | .365 | .391 | .417 | .469 | .495 | .521 | .573 | .625 | .677 | .781 | .885 | .990 | 1.09 | 1.20 | 1.30 | 1.56 | 1.88 |
| 26.0 | | | | .394 | .423 | .451 | .507 | .535 | .563 | .620 | .676 | .732 | .845 | .958 | 1.07 | 1.18 | 1.30 | 1.41 | 1.69 | 2.03 |
| 27.0 | | | | | .456 | .486 | .547 | .577 | .608 | .668 | .729 | .790 | .911 | 1.03 | 1.15 | 1.28 | 1.40 | 1.52 | 1.82 | 2.19 |
| 28.0 | | | | | .490 | .523 | .588 | .621 | .653 | .719 | .784 | .849 | .980 | 1.11 | 1.24 | 1.37 | 1.50 | 1.63 | 1.96 | 2.35 |
| 29.0 | | | | | .526 | .561 | .631 | .666 | .701 | .771 | .841 | .911 | 1.05 | 1.19 | 1.33 | 1.47 | 1.61 | 1.75 | 2.10 | 2.52 |
| 30.0 | | | | | .563 | .600 | .675 | .713 | .750 | .825 | .900 | .975 | 1.13 | 1.28 | 1.43 | 1.58 | 1.73 | 1.88 | 2.25 | 2.70 |
| 31.0 | | | | | | .641 | .721 | .761 | .801 | .881 | .961 | 1.04 | 1.20 | 1.36 | 1.52 | 1.68 | 1.84 | 2.00 | 2.40 | 2.88 |
| 32.0 | | | | | | .683 | .768 | .811 | .853 | .939 | 1.02 | 1.11 | 1.28 | 1.45 | 1.62 | 1.79 | 1.96 | 2.13 | 2.56 | 3.07 |
| 33.0 | | | | | | .726 | .817 | .862 | .908 | .998 | 1.09 | 1.18 | 1.36 | 1.54 | 1.72 | 1.91 | 2.09 | 2.27 | 2.72 | 3.27 |
| 34.0 | | | | | | .771 | .867 | .915 | .963 | 1.06 | 1.16 | 1.25 | 1.45 | 1.64 | 1.83 | 2.02 | 2.22 | 2.41 | 2.89 | 3.47 |
| 36.0 | | | | | | | .927 | 1.03 | 1.08 | 1.19 | 1.30 | 1.40 | 1.62 | 1.84 | 2.05 | 2.27 | 2.48 | 2.70 | 3.24 | 3.89 |
| 38.0 | | | | | | | 1.08 | 1.14 | 1.20 | 1.32 | 1.44 | 1.56 | 1.81 | 2.05 | 2.29 | 2.53 | 2.77 | 3.01 | 3.61 | 4.33 |
| 40.0 | | | | | | | 1.20 | 1.27 | 1.33 | 1.47 | 1.60 | 1.73 | 2.00 | 2.27 | 2.53 | 2.80 | 3.07 | 3.33 | 4.00 | 4.80 |
| 42.0 | | | | | | | 1.32 | 1.40 | 1.47 | 1.62 | 1.76 | 1.91 | 2.21 | 2.50 | 2.79 | 3.09 | 3.38 | 3.68 | 4.41 | 5.29 |
| 44.0 | | | | | | | | 1.53 | 1.61 | 1.77 | 1.94 | 2.10 | 2.42 | 2.74 | 3.07 | 3.93 | 3.71 | 4.03 | 4.84 | 5.81 |
| 46.0 | | | | | | | | 1.68 | 1.76 | 1.94 | 2.12 | 2.29 | 2.65 | 3.00 | 3.35 | 3.70 | 4.06 | 4.41 | 5.29 | 6.35 |
| 48.0 | | | | | | | | 1.82 | 1.92 | 2.11 | 2.30 | 2.50 | 2.88 | 3.26 | 3.65 | 4.03 | 4.42 | 4.80 | 5.76 | 6.91 |
| 50.0 | | | | | | | | 1.98 | 2.08 | 2.29 | 2.50 | 2.71 | 3.13 | 3.54 | 3.96 | 4.38 | 4.79 | 5.21 | 6.25 | 7.50 |

strength concretes. The value of $K_u$ is then substituted into Eq. (3.47), which is then solved for $b$ and $d$. Another parameter $F$, which is used to facilitate the selection of $b$ and $d$, is derived from Eq. (3.47) by multiplying $M_u$, which is expressed in foot-kips, by 12,000 to convert the moment $M_u$ into units of inch-pounds. Then Eq. (3.47) can be written

$$12{,}000 M_u = bd^2 K_u$$

Setting $bd^2/12{,}000 = F$, we can write

$$\frac{M_u}{K_u} = F \tag{3.48}$$

Values of $F$ are given in Table 3.4 as a function of $b$ and $d$. Once the designer evaluates $F$, Table 3.4 can be used to select combinations of $b$ and $d$ that provide the required proportions of the cross section. $F$, which is similar to a section modulus, is a measure of the size of the rectangular cross section required to carry a given value of moment $M_u$. A third design aid, Table 3.5, gives the moment capacity of 12-in-wide rectangular sections as a function of $d$ and $\rho$ for rectangular members reinforced with grade 60 steel. This design aid is used to proportion slabs.

**Example 3.19: The influence of concrete strength on moment capacity** For the cross section in Fig. 3.36 compare both the flexural capacity $M_u$ and the strain in the steel at failure for concrete with (a) $f'_c = 2.5$ kips/in$^2$ (17.24 MPa) and (b) $f'_c = 4$ kips/in$^2$ (27.58 MPa) $f_y = 50$ kips/in$^2$ (344.7 MPa), $A_s = 3$ in$^2$, $\phi = 0.9$.

SOLUTION (a) Using Eq. (3.24)

$$a = \frac{f_y A_s}{0.85 f'_c b} = \frac{150 \text{ kips}}{0.85(2.5)(12)} = 5.88 \text{ in}$$

$$c = \frac{a}{\beta_1} = \frac{5.88 \text{ in}}{0.85} = 6.92 \text{ in}$$

$$M_u = \phi A_s f_y \left( d - \frac{a}{2} \right)$$

$$M_u = 0.9(3)50 \left( 15.5 - \frac{5.88}{2} \right) = 1695.6 \text{ in} \cdot \text{kips} = 141.3 \text{ ft} \cdot \text{kips} \ (191.6 \text{ kN} \cdot \text{m})$$

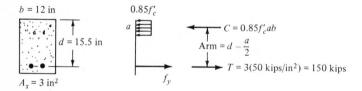

Figure 3.36

**Table 3.5  Resisting moments $M_u$, foot-kips, for sections 12 in wide,[6] $f_y = 60,000$**

| $\rho$ | Effective depth, in | | | | | | | | | | | | | |
|---|---|---|---|---|---|---|---|---|---|---|---|---|---|---|
|  | 3.0 | 3.5 | 4.0 | 4.5 | 5.0 | 5.5 | 6.0 | 6.5 | 7.0 | 8.0 | 9.0 | 10.0 | 11.0 | 12.0 |
| | | | | | | $f_c' = 3000\ \text{lb/in}^2$ | | | | | | | | |
| 0.002 | 0.9 | 1.3 | 1.7 | 2.1 | 2.6 | 3.2 | 3.8 | 4.5 | 5.2 | 6.7 | 8.5 | 10.5 | 12.8 | 15.2 |
| 0.003 | 1.4 | 1.9 | 2.5 | 3.2 | 3.9 | 4.7 | 5.6 | 6.6 | 7.7 | 10.0 | 12.7 | 15.6 | 18.9 | 22.5 |
| 0.004 | 1.9 | 2.5 | 3.3 | 4.2 | 5.1 | 6.2 | 7.4 | 8.7 | 10.1 | 13.2 | 16.7 | 20.6 | 24.9 | 29.6 |
| 0.005 | 2.3 | 3.1 | 4.1 | 5.1 | 6.4 | 7.7 | 9.1 | 10.7 | 12.4 | 16.3 | 20.6 | 25.4 | 30.7 | 36.6 |
| 0.006 | 2.7 | 3.7 | 4.8 | 6.1 | 7.5 | 9.1 | 10.8 | 12.7 | 14.8 | 19.3 | 24.4 | 30.1 | 36.4 | 43.4 |
| 0.007 | 3.1 | 4.2 | 5.5 | 7.0 | 8.7 | 10.5 | 12.5 | 14.7 | 17.0 | 22.2 | 28.1 | 34.7 | 42.0 | 49.9 |
| 0.008 | 3.5 | 4.8 | 6.3 | 7.9 | 9.8 | 11.8 | 14.1 | 16.5 | 19.2 | 25.0 | 31.7 | 39.1 | 47.3 | 56.3 |
| 0.009 | 3.9 | 5.3 | 7.0 | 8.8 | 10.9 | 13.1 | 15.6 | 18.4 | 21.3 | 27.8 | 35.2 | 43.4 | 52.6 | 62.6 |
| 0.010 | 4.3 | 5.8 | 7.6 | 9.6 | 11.9 | 14.4 | 17.1 | 20.1 | 23.3 | 30.5 | 38.6 | 47.6 | 57.6 | 68.6 |
| 0.011 | 4.7 | 6.3 | 8.3 | 10.5 | 12.9 | 15.6 | 18.6 | 21.8 | 25.3 | 33.1 | 41.9 | 51.7 | 62.5 | 74.4 |
| 0.012 | 5.0 | 6.8 | 8.9 | 11.3 | 13.9 | 16.8 | 20.0 | 23.5 | 27.3 | 35.6 | 45.1 | 55.6 | 67.3 | 80.1 |
| 0.013 | 5.3 | 7.3 | 9.5 | 12.0 | 14.9 | 18.0 | 21.4 | 25.1 | 29.1 | 38.0 | 48.1 | 59.4 | 71.9 | 85.6 |
| 0.014 | 5.7 | 7.7 | 10.1 | 12.8 | 15.8 | 19.1 | 22.7 | 26.7 | 30.9 | 40.4 | 51.1 | 63.1 | 76.4 | 90.9 |
| 0.015 | 6.0 | 8.2 | 10.7 | 13.5 | 16.7 | 20.2 | 24.0 | 28.2 | 32.7 | 42.7 | 54.0 | 66.7 | 80.7 | 96.0 |
| 0.016 | 6.3 | 8.6 | 11.2 | 14.2 | 17.5 | 21.2 | 25.2 | 29.6 | 34.3 | 44.9 | 56.8 | 70.1 | 84.8 | 100.9 |
| | | | | | | $f_c' = 4000\ \text{lb/in}^2$ | | | | | | | | |
| 0.002 | 1.0 | 1.3 | 1.7 | 2.1 | 2.7 | 3.2 | 3.8 | 4.5 | 5.2 | 6.8 | 8.6 | 10.6 | 12.8 | 15.3 |
| 0.003 | 1.4 | 1.9 | 2.5 | 3.2 | 3.9 | 4.8 | 5.7 | 6.7 | 7.7 | 10.1 | 12.8 | 15.8 | 19.1 | 22.7 |
| 0.004 | 1.9 | 2.6 | 3.3 | 4.2 | 5.2 | 6.3 | 7.5 | 8.8 | 10.2 | 13.3 | 16.9 | 20.8 | 25.2 | 30.0 |
| 0.005 | 2.3 | 3.2 | 4.1 | 5.2 | 6.5 | 7.8 | 9.3 | 10.9 | 12.6 | 16.5 | 20.9 | 25.8 | 31.2 | 37.2 |
| 0.006 | 2.8 | 3.8 | 4.9 | 6.2 | 7.7 | 9.3 | 11.0 | 13.0 | 15.0 | 19.6 | 24.9 | 30.7 | 37.1 | 44.2 |
| 0.007 | 3.2 | 4.3 | 5.7 | 7.2 | 8.9 | 10.7 | 12.8 | 15.0 | 17.4 | 22.7 | 28.7 | 35.5 | 42.9 | 51.1 |
| 0.008 | 3.6 | 4.9 | 6.4 | 8.1 | 10.0 | 12.1 | 14.5 | 17.0 | 19.7 | 25.7 | 32.5 | 40.1 | 48.6 | 57.8 |
| 0.009 | 4.0 | 5.5 | 7.2 | 9.1 | 11.2 | 13.5 | 16.1 | 18.9 | 21.9 | 28.6 | 36.2 | 44.7 | 54.1 | 64.4 |
| 0.010 | 4.4 | 6.0 | 7.9 | 10.0 | 12.3 | 14.9 | 17.7 | 20.8 | 24.1 | 31.5 | 39.9 | 49.2 | 59.6 | 70.9 |
| 0.011 | 4.8 | 6.6 | 8.6 | 10.9 | 13.4 | 16.2 | 19.3 | 22.7 | 26.3 | 34.3 | 43.4 | 53.6 | 64.9 | 77.2 |
| 0.012 | 5.2 | 7.1 | 9.3 | 11.7 | 14.5 | 17.5 | 20.9 | 24.5 | 28.4 | 37.1 | 46.9 | 57.9 | 70.1 | 83.4 |
| 0.013 | 5.6 | 7.6 | 9.9 | 12.6 | 15.5 | 18.8 | 22.4 | 26.2 | 30.4 | 39.8 | 50.3 | 62.1 | 75.2 | 89.5 |
| 0.014 | 6.0 | 8.1 | 10.6 | 13.4 | 16.6 | 20.0 | 23.8 | 28.0 | 32.5 | 42.4 | 53.6 | 66.2 | 80.1 | 95.4 |
| 0.015 | 6.3 | 8.6 | 11.2 | 14.2 | 17.6 | 21.2 | 25.3 | 29.7 | 34.4 | 45.0 | 56.9 | 70.2 | 85.0 | 101.2 |
| 0.016 | 6.7 | 9.1 | 11.9 | 15.0 | 18.5 | 22.4 | 26.7 | 31.3 | 36.3 | 47.5 | 60.1 | 74.2 | 89.7 | 106.8 |
| 0.017 | 7.0 | 9.6 | 12.5 | 15.8 | 19.5 | 23.6 | 28.1 | 33.0 | 38.2 | 49.9 | 63.2 | 78.0 | 94.4 | 112.3 |
| 0.018 | 7.4 | 10.0 | 13.1 | 16.5 | 20.4 | 24.7 | 29.4 | 34.5 | 40.0 | 52.3 | 66.2 | 81.7 | 98.9 | 117.7 |
| 0.019 | 7.7 | 10.5 | 13.7 | 17.3 | 21.3 | 25.8 | 30.7 | 36.1 | 41.8 | 54.6 | 69.1 | 85.3 | 103.3 | 122.9 |
| 0.020 | 8.0 | 10.9 | 14.2 | 18.0 | 22.2 | 26.9 | 32.0 | 37.6 | 43.6 | 56.9 | 72.0 | 88.9 | 107.5 | 128.0 |
| 0.021 | 8.4 | 11.5 | 15.0 | 19.0 | 23.4 | 28.3 | 33.7 | 39.5 | 45.9 | 59.0 | 75.8 | 93.6 | 113.3 | 134.8 |

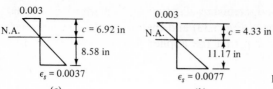

**Figure 3.37** Strains at failure: (*a*) $f_c' = 2.5\,\text{kips/in}^2$, (*b*) $f_c' = 4\,\text{kips/in}^2$.

From the strain diagram in Fig. 3.37*a*, we compute $\varepsilon_s$ by similar triangles

$$\frac{0.003}{c} = \frac{\varepsilon_s}{d - c} \qquad \text{and } \varepsilon_s = 0.0037$$

(*b*)

$$a = \frac{f_y A_s}{0.85 f_c' b} = \frac{150\ \text{kips}}{0.85(4)(12)} = 3.68\ \text{in}$$

$$c = \frac{a}{\beta_1} = \frac{3.68}{0.85} = 4.33\ \text{in}$$

$$M_u = \phi A_s f_y\left(d - \frac{a}{2}\right)$$

$$M_u = 0.9(3)50\left(15.5 - \frac{3.68}{2}\right) = 1844\ \text{in} \cdot \text{kips} = 153.7\ \text{ft} \cdot \text{kips}\ (208.4\ \text{kN} \cdot \text{m}).$$

From the strain diagram in Fig. 3.37*b* by similar triangles, we compute $\varepsilon_s$:

$$\frac{0.003}{c} = \frac{\varepsilon_s}{d - c} \qquad \text{and} \qquad \varepsilon_s = 0.0077$$

CONCLUSIONS For a 60 percent increase in concrete strength, the computations lead to two conclusions: (1) Increasing the strength of the concrete produces only an 8.8 percent increase in moment capacity. Since the tension force is the same in both beams because the area of steel is the same, the difference in moment capacity is due to the small increase in length of the arm between *T* and *C*. (2) The reduction in depth of the compression zone for the cross section made of higher-strength concrete results in a 108 percent increase in steel strain at failure. The larger strains result in a more ductile beam that can undergo much larger deflections before the secondary compression failure of the concrete occurs.

## 3.9 TRIAL METHOD

The *trial method*, a procedure for determining the area of steel required to reinforce a cross section for a particular value of moment, is based on estimating the arm of the internal couple. The procedure, which can be used for cross sections of any shape, produces approximate values of the required steel area. Since the procedure also permits the designer to compute an improved value of arm, an improved estimate of the required steel area can be made if the analysis is repeated.

In the trial method the designer estimates the length of the arm between the internal couple (see, for example, Fig. 3.33) and then solves for the tension force $T$ by using the basic relationship that the applied moment equals the flexural design strength; i.e.,

$$M_u = \phi T \times \text{arm}$$

and

$$T = \frac{M_u}{\phi \times \text{arm}} \qquad (3.49)$$

where $\phi = 0.9$, $M_u$ equals the factored moment, and the designer estimates the arm between the components of the internal couple $T$ and $C$. To start the procedure the arm may be estimated as some fraction of $d$, the effective depth. For the initial estimate of the arm, $0.85d$ for a rectangular section and $0.95d$ for a T-shaped section are recommended. After Eq. (3.49) has been solved for $T$, the approximate area of steel can be computed by dividing $T$ by $f_y$.

To produce a more accurate value of steel area, the components of the internal couple can be equated to provide a close estimate of the area $A_c$ of the stress block. In this step, $C$, is expressed as the product of the area of the stress block and $0.85f'_c$

$$C = T$$

$$0.85f'_c A_c = T$$

$$A_c = \frac{T}{0.85f'_c} \qquad (3.50)$$

where $A_c$ = approximate area of stress block
$\quad\ T$ = approximate value of tension force predicted by Eq. (3.49)
$\quad\ f'_c$ = compressive strength of concrete

Once $A_c$ has been evaluated, the designer can locate the position of $C$ (it passes through the centroid of $A_c$) and recompute the distance between $T$ and $C$. If the beam has a rectangular section, the designer can express the arm as $d - a/2$ and select values of the arm by guessing values of $a$. An initial value of $a = 0.15d$ is recommended. Regardless of the first assumption for the arm, two cycles of the trial method should be adequate to produce a close estimate of the required steel area. Examples 3.20 and 3.21 illustrate the method for a nonrectangular and a rectangular cross section. For these problems it will be assumed that the area of steel from the analysis is less than three-fourths of the balanced area. If this information is not given, the designer will also be required to compute the area of balanced steel to ensure that the cross section will be underreinforced as required by the ACI Code.

**Example 3.20: Use of the trial method to determine the required area of flexural steel**
If the cross section in Fig. 3.38 carries a factored moment $M_u$ of 115 ft · kips (155.94 kN · m), what area of steel is required? $f'_c = 3$ kips/in$^2$ (20.68 MPa), and $f_y = 60$ kips lb/in$^2$ (413.7 MPa).

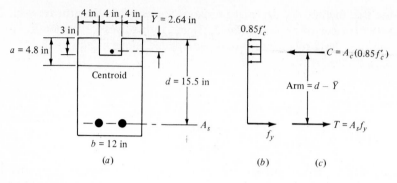

**Figure 3.38** (*a*) Cross section, (*b*) stresses, (*c*) internal couple.

SOLUTION In the initial analysis we estimate the arm of the internal couple. Try arm = 0.85*d*

$$\text{Arm} = 0.85(15.5) = 13.2 \text{ in } (335.3 \text{ mm})$$

Compute *T* using Eq. (3.49)

$$T_1 = \frac{M_u}{\phi \times \text{arm}} = \frac{115(12)}{0.9(13.2)} = 116.2 \text{ kips } (516.86 \text{ kN})$$

Approximate value of $A_s$

$$A_s = \frac{T_1}{f_y} = \frac{116.2 \text{ kips}}{60 \text{ kips/in}^2} = 1.94 \text{ in}^2 \ (1251.7 \text{ mm}^2)$$

To improve the value of $A_s$, repeat the analysis with a revised value of arm. Use the distance between the centroid of the tension steel and the centroid of the compression zone. Compute $A_c$ using Eq. (3.50) with $T = T_1$ from the initial analysis

$$A_c = \frac{T_1}{0.85f'_c} = \frac{116.2 \text{ kips}}{0.85(3 \text{ kips/in}^2)} = 45.6 \text{ in}^2 \ (29421 \text{ mm}^2)$$

Establish the depth of the stress block from the geometry of the compression zone. The distance that the stress block extends below the notch equals

$$\frac{45.6 - (3)(4)(2)}{12} = 1.8 \text{ in } (45.7 \text{ mm})$$

$$a = 3 + 1.8 = 4.8 \text{ in } (242 \text{ mm})$$

Locate the centroid of the stress block. Sum moments of the stress-block area about a horizontal axis through the top of the cross section

$$A\bar{Y} = \Sigma A_n y_n$$

$$\bar{Y} = \frac{3(4)(1.5)(2) + 21.6(3 + 0.9)}{45.6} = 2.64 \text{ in } (67 \text{ mm})$$

Repeat the analysis for $A_s$ using the new value of arm, where the arm is

$$d - \bar{Y} = 15.5 - 2.64 = 12.86 \text{ in } (326.6 \text{ mm})$$

$$T_2 = \frac{M_u}{\phi \times \text{arm}} = \frac{115(12)}{0.9(12.86)} = 119.23 \text{ kips}$$

$$A_s = \frac{T_2}{f_y} = \frac{119.23}{60} = 1.99 \text{ in}^2 \ (1284 \text{ mm}^2)$$

Use two no. 9 bars; $A_{s,\,\text{sup}} = 2 \text{ in}^2 \ (1290 \text{ mm}^2)$.

**Example 3.21:** Use the trial method to determine the area of tension steel required by the cross section in Fig. 3.39 to carry a factored moment $M_u = 180 \text{ ft} \cdot \text{kips} \ (244.1 \text{ kN} \cdot \text{m})$; $f'_c = 3 \text{ kips/in}^2 \ (20.68 \text{ MPa})$, and $f_y = 40 \text{ kips/in}^2 \ (275.8 \text{ MPa})$.

SOLUTION For the initial analysis estimate the arm of the internal couple by guessing $a \approx 0.15d \approx 3 \text{ in}$

$$\text{Arm} = d - \frac{a}{2} = 20.4 - \tfrac{3}{2} = 18.9 \text{ in } (480 \text{ mm})$$

Then use Eq. (3.49) to compute $T$

$$T_1 = \frac{M_u}{\phi(d - a/2)} = \frac{180(12)}{0.9(18.9)} = 127 \text{ kips}$$

The initial estimate of $A_s$ is

$$\frac{T_1}{f_y} = \frac{127 \text{ kips}}{40 \text{ kips/in}^2} = 3.18 \text{ in}^2 \ (2051.7 \text{ mm}^2)$$

Repeat the analysis using an arm based on an improved value of $a$. Equate $T_1 = C$

$$127 \text{ kips} = 0.85 f'_c A_c = 0.85(3a)(12)$$

$$a = 4.15 \text{ in}$$

$$\text{Arm} = d - \frac{a}{2} = 20.4 - \frac{4.15}{2} = 18.33 \text{ in}$$

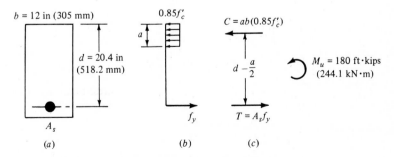

**Figure 3.39** (*a*) Cross section, (*b*) stresses, (*c*) internal couple.

Recompute $T$

$$T_2 = \frac{M_u}{\phi(d - a/2)} = \frac{180(12)}{0.9(18.33)} = 131 \text{ kips}$$

The second estimate of $A_s$ is

$$A_s = \frac{T_2}{f_y} = \frac{131 \text{ kips}}{40 \text{ kips/in}^2} = 3.28 \text{ in}^2 \ (2116.3 \text{ mm}^2)$$

Since the result of the second analysis is close to that of the first, the design is complete. Use two no. 10 and one no. 9; $A_s$ supplied $= 3.53 \text{ in}^2$.

**Example 3.22:** A continuous beam of constant depth and with a rectangular cross section is to be designed for a uniform factored load of $w_u = 3$ kips/ft (43.8 kN/m) (see Fig. 3.40). Using a $\rho$ of approximately $\frac{3}{4}\rho_b$ at the point of maximum moment, establish the depth of the cross section if $b = 14$ in. Round the depth off to the nearest even inch and compute the required area of steel at all points of maximum positive and negative moment. $f'_c = 3 \text{ kips/in}^2$, and $f_y = 50 \text{ kips/in}^2$.

SOLUTION Use Eq. (3.37) to compute or read from Table 3.3 that $\frac{3}{4}\rho_b = 0.0204$. Using Eq. (3.46), base dimensions on $M_u = 283.5$ ft · kips at $B$

$$M_u = \phi\rho f_y bd^2\left(1 - \frac{\rho f_y}{1.7f'_c}\right)$$

$$283.5(12) = 0.9(0.0204)(50)(14d^2)\left[1 - \frac{0.0204(50)}{1.7(3)}\right]$$

$$d = 18.19 \text{ in}$$

$$\text{Depth } h = 18.19 + 2.6 \text{ cover} = 20.79 \text{ in} \qquad \text{use 22 in}$$

Recompute

$$d = h - 2.6 \text{ in} = 22 - 2.6 = 19.4 \text{ in}$$

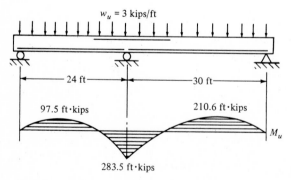

Figure 3.40

Recompute $\rho$ by solving Eq. (3.46) with $b = 14$ in and $d = 19.4$

$$283.5(12) = 0.9\rho(50)(14)(19.4^2)\left[1 - \frac{\rho(50)}{1.7(3)}\right]$$

$$\rho = 0.0173$$

$$A_s = \rho bd = 0.0173(14)(19.4) = 4.7 \text{ in}^2$$

Solving Eq. (3.46) for $\rho$, compute $A_s$ for $M_u = 97.5$ ft · kips, $b = 14$ in, and $d = 19.4$ in

$$97.5(12) = 0.9\rho(50)(14)(19.4^2)\left[1 - \frac{\rho(50)}{1.7(3)}\right]$$

$$\rho = 0.0052 > \rho_{\min} = \frac{200}{f_y} = 0.004 \qquad \text{OK}$$

$$A_s = \rho bd = 0.0052(14)(19.4) = 1.41 \text{ in}^2$$

Compute $A_s$ for $M_u = 210.6$ ft · kips. Use Tables 3.3 and 3.4. Enter Table 3.4 with $b = 14$ in and $d = 19.4$, interpolate, and read $F = 0.44$. Using Equation (3.48), solve for $K_u$

$$K_u = \frac{M_u}{F} = \frac{210.6}{0.44} = 479$$

Enter Table 3.3 with $K_u = 479$. In column headed by $f_y = 50,000$ interpolate to find $\rho = 0.0121$; then

$$A_s = \rho bd = 0.0121(14)(19.4) = 3.29 \text{ in}^2$$

## 3.10 BALANCED STEEL FOR BEAMS WITH NONRECTANGULAR COMPRESSION ZONES

In this section we establish a general procedure for the computation of the balanced steel area $A_{sb}$ for a cross section of any shape which is symmetrical with respect to a vertical axis or which is constrained so that under load it deflects vertically without twisting. To make the procedure completely general, steel may also be present in the compression zone; see Fig. 3.41. $C_c$ is not located at $a/2$ because the stress block is not a rectangle. The step-by-step procedure for computing $A_{sb}$ is detailed below.

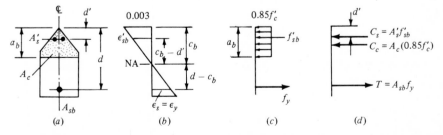

**Figure 3.41** Balanced steel for a beam with a nonrectangular compression zone: (a) cross section, (b) strains at failure, (c) stresses, (d) internal couple. $\varepsilon_{sb}' =$ strain in compression steel for a beam reinforced with balanced steel, $f_{sb}' =$ stress in compression steel for a beam reinforced with balanced steel, $A_c =$ area of stress block, and $C_c =$ resultant of concrete compressive stresses.

## Procedure for Determining Balanced Steel

**Step 1**  As in the case of a rectangular beam, the balanced area of steel $A_{sb}$ positions the neutral axis so that the maximum strain in the concrete reaches a value of 0.003 just as the tension steel yields.

**Step 2**  $c_b$, the distance of the compression surface from the neutral axis, is computed from the geometry of the strain curve using similar triangles

$$c_b = \frac{87,000}{f_y + 87,000} d \tag{3.34}$$

**Step 3**  Tests show that for nonrectangular sections[5] $a_b = \beta_1 c_b$.

**Step 4**  $\epsilon'_{sb}$ is also computed from the geometry of the compression-zone strain distribution (Fig. 3.41b)

$$\frac{0.003}{c_b} = \frac{\epsilon'_{sb}}{c_b - d'}$$

$$\epsilon'_{sb} = \frac{c_b - d'}{c_b} (0.003) \tag{3.51}$$

Once $\epsilon'_{sb}$ has been computed, $f'_{sb}$ can be evaluated. To speed the computation of $f'_{sb}$, the design aid given in Fig. 3.42 may be used.

**Step 5**  The compression forces in the steel and concrete are next computed by multiplying the respective areas of steel and concrete by the associated stresses (see Fig. 3.41d).

**Step 6**  Since the total compression force must equal the tension force (these forces constitute the internal couple), we can write

$$T = C_c + C_s$$

Substituting $T = A_{sb} f_y$ and solving for $A_{sb}$ gives

$$A_{sb} = \frac{C_c + C_s}{f_y} \tag{3.52}$$

If no compression steel is present, $C_s = 0$. To ensure ductile behavior, the maximum steel permitted by the 1977 ACI Code equals

$$A_{s,\,\text{max}} = \frac{\frac{3}{4} C_c}{f_y} + \frac{C_s}{f_y} \tag{3.53}$$

Since the steel in the compression zone adds to the member's ductility, the $\frac{3}{4}$ factor is applied only to the compression force $C_c$.

The use of Eqs. (3.52) and (3.53) are illustrated in Examples 3.23 and 3.24.

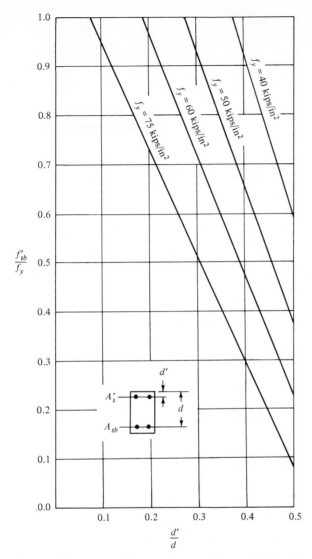

**Figure 3.42** Variation of $f'_{sb}$ with $d'/d$ for a balanced failure. Enter the chart with $f_y$ and $d'/d$. Read $f'_{sb}/f_y$. If a point falls above the top of the chart, the compression steel has yielded and $f'_{sb} = f_y$. The graph may be used for any shape of cross section.

**Example 3.23:** (*a*) Determine balanced steel $A_{sb}$ for the symmetrical cross section in Fig. 3.43. (*b*) What is the maximum area of steel permitted in the cross section by the ACI Code? $f'_c = 3.5$ kips/in² (24.13 MPa), and $f_y = 70$ kips/in² (482.7 MPa).

SOLUTION (*a*) Consider the beam to be reinforced with balanced steel $A_{sb}$. Use Eq. (3.34) to compute $c_b$

$$c_b = \frac{87{,}000}{87{,}000 + f_y}\,d = \frac{87{,}000}{87{,}000 + 70{,}000(27.25)} = 15.24 \text{ in}$$

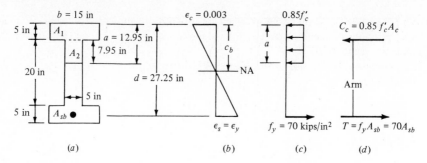

**Figure 3.43** (a) Cross section, (b) strains at balanced failure, (c) stresses, (d) internal couple.

Using Eq. (3.21), compute $a$

$$a = \beta_1 c_b = 0.85(15.24) = 12.95 \text{ in } (329 \text{ mm})$$

Compute the resultant compression force $C_c$. Break the compression zone into two rectangular areas (see Fig. 3.43a)

$$C_c = 0.85f'_c A_c$$
$$= 0.85f'_c(A_1 + A_2) = 0.85(3.5 \text{ kips/in}^2)[15(5) + 7.95(5)]$$
$$= 341.38 \text{ kips } (1518.5 \text{ kN})$$

Equate $T = C_c$, and solve for $A_{sb}$

$$70A_{sb} = 341.38 \text{ kips}$$
$$A_{sb} = 4.88 \text{ in}^2$$

(b)

$$A_{s,\text{max}} = \tfrac{3}{4}A_{sb} = 0.75(4.88 \text{ in}^2) = 3.66 \text{ in}^2 (2361.4 \text{ mm}^2)$$

**Example 3.24:** If $A'_s$ equals 2 in$^2$, determine (a) the area of balanced steel $A_{sb}$ for the cross section in Fig. 3.44 and (b) the maximum area of flexural steel permitted in the cross section by the ACI Code. $f'_c = 3 \text{ kips/in}^2$ (20.68 MPa), and $f_y = 58 \text{ kips/in}^2$ (400 MPa).

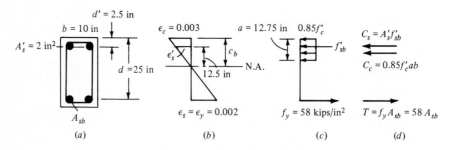

**Figure 3.44** (a) Cross section, (b) strains, (c) stresses, (d) internal couple.

SOLUTION (a) Assume that the cross section is reinforced with balanced steel. Balanced steel positions the neutral axis so that the tension steel yields just as the strain in the concrete reaches 0.003.

Using the strain diagram in Fig. 3.44b, locate the neutral axis by use of similar triangles

$$\frac{0.003}{c_b} = \frac{0.002}{25 - c_b}$$

$$c_b = 15 \text{ in}$$

Determine $\epsilon'_s$ by similar triangles

$$\frac{\epsilon'_s}{12.5 \text{ in}} = \frac{0.003}{15 \text{ in}}$$

$$\epsilon'_s = 0.0025$$

Since $\epsilon'_s > \epsilon_y = 0.002$, the compression steel yields.

Compute the total compression force $C_{tot} = C_s + C_c$

$$C_c = 0.85f'_c ab = 0.85(3 \text{ kips/in}^2)(12.75 \text{ in})(10 \text{ in}) = 325.1 \text{ kips†}$$

where

$$a = \beta_1 c_b = 0.85(15 \text{ in}) = 12.75 \text{ in}$$

$$C_s = f_y A'_s = (58 \text{ kips/in}^2)(2) = 116 \text{ kips}$$

Equate $T = C_c + C_s$, and solve for $A_{sb}$, where $T = A_{sb} f_y$

$$A_{sb}(58) = 325.1 + 116 = 441.1 \text{ kips}$$

$$A_{sb} = 7.61 \text{ in}^2$$

(b) Use Eq. (3.53) to compute the maximum area of flexural steel

$$A_{s,\,max} = \frac{0.75C_c}{f_y} + \frac{C_s}{f_y}$$

$$= \frac{0.75(325.1 \text{ kips})}{58 \text{ kips/in}^2} + \frac{116 \text{ kips}}{58 \text{ kips/in}^2}$$

$$= 6.20 \text{ in}^2 \ (3876.2 \text{ mm}^2)$$

# 3.11 DESIGN OF BEAMS WITH COMPRESSION STEEL

If a beam designed in accordance with the ACI Code is reinforced with tension steel only, the maximum flexural capacity the cross section can develop is achieved when an area of steel equal to three-fourths of the balanced steel area is used. When restrictions are placed on the dimensions of a cross section, the moment capacity of a member (even when reinforced with three-fourths of the balanced steel area)

---

† If $A'_s$ is large, it should be deducted from the area of the concrete compression zone. If this adjustment is made, $C_c = 320$ kips, and $A_{s,\,max} = 6.14 \text{ in}^2$.

may not be adequate to supply the required moment capacity. Under such conditions, additional moment capacity can be created without producing a brittle overreinforced beam by adding additional reinforcement to both the tension and compression sides of the cross section. As shown by Eq. (3.53), the maximum area of tension steel that can be used to reinforce a cross section is a direct function of both the strength of the concrete compression zone and the area of the compression steel $A_s'$.

Figure 3.45 illustrates two situations in which compression steel can be used advantageously. In Fig. 3.45a compression steel is used to increase the flexural capacity of the compression zone of a prefabricated beam whose sides have been cut back to provide a seat to support beams framing in from each side. Figure 3.45b shows a common design situation in which compression steel is used to reduce the size of a continuous T beam of constant cross section by adding flexural capacity in the region where the effective cross section is smallest and the moment greatest. Near midspan of a continuous beam (see section 1 of Fig. 3.45b), where the positive moment creates compression in the flange, the beam behaves as if it were a rectangular beam with a width equal to that of the flange. Even if the beam is shallow, the large flange provides the potential for a large moment capacity.

If the moment produces tension in the flange and compression in the web (the situation at the supports where negative bending occurs), the beam, which now behaves like a narrow rectangular beam with a width equal to that of the web, has a much smaller flexural capacity than the flanged section at midspan. If compression steel is added to the compression zone (see section 2 of Fig. 3.45b),

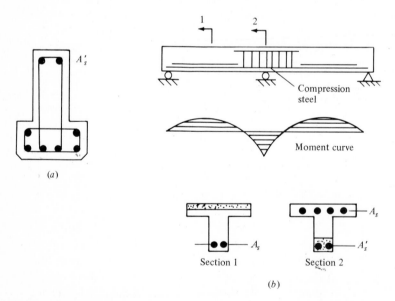

**Figure 3.45** Examples of beams reinforced with compression steel: (a) precast inverted T beam, (b) continuous beam with a portion of the positive steel extended into the supports to be used as compression steel.

the flexural strength can be substantially raised without increasing the width of the web or the depth of the cross section. By using compression steel to raise the capacity of the compression zone the dead weight can be reduced and the headroom increased.

To be most effective, compression steel should be placed where the compressive strains are greatest, i.e., as far as possible from the neutral axis. If compression steel is positioned near the neutral axis, the compressive strains may be too small to stress the steel to its full capacity and the compression steel has little influence on the flexural strength or behavior of the member.

Besides increasing the flexural capacity of a cross section, compression steel produces a marked improvement in behavior by raising the amount of compressive strain the concrete can sustain before crushing and by reducing the tendency of the concrete to break down at high levels of strain.[7] Stabilizing the compression zone of a highly stressed beam reduces creep and increases ductility. Comparing the load-deflection curves of two underreinforced beams of identical proportions (except for the presence of compression steel in one), Fig. 3.46 illustrates the improvement in ductility afforded by the addition of compression steel. As indicated in Fig. 3.46, the flexural capacity is not increased significantly by the addition of compression steel to an underreinforced beam because the magnitude of the internal couple is controlled by the area of the tension steel.

Recognizing the beneficial effect of compression steel on bending behavior, many building codes require that all flexural members of structures located in seismic zones be reinforced with a minimum area of compression steel, even when the design calculations indicate that compression steel is not required for strength. The addition of compression steel produces tough ductile members that can withstand the large bending deformations and repeated reversals of stress produced in building members by cyclic earthquake-induced ground motions.

Since beams reinforced with compression steel (used to increase flexural capacity) have a high reinforcement ratio, their ductility is limited. As the reinforcement ratio increases, the distance between the neutral axis and the top of the compression zone increases, so that less rotation is required to bring the strain in

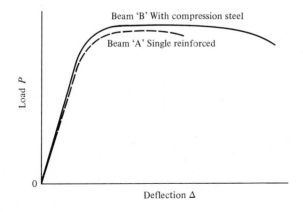

**Figure 3.46** Load-deflection curves showing the improvement in ductility and toughness produced by compression steel.

the concrete at the top of the compression zone to failure. Recognizing that an improvement in the ductility of concrete in compression can be achieved by providing lateral confinement of the concrete, ACI Code §7.11.1 requires compression steel to be enclosed by closely spaced ties throughout the region in which it is used. By providing a certain limited amount of lateral confinement of the concrete in the compression zone, ties increase the ultimate strain required to produce a compression failure and also reduce the rate at which heavily compressed concrete—strained into the inelastic region—breaks down.[7]

When no. 10 or smaller bars are used as compression steel, ACI Code §7.10.5.1 specifies that ties be at least $\frac{3}{8}$ in (9.5 mm) in diameter. If bundled bars or no. 11 or larger bars are used as compression steel, ties must be at least $\frac{1}{2}$ in (13 mm) in diameter. In accordance with ACI Code §7.10.5.2, the maximum spacing of ties is not to exceed the smallest of the following distances:

1. Sixteen bar diameters of the compression steel
2. Forty-eight tie diameters
3. The least dimension of the cross section

Although compression steel permits the use of large areas of tension steel, the designer must verify (1) that the steel can be fitted into the tension zone while maintaining the required spacing between bars and the minimum concrete cover specified by the ACI Code and (2) that the limit on crack width as measured by the ACI expression $z = 0.6f_y \sqrt[3]{d_c A}$ can be satisfied. While the use of a small number of large-diameter bars increases the spacing between bars, the second requirement, the control of crack width, is most easily satisfied by specifying a large number of small-diameter bars.

When designing a beam with compression steel, it is convenient to break the total internal moment into two couples. The first couple $M_1$ (Fig. 3.47c) represents the moment capacity of the cross section reinforced with $\frac{3}{4}A_{sb}$, where $A_{sb}$ applies to the section without compression steel. The second couple $M_2$ represents the moment capacity produced by the forces in the compression steel and in the additional tension steel, which is added to balance the force in the compression steel (Fig. 3.47d). The total moment capacity $M_u$ of the cross section can then be expressed as

$$M_u = \phi(M_1 + M_2)$$

where $\phi = 0.9$, the concrete couple is $M_1 = T_1(d - a/2)$, and the steel couple is $M_2 = T_2(d - d')$.

### Design Procedure

**Step 1**   Determine the moment $M_1$ which the beam can carry using $A_{s_1} = \frac{3}{4}A_{sb}$, where $A_{sb}$ represents balanced steel for the cross section *without* compression steel (see Fig. 3.47c).

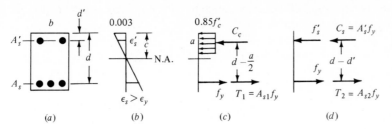

**Figure 3.47** Moment capacity of a beam with compression steel: (a) cross section with $A_s = A_{s1} + A_{s2}$, where $A_{s1} = \frac{3}{4}A_{sb}$, (b) strain, (c) concrete couple $M_1$, (d) steel couple $M_2$.

**Step 2** The excess moment $M_2$, the difference between the total moment and the moment $M_1$, is to be carried by a couple composed of compression steel and additional bottom steel. To compute these areas of steel, establish stresses. For tension steel $f_s = f_y$, tension steel will always yield at failure since the beam is underreinforced. For compression steel $f'_s \leq f_y$. To establish $f'_s$ compute the strain $\epsilon'_s$ in the compression steel using the strain distribution at failure. The location of the neutral axis can be closely approximated by using the neutral axis associated with $\frac{3}{4}A_{sb}$ in step 1 (see Fig. 3.47b).

**Step 3** Determine the magnitude of the internal tension force $T_2$ of the compression-steel couple by summing moments about the compression steel and equating to $M_2$ (see Fig. 3.47d)

$$T_2 = \frac{M_2}{\phi(d - d')} \qquad \text{where } M_2 = \frac{M_u}{\phi} - M_1$$

**Step 4** Compute the areas of additional bottom steel and the compression steel

At bottom:

$$A_{s2} = \frac{T_2}{f_y}$$

At top:

$$A'_s = \frac{C_s}{f'_s} \qquad \text{where } C_s = T_2$$

## Investigation of Moment Capacity

**Step 1** Verify that the beam is underreinforced

$$A_{s,\,max} \leq \frac{\frac{3}{4}C_c + C_s}{f_y} \tag{3.53}$$

where $C_c$ is the force in the concrete and $C_s$ is the force in the compression steel when the cross section is reinforced with $A_{sb}$.

**Step 2** Since horizontal equilibrium requires the total compression force to equal the total tension force,

$$T = C_s + C_c = A'_s f'_s + ab(0.85f'_c) \tag{1}$$

Equation (1) contains two unknowns, $f'_s$ and $a$. Because the neutral axis must position itself so that the forces in the compression zone $C_s$ and $C_c$ must equal $T$, the trial procedure below can be established to provide a simple solution for the unknowns in Eq. (1).

## Trial Procedure

**Step 1** Guess a value for $c$, the distance between the neutral axis and the outside compression surface.

**Step 2** Compute the depth of the stress block using $a = \beta_1 c$. Next compute $\epsilon'_s$ from the geometry of the strain distribution. With $\epsilon'_s$ known, evaluate $f'_s$.

**Step 3** Compute

$$C_s = f'_s A_s \qquad \text{and} \qquad C_c = A_c(0.85f'_c)$$

where $A_c$ is the area of concrete in the stress block.

**Step 4** Compare $T$ and the sum of $C_s + C_c$. As components of the internal couple, they must be equal.

If $C_s + C_c < T$, $c$ was assumed too small (to increase the compression force, the neutral axis must drop; therefore guess a larger value for $c$ and repeat steps 2 to 4).

If $C_s + C_c > T$, $c$ was assumed too large; therefore guess a smaller value of $c$ and repeat.

If $C_s + C_c = T$, $c$ was assumed correctly and the position of the neutral axis was guessed correctly.

Once the location of the neutral axis has been established, the moment capacity can be evaluated by summing moments of $C_s$ and $C_c$ about $T$.

Since $C_s$ and $C_c$ are often located at approximately the same elevation, the moment capacity of the cross section is not sensitive to the magnitude of the two forces as long as they equal the tension force $T$; therefore, the trial method need be carried only to the point where the sum of $C_s$ and $C_c$ is within 4 or 5 percent of the value of $T$. At this point, the value of $C_s$ can be fixed and the force in the concrete $C_c$ computed from $C_c = T - C_s$. The flexural capacity is then determined by establishing the depth of the stress block to evaluate the moment arm between $C_c$ and $T$ and then summing moments of the forces $C_c$ and $C_s$ about $T$.

**Alternate procedure** Since the compression steel is normally strained to its yield point or slightly beyond, the number of trials can often be reduced by assuming $f_s' = f_y$ initially. After $C_s$ has been established, the force in the concrete can be computed from $C_c = T - C_s$. After $C_c$ has been determined, $a$, the depth of the stress block, can be established. Then the position of the neutral axis can be calculated from $c = a/\beta_1$. Once the position of the neutral axis has been established, the strain in the compression steel can be computed to verify the initial assumption that the compression steel has yielded.

**Example 3.25: Design of a double-reinforced concrete beam** Factored loads produce a maximum design moment $M_u = 200$ ft · kips (271.2 kN · m) in a reinforced concrete beam. If the width of the beam is limited to 10 in (254 mm) and the effective depth must not exceed 16 in (406 mm), select the reinforcing steel (see Fig. 3.48a). $f_c' = 3$ kips/in² (20.68 MPa), and $f_y = 50$ kips/in² (344.7 MPa).

SOLUTION If the cross section in Fig. 3.48a is reinforced with the maximum area of tension steel permitted by the ACI Code ($\frac{3}{4}A_{sb}$), computations show that its capacity is not adequate to carry the design moment of 200 ft · kips; therefore, compression steel is required.

Compute the moment capacity $M_1$ of the cross section reinforced with an area of tension steel $A_{s_1} = \frac{3}{4}A_{sb}$. From Table 3.3, $\frac{3}{4}\rho_b = 0.0204$ and $K_u = 734$

$$A_{s_1} = \tfrac{3}{4}\rho_b bd = 0.0204(10)(16) = 3.26 \text{ in}^2$$

From Eq. (3.47)

$$M_1 = K_u bd^2 = \frac{734(10)(16)^2}{1000} = 1879 \text{ in · kips (212.33 kN · m)}$$

Since $M_1$ is less than $M_u = 200$ ft · kips (2400 in · kips), compression steel $A_s'$ and additional tension steel $A_{s_2}$ will be added to form a second couple $M_2$ to carry the balance of the moment

$$M_2 = 2400 \text{ in · kips} - 1879 \text{ in · kips} = 521 \text{ in · kips}$$

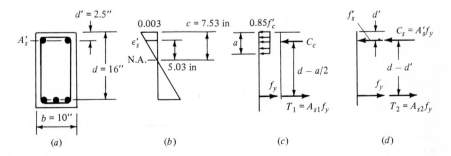

**Figure 3.48** (a) Cross section, (b) strains at failure, (c) concrete couple $M_1$, (d) steel couple $M_2$.

Compute the additional tension steel $A_{s_2}$ by summing moments about $A'_s$ (Fig. 3.48d)

$$M_2 = \phi T_2(d - d')$$
$$521 \text{ in} \cdot \text{kips} = 0.9 A_{s_2}(50 \text{ kips/in}^2)(16 - 2.5)$$
$$A_{s_2} = 0.86 \text{ in}^2$$

The total area of tension steel is

$$A_s = A_{s_1} + A_{s_2} = 3.26 + 0.86 = 4.12 \text{ in}^2$$

Use three no. 9 and two no. 7; $A_s$ supplied $= 4.2 \text{ in}^2$.

To compute the area of compression steel we must first establish the stress in the compression steel at failure from the geometry of the strain curve in Fig. 3.48b. The approximate position of the neutral axis can be taken as that associated with $A_{s_1}$. We solve for $c$ by first determining $a$, the depth of the stress block

$$C_c = T_1 = A_{s_1} f_y$$

$$0.85 f'_c b a = 3.26 \text{ in}^2 \ (50 \text{ kips/in}^2)$$

$$a = \frac{163}{0.85(3)(10 \text{ in})} = 6.4 \text{ in} \qquad c = \frac{a}{\beta_1} = \frac{6.4}{0.85} = 7.53 \text{ in}$$

From similar triangles in Fig. 3.48b, establish $\epsilon'_s$

$$\frac{0.003}{7.53} = \frac{\epsilon'_s}{5.03}$$

$$\epsilon'_s = 0.002$$

Since $\epsilon'_s > \epsilon_y = 0.00172$, compression steel yields $f'_s = 50 \text{ kips/in}^2$.

Compute $A'_s$; equate $T_2 = C_s$, where $T_2 = 0.86(50 \text{ kips/in}^2) = 42.8 \text{ kips}$

$$A'_s f_y = 42.8 \text{ kips}$$

$$A'_s = \frac{42.8}{50} = 0.86 \text{ in}^2$$

Use two no. 6 to give $A'_s = 0.88 \text{ in}^2$.

Design ties for compression steel. To add ductility to the beam ACI Code §7.11.1 requires that closed ties be placed around compression steel. Use no. 3 ties. The required spacing is

$$\text{16 diameters of longitudinal steel} = 16(\tfrac{6}{8}) = 12 \text{ in}$$
$$\text{48 tie diameters} = 48(\tfrac{3}{8}) = 18 \text{ in}$$
$$\text{Least dimension of cross section} = 10 \text{ in} \qquad \text{Controls}$$

A detailed sketch is shown in Fig. 3.49.

**Example 3.26:** Investigate the flexural capacity of the member designed in Example 3.25 using the theoretical areas of steel previously computed (see Fig. 3.50); $f'_c = 3 \text{ kips/in}^2$ (20.68 MPa), and $f_y = 50 \text{ kips/in}^2$ (344.75 MPa).

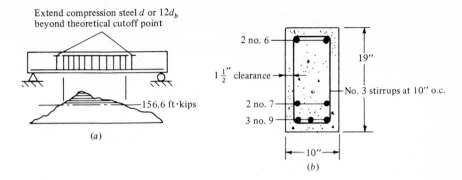

**Figure 3.49** Reinforcing details: (*a*) compression steel required in region where $M_u$ exceeds 156 ft · kips, (*b*) cross section.

SOLUTION Compute $T$ (bottom steel yields at failure)

$$T = A_s f_y = 4.12(50 \text{ kips/in}^2) = 206 \text{ kips}$$

Equate $T = C$, where $C = C_c + C_s$

$$206 \text{ kips} = C_c + C_s$$

Assuming that $f_s' = f_y$ leads to

$$C_s = A_s' f_y = 0.86(50) = 43 \text{ kips}$$

$$206 \text{ kips} = C_c + 43$$

$$C_c = 163 \text{ kips}$$

Solve for the area of the stress block and compute $a$

$$C_c = 0.85 f_c' ab$$

$$163 = 0.85(3)(10a)$$

$$a = 6.39 \text{ in}$$

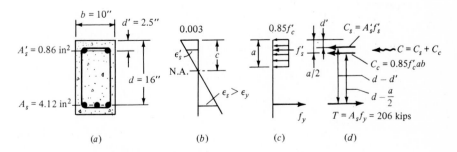

**Figure 3.50** (*a*) Cross section, (*b*) strains at failure, (*c*) stresses, (*d*) internal forces.

Compute

$$c = \frac{a}{\beta_1} = \frac{6.39 \text{ in}}{0.85} = 7.52 \text{ in}$$

Verify the stress in compression steel using the strain diagram (Fig. 3.51)

$$\frac{\epsilon_s'}{5.02 \text{ in}} = \frac{0.003}{7.52 \text{ in}}$$

Since $\epsilon_s' = 0.002 > \epsilon_y$, $f_s' = f_y$, as assumed above.

Sum moments about $T$ to calculate $M_u$

$$M_u = \phi(C_s \times \text{arm} + C_c \times \text{arm})$$

$$= 0.9\left[(43 \text{ kips})(16 - 2.5) + (163 \text{ kips})\left(16 - \frac{6.39}{2}\right)\right]$$

$$= 2400.9 \text{ in} \cdot \text{kips}$$

**Example 3.27:** Compare the flexural design strengths of the two cross sections in Fig. 3.52, both having the same area of tension steel; $f_c' = 3 \text{ kips/in}^2$, and $f_y = 50 \text{ kips/in}^2$.

SOLUTION Compute $M_u$ for the cross section in Fig. 3.52a (no compression steel)

$$\rho = \frac{A_s}{bd} = \frac{3}{12(17.5)} = 0.0143$$

In Table 3.3, read $K_u = 552.8$. Using Eq. (3.47)

$$M_u = K_u bd^2 = \frac{552.8(12)(17.5^2)}{12,000} = 169.3 \text{ ft} \cdot \text{kips}$$

Compute $M_u$ for cross section with compression steel (Fig. 3.52b) using the trial procedure in Sec. 3.11. Guess a trial value of $c = 4$ in; then $a = \beta_1 c = 0.85(4) = 3.4$ in

$$\epsilon_s' = \frac{1.5}{4}(0.003) = 0.001125$$

$$f_s' = \epsilon_s' E_s = 0.001125(29 \times 10^3) = 32.6 \text{ kips/in}^2$$

Compute $C = C_c + C_s$ and compare with $T = A_s f_y = 150$ kips

$$C = 0.85f_c' ab, + A_s' f_y = (2.55 \text{ kips/in}^2)(3.4)(12) + (2)(32.6 \text{ kips/in}^2)$$

$$= 169.2 \text{ kips}$$

Since $C$ exceeds $T$, $c$ was assumed too large initially; therefore, assume a smaller value of $c$, say 3.7 in and repeat the computations. For the new value of $c$, we compute

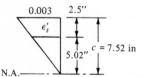

Figure 3.51 Strain distribution in compression zone.

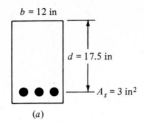

(a)

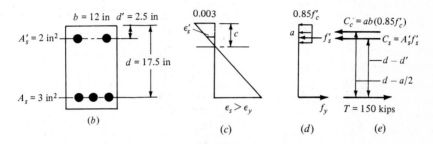

(b)  (c)  (d)  (e)

**Figure 3.52** (a) Tension steel only, (b) doubly reinforced, (c) strains, (d) stresses, (e) forces.

$f'_s = 28.2$ kips/in² and $C_c + C_s = 152.6$ kips. Because the value of $C_c + C_s$ is only slightly greater than $T(150 \text{ kips})$, the analysis is sufficiently accurate to establish $f'_s$. If $f'_s$ is rounded to 28 kips/in², we can compute $C_s$ as

$$C_s = A'_s f'_s = (2 \text{ in}^2)(28 \text{ kips/in}^2) = 56 \text{ kips}$$

Then $$C_c = T - C_s = 150 - 56 = 94 \text{ kips}$$

Next we solve for $a$

$$C_c = ab(0.85f'_c)$$

$$94 = a(12)(2.55)$$

$$a = 3.07 \text{ in}$$

Evaluate $M_u$ by summing moments of forces about $T$ to give

$$M_u = \phi\left[C_c\left(d - \frac{a}{2}\right) + C_s(d - d')\right]$$

$$= 0.9\left[94\left(17.5 - \frac{3.07}{2}\right) + 56(17.5 - 2.5)\right]\frac{1}{12} = 175.6 \text{ ft} \cdot \text{kip}$$

Example 3.27 indicates that compression steel in an underreinforced beam does not significantly increase the moment capacity. Since the moment capacity of an underreinforced beam is controlled by the tensile strength of the steel, to increase compressive strength by adding compression steel adds strength where it cannot be used. Compression steel does increase the beam's ductility and reduces creep.

## 3.12 DESIGN OF T BEAMS

### Introduction

Beams with T-shaped cross sections are used extensively as components of concrete structures. They occur most frequently when concrete beams are poured monolithically with slabs to form the floors of buildings and the roadways of bridges. Rigidly joined together by reinforcement, a portion of the slab acts with the beam to produce a T-shaped flexural member. The slab is termed the flange, and the portion of the beam that projects below the slab is called the stem (Fig. 3.53).

Although isolated T beams of poured-in-place concrete are uncommon, large quantities of T beams and double T beams are produced by the precast concrete industry for use as components of prefabricated buildings (see Fig. 3.54). These members are typically placed side by side with their flanges joined to form a floor. Since precast beams of the same nominal depth differ slightly in height as a result of the manufacturing process, several inches of concrete topping are often placed on top of the flanges to form a level surface. Light reinforcement, such as welded wire mesh or small-diameter deformed reinforcing bars, is added to the topping to provide continuity and reduce cracking.

### State of Stress at Failure

A T-shaped cross section is most efficiently used when the flange is placed in compression. The wide flange not only permits a large compression force to develop but also maximizes the arm of the internal couple by positioning the resultant of the compression stresses near the compression surface (see Fig. 3.55). The elimination of concrete from the tension zone, where only the steel reinforcement is effective in carrying tension, reduces the dead weight but does not influence the bending strength of the cross section. For long-span beams, where a large percentage of the design moment is produced by the dead weight of the member, use of the T-shaped section will result in a considerable reduction in weight, which in turn will permit the design of smaller and lighter members.

Since the flange of the typical T beam is wide, the depth of the stress block will normally be small. As a result, when failure occurs, the position of the neutral axis will usually be located in the flange near the compression surface. As shown in Fig. 3.55b, the strains in the steel at failure will be many times greater than those in the concrete because of the location of the neutral axis; therefore a ductile mode of failure associated with large deflections and extensive stretching of the steel is assured.

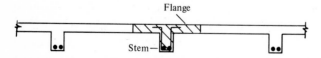

**Figure 3.53** Floor system with T beams.

(b)

(a)

**Figure 3.54** Precast beams: (a) T beam, (b) double T beam.

## Effective Width of Flange

In beams with a compact cross section, stress in the compression zone is assumed to be constant in magnitude across the width of the beam (Fig. 3.56). In T beams with *long thin flanges* the stresses vary across the flange width because of the shear deformations of the flange. The approximate variation of stress in the flange is shown in Fig. 3.57.

To simplify the design of T beams, the variable stress distribution acting over the full width of flange is replaced by an equivalent uniform stress, which is assumed to act over a reduced width $b_{eff}$, selected so that the uniform stress acting over the reduced width produces the same resultant compression force in the flange as the actual stress, which varies over the full width $b$. To establish the effective width of slab that acts as the compression zone for a beam that is a component of T-beam-and-slab construction, ACI Code §8.10 gives the following criteria.

**Case 1: Symmetrical cross section** The effective width (see Fig. 3.58a) is given by the smallest value of

1. One fourth of the beam's span length
2. The stem width plus a flange overhang of 8 times the slab thickness on each side of the stem
3. The stem width plus a flange overhang not greater than half the clear distance to the next beam

**Case 2: Beam with an L-shaped flange** The width of flange (see Fig. 3.58b) is to be taken as the stem width plus a flange *overhang* equal to the smallest of

1. One-twelfth the beam's span length
2. Six times the thickness of the slab
3. One-half the clear distance to the next beam

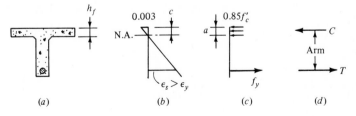

(a)          (b)          (c)          (d)

**Figure 3.55** State of stress in a T beam at failure: (a) cross section, (b) strains, (c) stresses, (d) internal couple.

**Figure 3.56** Variation of stresses in a compact section.

## Distribution of Flexural Reinforcement in the Flanges

**Longitudinal reinforcement**   When the flanges of T beams carry tensile stresses, ACI Code §10.6.6 requires that part of the main reinforcement be spread over a width equal to the smaller of the effective flange width (Fig. 3.45b) or a width equal to one-tenth of the span. Further, if the effective flange width exceeds one-tenth of the span, some reinforcement should be placed in the outer sections of the flange. This provision will ensure that many fine cracks rather than a few wide cracks will develop on the top surface perpendicular to the span of the beam.

**Transverse reinforcement**   Load applied directly to the flange of a T beam will cause the flanges to bend downward (Fig. 3.59). To prevent a bending failure of the flange, transverse reinforcement must be added to the top of the flange overhangs. This reinforcement can be sized by treating the flange overhangs as cantilevers fixed at the face of the stem and having a span equal to the length of the flange overhang (Fig. 3.59b). ACI Code §8.10.5.2 requires that the spacing of the transverse reinforcement not exceed 5 times the slab thickness or 18 in (457 mm). Additional longitudinal steel will be required in the flange to hold the transverse steel in position when concrete is poured.

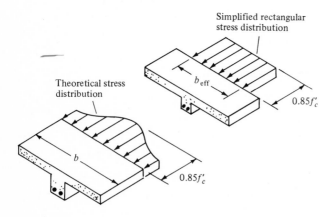

**Figure 3.57** Variation of compression stresses in the flange of a T beam.

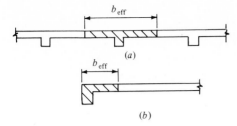

**Figure 3.58** Effective flange width.

## T-Beam Design

Most T beams occur as part of continuous floor systems (see Fig. 3.53). The dimensions of these beams are normally determined by the strength required to carry the shear and moment at the supports, where the compression zone is at the bottom of the web and the member acts as a rectangular beam whose width is equal to that of the web (see Fig. 3.45b). In regions of positive moment, where the flange is in compression, the designer has only to select the area of the flexural steel and verify that it can be placed in the web with the required spacing between bars. The minimum area of flexural steel equals $200b_w d/f_y$. In T-beam design, as in rectangular-beam design, the maximum area of steel to be used as flexural reinforcement is equal to $\frac{3}{4}A_{sb}$ (the procedure for computing $A_{sb}$ was given in Sec. 3.10).

**Case 1: Stress block confined to flange**  The design procedure for determining the moment capacity of a particular T-shaped cross section will depend on the position of the bottom of the stress block. If the stress block lies completely in the flange, the most common case, the beam is designed exactly like a rectangular beam (Fig. 3.60). On the other hand, if the bottom of the stress block falls in the web, the stress block must be divided into known areas, the forces on these areas computed, and the moment capacity of the cross section established by summing the forces in the compression zone about the centroid of the tension steel (Fig. 3.61). As an alternative, the trial method discussed in Sec. 3.9 can be used to select the reinforcement.

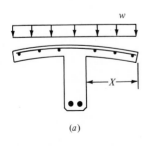

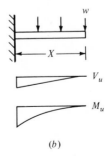

**Figure 3.59** (a) Transverse bending of T-beam flange; (b) shear and moment curves for flange overhang.

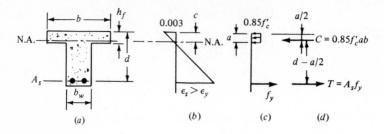

**Figure 3.60** (a) Cross section, (b) strains at failure, (c) stresses, (d) internal couple.

**Case 2: Stress block extended into stem** Break the total internal moment capacity $\phi M_n$ into two couples (Fig. 3.61). One couple represents the moment capacity of the flange overhangs, and the second represents the moment capacity of the rectangular beam portion. The total moment capacity is

$$\phi M_n = \phi(M_1 + M_2) \tag{1}$$

where

$$M_1 = 85 f'_c A_f \left(d - \frac{h_f}{2}\right) \tag{2}$$

$$M_2 = 0.85 f'_c A_w \left(d - \frac{a}{2}\right) \tag{3}$$

Substituting Eqs. (2) and (3) into (1) gives

$$\phi M_n = \phi\left[0.85 f'_c A_f\left(d - \frac{h_f}{2}\right) + 85 f'_c A_w\left(d - \frac{a}{2}\right)\right]$$

**Example 3.28: T-beam investigation** Determine the moment capacity of the T beam in Fig. 3.62; $f'_c = 2.5$ kips/in² (17.24 MPa), and $f_y = 60$ kips/in² (413.7 MPa).

SOLUTION Determine the area of the compression zone by equating $C = T$

$$0.85 f'_c A_c = A_s f_y = 280.8 \text{ kips}$$

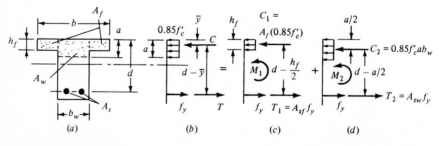

**Figure 3.61** (a) Cross section with $A_s = A_{sf} + A_{sw}$, where $A_{sf}$ is the portion of $A_s$ used to balance compression force in flange overhangs and $A_{sw}$ is the portion of $A_s$ used to balance compression force in web; (b) stresses; (c) moment: flange overhangs and $A_{sf}$; (d) moment: web and $A_{sw}$.

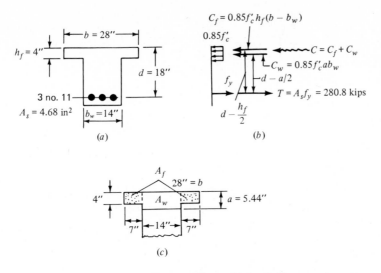

**Figure 3.62** (*a*) Cross section, (*b*) internal forces, (*c*) detail of compression zone.

where $A_c$, the area of the stress block, is

$$A_c = \frac{280.9}{2.5(0.85)} = 132.1 \text{ in}^2$$

Break $A_c$ into two parts, the area of flange overhangs $A_f$ and the area of web $A_w$ (see Fig. 3.62*c*)

$$A_f = 14(4) = 56 \text{ in}^2$$

$$A_w = 132.1 - 56 = 76.1 \text{ in}^2$$

$$a(14 \text{ in}) = 76.1$$

$$a = 5.44 \text{ in}$$

Represent the compression force in the flange overhangs by $C_f$ and the compression force in the web by $C_w$ (Fig. 3.62*b*)

$$C_f = A_f 0.85 f'_c \qquad\qquad C_w = A_w 0.85 f'_c$$
$$C_f = 56(0.85)(2.5) = 119 \text{ kips} \qquad C_w = 76.1(0.85)(2.5) = 161.8 \text{ kips}$$

Compute $\phi M_n$ by summing the compression forces about the centroid of the tension steel

$$\phi M_n = \phi \left[ C_f \left( d - \frac{h_f}{2} \right) + C_w \left( d - \frac{a}{2} \right) \right]$$

$$= 0.9[(119 \text{ kips})(16 \text{ in}) + (161.8 \text{ kips})(15.28 \text{ in})] = 3939 \text{ in} \cdot \text{kips} = 328.25 \text{ ft} \cdot \text{kips}$$

**Example 3.29: T-beam design by method of trials**   What area of steel is required in the T beam in Fig. 3.63 to carry $M_u = 100 \text{ ft} \cdot \text{kips}$? $f'_c = 3 \text{ kips/in}^2$, and $f_y = 60 \text{ kips/in}^2$.

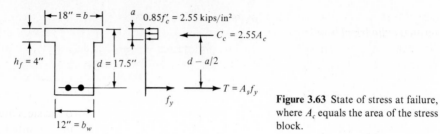

**Figure 3.63** State of stress at failure, where $A_c$ equals the area of the stress block.

SOLUTION Compute an approximate value of $A_s$ by assuming a value of $a$. A good estimate can be made by assuming $a = h_f = 4$ in. Sum moments about $C_c$

$$M_u = \phi T\left(d - \frac{a}{2}\right)$$

$$100(12) = 0.9T(17.5 - 2)$$

$$T = 86 \text{ kips}$$

(T is approximate because $a$ is not exact.)

$$A_s \approx \frac{T}{f_y} \approx \frac{86}{60} = 1.43 \text{ in}^2$$

Recompute an improved value of $a$ by equating $C_c = T$

$$86 \text{ kips} = A_c(0.85)(3)$$

Since $A_c = 33.7 \text{ in}^2$, $a$ lies in the flange

$$a(18) = 33.7 \quad \text{and} \quad a = 1.87 \text{ in}$$

Repeat the analysis using the computed value of $a$

$$M_u = \phi T\left(d - \frac{a}{2}\right)$$

$$100(12) = 0.9T\left(17.5 - \frac{1.87}{2}\right)$$

$$T = 80.5 \text{ kips}$$

$$A_s \approx \frac{T}{f_y} = \frac{80.5}{60} = 1.34 \text{ in}^2$$

Use 2 no. 8 bars, $A_s$ supplied $= 1.57 \text{ in}^2$.

## 3.13 ONE-WAY SLABS

Slabs, used extensively in roof and floor systems, are structural elements whose width and length are large compared with their thickness. Except for heavily loaded slabs, the minimum thickness of slabs is controlled by deflection limitations (see Table 3.1). If the slab is supported so that it bends in one direction only, termed *cylindrical bending*, it is classified as a *one-way slab*. In the design of one-way slabs the assumption is made that the slab behaves like a series of individual beams placed side by side. Typically, a 1-ft (305-mm)-wide strip of slab is analyzed and the reinforcement required for that strip is used throughout the width of the slab. Normally, uniformly spaced bars of a single size are used as slab reinforcement. If a single bar were available with an area $A_b$ that was equal to the steel area $A_s$ required for a 12-in strip of slab, bars spaced 12 in on center would supply the correct area of steel. Since a limited number of bar sizes are available, the spacing of bars must be varied to supply the required area of steel. An equation for the required spacing can be derived by expressing in terms of steel area per 12 in of slab width (1) the area of steel $A_s$ required for a 12-in strip and (2) the area of steel supplied by bars of area $A_b$ spaced $s$ inches apart. Equating these quantities we may write

$$\frac{A_s}{12} = \frac{A_b}{s}$$

and

$$s = \frac{12A_b}{A_s} \quad \text{in} \tag{3.54}$$

As an alternate to Eq. (3.54), the area of steel supplied in a 12-in width of slab by uniformly spaced bars can be read directly from Table B.2 in the Appendix.

If the slab is continuous over several supports (see Fig. 3.64b), it must be analyzed as a continuous beam. As indicated by the curvature, continuous slabs require top steel over the supports and bottom steel in the midspan sections to carry tensile stresses. The analysis of floor systems composed of continuous beams and slabs is covered in Chap. 9.

Although one-way slabs are assumed to bend in one direction only, reinforcement must also be placed perpendicular to the flexural steel to control crack width due to dimensional changes in the slab produced by a drop in temperature and shrinkage (see Sec. 3.14 for details).

### T-Beam–and–Slab Construction

T-beam–and–slab construction is a common structural system that results when slabs are poured monolithically with beams (see Fig. 3.65). In this system the slabs contribute to the load transmission in two directions. In the direction in which the

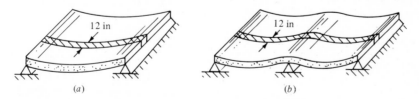

**Figure 3.64** One-way slabs: (*a*) simply supported, (*b*) continuous.

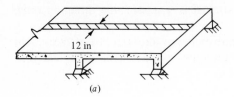

12 in

(a)

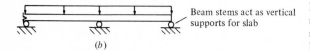

Beam stems act as vertical
supports for slab

(b)

**Figure 3.65** T beam and slab
system: (a) floor system; (b) 1-ft
strip of slab idealized as continuous
beam.

beams span, the slab acts with the stem to form a T beam in regions of positive
moment. In the direction perpendicular to the span of the beams, the slab is assumed
to span as a continuous beam between the beam stems, which are assumed to act
as vertical supports for the slab. The design of T-beam–and–slab systems is
discussed in detail in Chap. 9.

## 3.14 TEMPERATURE AND SHRINKAGE STEEL

When a slab is restrained along the boundaries, as most slabs are by edge beams
or by adjacent slabs, longitudinal tensile stresses are induced in the slab by a drop
in temperature or by shrinkage. These longitudinal stresses, which resist the contraction of the slab, produce cracks. If the area of slab steel is too small, cracks
appear in a zigzag irregular pattern wherever the slab is weakest. To minimize
their effect, the designer must add reinforcement called *shrinkage* and *temperature
steel* or simply *temperature steel.* This reinforcement is placed perpendicular to the
main flexural steel and is wired to it to form a rigid mat. The Code requires that
the area of steel supplied for moment be not less than that required for temperature.

If temperature steel is used, a large number of narrow cracks develop. Since
they are not visible to the naked eye, unsightly cracking that detracts from the
appearance of the slab is avoided. Temperature steel also serves the secondary
function of holding the main steel securely in place during construction while the
concrete is poured into the forms and provides some transverse bending strength
to distribute concentrated loads laterally so that the possibility of local overstress
is reduced.

Temperature steel is designed on an empirical basis. The minimum required
ratio of temperature steel area to the gross area of the concrete slab for various
grades of deformed bars is specified by ACI Code §7.12 as

$$\frac{A_s}{A_g} = \begin{cases} 0.002 & \text{for } f_y = 40 \text{ to } 50 \text{ kips/in}^2 \\ 0.0018 & \text{for } f_y = 60 \text{ kips/in}^2 \\ \dfrac{0.0018(60,000)}{f_y \text{ lb/in}^2} & \text{for } f_y > 60 \text{ kips/in}^2 \end{cases}$$

where $A_s$ is the required area of the temperature steel and $A_g$ the gross area of the concrete.

In any case, the ratio $A_s/A_g$ must not be less than 0.0014, and the spacing between bars is not to exceed 18 in or 5 times the slab thickness.

**Example 3.30:** Compute the required spacing of temperature steel for the one-way slab in Fig. 3.66; $f_y = 50$ kips/in². Use no. 4 bars.

SOLUTION

$$A_s = 0.002 \, A_g = 0.002(8)(12) = 0.192 \text{ in}^2$$

Use no. 4 bars at 12 in on center; $A_s$ supplied $= 0.2$ in²

**Example 3.31: Design of a one-way slab** Determine the required depth and design the reinforcement for the uniformly loaded slab spanning 16 ft between simple supports in Fig. 3.67. In addition to its dead weight, the slab carries a live load of 100 lb/ft²; $f'_c = 3$ kips/in², and $f_y = 60$ kips/in².

SOLUTION To establish the flexural reinforcement, analyze a 1-ft-wide strip of slab as a simply supported beam. Using Table 3.1, establish the depth of slab required to control deflections

$$h \geqslant \frac{l}{20} = \frac{16(12)}{20} = 9.6 \text{ in} \qquad \text{use 10 in}$$

Providing $\frac{3}{4}$ in (19 mm) of clear cover to protect the reinforcement and allowing $\frac{1}{4}$ in for half a bar diameter gives $d = 9$ in. The weight of slab per square foot is

$$150(\tfrac{10}{12}) = 124.5 \text{ lb/ft}^2$$

The factored design load is

$$w_u = 1.7(100) + 1.4(124.5) = 344 \text{ lb/ft}^2$$

When a 1-ft strip is analyzed as a uniformly loaded simple beam, the moment $M_u$ at midspan is

$$M_u = \frac{w_u l^2}{8} = \frac{0.344(16^2)}{8} = 11.02 \text{ ft} \cdot \text{kips}$$

Solve for the area of flexural steel $A_s$ using the trial method

$$M_u = \phi A_s f_y \left( d - \frac{a}{2} \right)$$

Guess $a = 1$ in

$$11.02(12) = 0.9 A_s 60(9 - \tfrac{1}{2}) \qquad \text{and} \qquad A_s = 0.288 \text{ in}^2$$

Equating $T = C$, gives $a = 0.56$ in; repeating the analysis with $a = 0.56$ in, we compute $A_s = 0.281$ in².

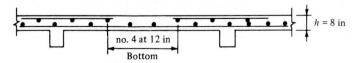

no. 4 at 12 in

Bottom

$h = 8$ in

**Figure 3.66**

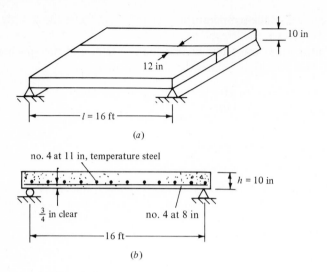

**Figure 3.67** (*a*) Simply supported one-way slab; (*b*) reinforcing details.

Alternately, $\rho$ may be determined from Table 3.5. Interpolating between $\rho = 0.002$ and 0.003, we compute $\rho = 0.0026$; then $A_s = \rho bd = 0.0026(12)9 = 0.281$ in$^2$.
Compute spacing of reinforcement by Eq. (3.54) if no. 4 bars ($A_b = 0.2$ in$^2$) are used

$$s = \frac{12A_b}{A_s} = \frac{12(0.2)}{0.281} = 8.5 \text{ in} \qquad \text{round to 8 in}$$

Alternatively the required spacing can be determined from Table B.2, which indicates that no. 4 bars at 8 in supply 0.29 in$^2$. To tie the slab together in the transverse direction and to prevent the formation of a small number of wide cracks add temperature and shrinkage steel

$$\frac{A_s}{A_g} = 0.0018 \qquad A_s = 0.0018(10)(12) = 0.216 \text{ in}^2$$

Use no. 4 bars at 11 in on center placed on top of flexural steel. See Fig. 3.67*b* for reinforcing details.

## PROBLEMS

**3.1** Determine the smallest value of uniform load that will cause the reinforced concrete beam in Fig. P3.1 to crack: $f_c' = 3000$ lb/in$^2$.

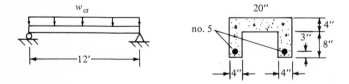

**Figure P3.1**

(a) Beam is constructed of stone concrete.

(b) Beam is constructed of lightweight concrete with $f_{ct} = 320$ lb/in$^2$.

**3.2** Determine the cracking moment $M_{cr}$ of the T-beam cross section in Fig. P3.2 if $f'_c = 3.6$ kips/in$^2$.

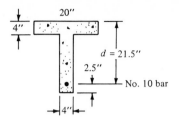

**Figure P3.2**

**3.3** The beam is constructed of stone concrete with $f'_c = 3$ kips/in$^2$. Will it crack if the loads produce the moment curve shown in Fig. P3.3? If the beam cracks, indicate the location of the cracks.

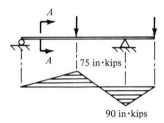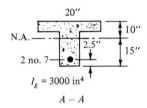

**Figure P3.3**

**3.4** The beam in Fig. P3.4 carries two equal loads. What value of these loads will just cause initial cracking of the beam? Neglect the weight of the beam; $f'_c = 3.6$ kips/in$^2$ and $f_y = 60$ kips/in$^2$.

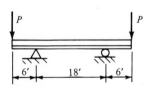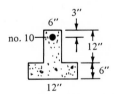

**Figure P3.4**

**3.5** For each of the cross sections in Fig. P3.5 compute the moment of inertia of the cracked transformed section. All members are constructed of stone concrete. Figure P3.5 continues on page 130.

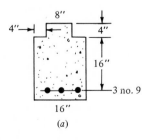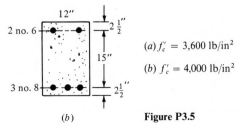

(a) $f'_c = 3,600$ lb/in$^2$

(b) $f'_c = 4,000$ lb/in$^2$

**Figure P3.5**

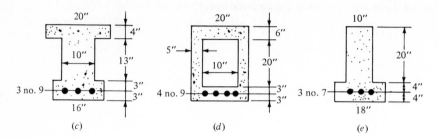

(c)    (d)    (e)

**Figure P3.5 (continued)**

(c) $f'_c = 2500 \text{ lb/in}^2$     (d) $f'_c = 4000 \text{ lb/in}^2$     (e) $f'_c = 3000 \text{ lb/in}^2$

**3.6** What is the minimum depth required for the continuous beam in Fig. P3.6 to control deflections if $f'_c = 4 \text{ kips/in}^2$, $f_y = 50 \text{ kips/in}^2$, and $w = 120 \text{ lb/ft}^3$?

**Figure P3.6**

**3.7** The floor beam in Fig. P3.7 supports nonstructural elements not likely to be damaged by large deflections. If the dead load equals 1.4 kips/ft and the live load equals 0.8 kip/ft, does the beam satisfy ACI Code requirements for allowable deflections? $f'_c = 3.5 \text{ kips/in}^2$, $f_y = 60 \text{ kips/in}^2$, sand-lightweight concrete $w = 135 \text{ lb/ft}^3$. Assume live load acts continuously.

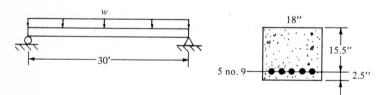

**Figure P3.7**

**3.8** Under what conditions is $I_{cr}$ the appropriate value of moment of inertia to use in deflection computations?

**3.9** For the beam in Fig. P3.9 compute (a) the immediate deflection produced by total load, and (b) estimate the additional long-term deflection if the live load acts continuously. $f'_c = 4 \text{ kips/in}^2$ and $f_y = 50 \text{ kips/in}^2$.

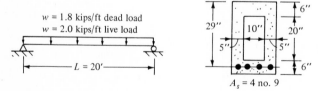

**Figure P3.9**

**3.10** The dead weight of the beam in Fig. P3.10 is 384 lb/ft. Compute (*a*) the midspan deflection due to dead load and (*b*) the midspan deflection due to total load. $f'_c = 3$ kips/in².

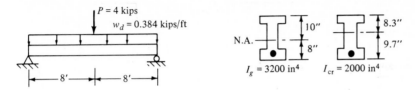

**Figure 3.10**

**3.11** For each of the cross sections in Fig. P3.11 determine whether the reinforcement satisfies the requirements of ACI Code §10.6.4 for controlling crack width. Beams subject to exterior exposure. Also estimate the width of cracks on the tension surface. $f'_c = 3.6$ kips/in², $f_y = 60$ kips/in².

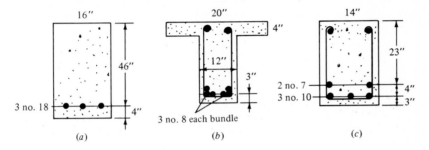

**Figure P3.11** Note that a beam over 3 ft also needs additional steel in the upper region of the tension zone.

**3.12** Determine the area and spacing of the additional reinforcement required by the code to control cracking in the web of the deep beam in Fig. P3.12.

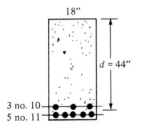

**Figure P3.12**

**3.13** Determine the flexural strength $\phi M_n$ and the strain in the tension steel at failure for the cross section in Fig. P3.13 if $f_y = 50$ kips/in² and (*a*) $f'_c = 2.5$ kips/in² and (*b*) $f'_c = 4$ kips/in². Which strength concrete produces the most ductile member?

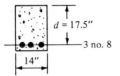

**Figure P3.13**

**3.14** Each leg of the cross section in Fig. P3.14 is reinforced with a no. 9 bar. Determine (a) the flexural capacity $M_u$ of the cross section and (b) the strain in the steel at failure. $f'_c = 3.6$ kips/in$^2$, and $f_y = 50$ kips/in$^2$.

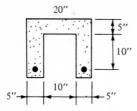

**Figure P3.14**

**3.15** Based on the design strength $\phi M_n$ of the cross section, determine the maximum distance a simply supported beam can span if it has the cross section shown in Fig. P3.15. The beam must carry a uniform service dead load of 0.8 kip/ft and a uniform service live load of 0.9 kip/ft. $f'_c = 3.5$ kips/in$^2$, and $f_y = 60$ kips/in$^2$.

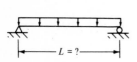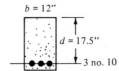

**Figure P3.15**

**3.16** Determine the flexural design strength of the triangular cross section in Fig. P3.16. $f'_c = 3.6$ kips/in$^2$, and $f_y = 50$ kips/in$^2$.

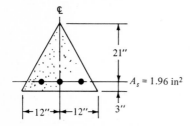

**Figure P3.16**

**3.17** In addition to its own dead weight, which must be estimated, the beam in Fig. P3.17 carries uniform service loads of 800 lb/ft dead load and 1500 lb/ft live load. Exterior exposure. Design the beam for (a) a 24-ft span, $d = 2b$; (b) a 60-ft span, $d = 2.5b$. Final dimensions should be rounded to the even inch. $f'_c = 3$ kips/in$^2$, $f_y = 60$ kips/in$^2$.

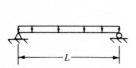

**Figure P3.17**

**3.18** Design a 10-in-wide beam of constant depth with a rectangular cross section to carry the service loads shown in Fig. P3.18. The dead load contains an allowance for the weight of the beam. Establish the proportions of the cross section at the point of maximum moment using $\rho = 5/8\rho_b$. Then select the area of steel at the points of maximum moment. Neglect consideration of deflections when setting the depth but check $z$.

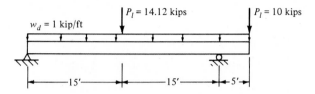

**Figure P3.18**

**3.19** If the beam in Fig. P3.19 is reinforced with the minimum area of steel permitted by the ACI Code, what is its flexural design strength $\phi M_n$? $f'_c = 3$ kips/in$^2$, and $f_y = 60$ kips/in$^2$.

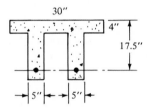

**Figure P3.19**

**3.20** Estimate the flexural design strength $\phi M_n$ of the cross section in Fig. P3.20 within 5 percent of the exact value. Also estimate the stress in the compression steel at failure. $f'_c = 3$ kips/in$^2$, and $f_y = 60$ kips/in$^2$.

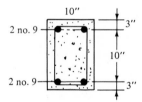

**Figure P3.20**

**3.21** A precast beam with the cross section shown in Fig. P3.21 must carry a factored moment $M_u = 280$ ft · kips. Determine the required reinforcement (compression steel required). $f'_c = 4$ kips/in$^2$, and $f_y = 60$ kips/in$^2$.

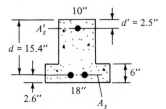

**Figure P3.21**

**3.22** The T-shaped section in Fig. P3.22 is loaded by a negative moment $M_u = 180$ ft · kips that produces tension in the flange. (*a*) What is required area of flexural steel? (*b*) Verify that the beam is underreinforced. $f'_c = 3$ kips/in², and $f_y = 72$ kips/in².

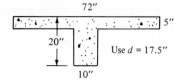

Use $d = 17.5''$

**Figure P3.22**

**3.23** Determine the balanced area of steel for the cross section in Fig. P3.23; $f'_c = 3$ kips/in², and $f_y = 60$ kips/in².

**Figure P3.23**

**3.24** The hollow box beam in Fig. P3.24 must carry a factored moment of 400 ft · kips. (*a*) Select the area of flexural steel and (*b*) determine balanced steel for the cross section. $f'_c = 4$ kips/in², and $f_y = 50$ kips/in².

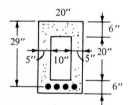

**Figure P3.24**

**3.25** Determine the flexural design strength $\phi M_n$ of the cross section in Fig. P3.25. $f'_c = 3$ kips/in², and $f_y = 60$ kips/in².

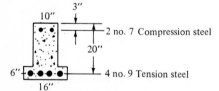

2 no. 7 Compression steel

4 no. 9 Tension steel

**Figure P3.25**

**3.26** What area of tension steel is required if the T beam in Fig. P3.26 is to carry a factored moment $M_u = 400$ ft · kips? $f'_c = 3.6$ kips/in², and $f_y = 50$ kips/in².

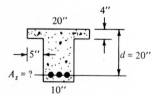

$A_s = ?$

$d = 20''$

**Figure P3.26**

**3.27** A factored moment $M_u = 24$ ft · kips acts on the cross section in Fig. P3.27. What area of flexural steel is required by the ACI Code? Check the area required for strength as well as the minimum steel area. $f'_c = 3.5$ kips/in$^2$, and $f_y = 50,000$ lb/in$^2$.

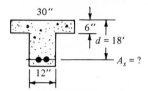

**Figure P3.27**

**3.28** Considering only moment, design a double T beam (Fig. P3.28) to carry its own weight and a service live load of 150 lb/ft$^2$. Exterior exposure. $f'_c = 3.5$ kips/in$^2$, and $f_y = 75$ kips/in$^2$.

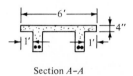
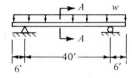

Section A–A

**Figure P3.28**

**3.29** Determine (a) the flexural design strength $\phi M_n$ of the T beam in Fig. P3.29 and (b) balanced steel for the cross section. $f'_c = 3.5$ kips/in$^2$, and $f_y = 50$ kips/in$^2$.

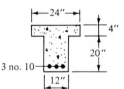

**Figure P3.29**

# REFERENCES

1. ACI Committee 435: Variability of Deflections of Simply Supported Reinforced Concrete Beams, *J. ACI*, vol. 69, no. 1, pp. 29–35, January 1972.
2. ACI Committee 435: Allowable Deflections, *J. ACI*, vol. 65, no. 6, pp. 433–44, June 1968.
3. ACI Committee 224: Control of Cracking in Concrete Structures, *J. ACI*, vol. 69, no. 12, pp. 717–752, December 1972.
4. A. H. Mattock, L. B. Kriz, and E. Hognestad: Rectangular Concrete Stress Distribution in Ultimate Strength Design, *J. ACI*, vol. 57, pp. 875–929, February 1961.
5. A. H. Mattock and L. B. Kriz: Ultimate Strength of Nonrectangular Structural Concrete Members, *J. ACI*, vol. 57, pp. 737–766, January 1961.
6. Ultimate Strength Design Handbook, vol. 1, *ACI Spec. Publ.* 17, Detroit, 1971.
7. P. Park and T. Pauly: "Reinforced Concrete Structures," Wiley-Interscience, New York, 1975.
8. P. Gergely and L. Lutz: "Maximum Crack Width in Reinforced Concrete Flexural Members," *Causes, Mechanism, and Control of Cracking in Concrete*, SP-20, ACI, Detroit, 1968.

East Fork Chowchilla River Bridge, California; a posttensioned cast-in-place box girder bridge. (*Photograph by the Post-tensioning Institute.*)

# FOUR

## SHEAR AND DIAGONAL TENSION

### 4.1 INTRODUCTION

Loads applied to beams produce not only the bending moments considered in Chap. 3 but also internal shear forces. The free-body diagrams in Fig. 4.1*b*, cut by section 1-1, show the internal shear force *V* acting perpendicular to the longitudinal axis of the beam. Shear is the internal force required to produce vertical equilibrium of the free bodies.

In most cases, stresses produced by shear forces are much smaller than those produced by moment; therefore most beams are routinely proportioned for moment rather than shear. Once the beam has been sized, shear is checked to determine whether shear reinforcing is required. The only significant exception is a very short beam carrying heavy loads, where the shear stresses may be high enough to govern the proportions of the cross section.

Although reinforced concrete beams, as a standard procedure, are proportioned for moment without considering the interaction of shear, shear in certain parts of the beam will often exceed the beam's shear strength and in those locations shear reinforcement will be required to offset the possibility of a failure.

In reinforced concrete beams shear does not produce failure directly on the vertical plane on which it acts, as one might anticipate. The major effect of shear is to induce tensile stresses on diagonal planes oriented at 45° to the plane on which the shear acts. Since concrete has a relatively low tensile strength compared with its shear and compressive strengths, overstress will always be initiated by tension stresses. When these diagonal tension stresses in combination with bending stresses created by moment exceed the tensile strength of the concrete, diagonal cracking develops that can split the beam (Fig. 4.2). These *diagonal tension failures*

137

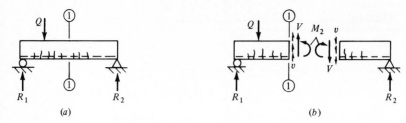

**Figure 4.1** Shear force and bending moment: (a) loaded beam; (b) free-body diagrams of the beam, cut by section 1-1, showing internal forces; $V$ = shear.

occur without warning; since they frequently cause the beam to collapse completely, they are dangerous and must be prevented by the design process.

Shear produces two types of diagonal cracks in reinforced concrete beams. *Flexure-shear cracks*, the most common type, develop from the tip of a flexural crack. For these cracks to form, the bending moment must exceed the cracking moment of the cross section and a significant shear must exist. A second type, a *web-shear crack*, can open near middepth at the level of the centroid of an uncracked section when the diagonal tension stresses produced by shear exceed the tensile strength of the concrete. This type of crack moves on a diagonal path to the tension surface. Although uncommon, web-shear cracks occasionally develop in beams with thin webs in regions of high shear and low moment, a combination of internal forces that exists adjacent to simple supports or at points of inflection in continuous beams (Fig. 4.3).

As has been pointed out, concrete beams crack under normal service loads due to moment; such cracks do not lead to failure because the bending steel is positioned and proportioned to carry the tension created by moment. However, this normal type of flexural cracking reduces the effective cross section available to carry the shear forces. As a result, flexural cracking increases the intensity of the shear stresses on the uncracked section above the tip of the crack and increases the likelihood of a diagonal tension failure.

Since the objective of the ACI Code is to produce ductile beams that do not rupture at failure, the designer must provide a greater factor of safety against a shear failure than against a bending failure, thereby ensuring that a bending failure will always precede a shear failure.

In regions of the beam where shear is low, the concrete can normally sustain the diagonal tension stresses, but in regions of high shear, beams require reinforcement for diagonal tension. When reinforcement is required, hoops of steel, called *stirrups*, or bent bars are provided in regions where shear reinforcement is necessary (Fig. 4.4).

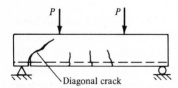

Diagonal crack

**Figure 4.2** Diagonal tension failure.

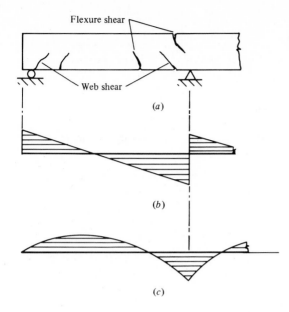

Figure **4.3** Web-shear and flexure-shear cracks: (*a*) types of cracks, (*b*) shear curve, (*c*) moment curve.

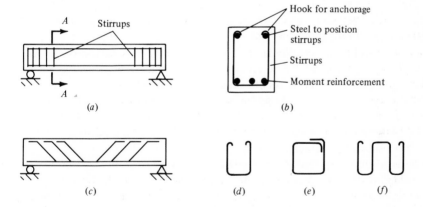

Figure **4.4** Types of shear reinforcement: (*a*) vertical steel, (*b*) section *A-A*, (*c*) bent bars; types of stirrups: (*d*) open, (*e*) closed, (*f*) multileg.

## 4.2 DESCRIPTION OF A SHEAR FAILURE IN A BEAM REINFORCED FOR MOMENT

To establish the effect of shear stresses on the behavior of a beam designed and reinforced for moment, the pattern of cracking in the beam of Fig. 4.5 will be examined as it is loaded to failure by concentrated loads acting at the third points. In this discussion, the contribution of the beam's own weight to the shear and moment will be neglected. This assumption limits shear to the outer regions of

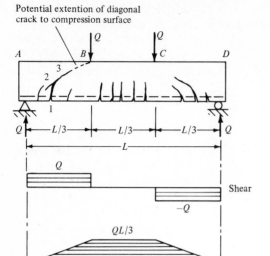

**Figure 4.5** Influence of shear on shape of crack.

the beam and eliminates its influence on the center section. In a real design situation, however, the effect of the beam's weight on the design loads must be taken into account.

Between points $B$ and $C$ of the beam under consideration, i.e., in the middle third of the beam, the shear is zero, and only moment exists. In this center region vertical cracks develop perpendicular to the direction of the maximum tensile stresses produced by moment. These cracks initiate at the bottom surface, where the tensile stresses are greatest. As the load increases, the cracks continue to propagate, always in a vertical direction. (The response of a beam to moment was covered in detail in Chap. 3.) However, at both ends of the beam, where both moment and shear exist (between $A$ and $B$ and between $C$ and $D$), a different mode of cracking occurs. These cracks also initiate at the bottom surface and extend first in a vertical direction, but as they move upward, their slope decreases progressively, reaching 45° near middepth. As the cracks move into the upper part of the compression zone, the slope gradually approaches the horizontal. Although most cracks stop when they reach the heavily stressed region of the upper compression zone, at a particular value of load a critical diagonal crack may suddenly tear through the beam to the compression surface. As the beam splits into two sections, a sudden brittle failure occurs. If crack patterns produced by moment in the center region are compared with those produced by shear and moment in the end regions, it becomes evident that shear has combined with moment to produce diagonal cracking, which under certain conditions can lead to an undesirable mode of failure.

Since concrete has a low tensile strength compared with its shear and compressive strengths, crack patterns must reflect the direction of the maximum or principal tensile stresses. To verify that the shape of the major diagonal crack between points $A$ and $B$ in Fig. 4.5 is produced by tensile stresses, the direction of the maximum tensile stress at the bottom surface (point 1), at the neutral axis (point 2), and at the center of the compression zone (point 3) will be established. At the points selected it will be seen that the slopes of the diagonal cracks are distinctly different because different stresses dominate. Although both shear and bending stresses influence the state of stress at all points along a diagonal crack, for clarity the following discussion will simplify the state of stress by neglecting the influence of stresses that are small.

Along the bottom edge of the uncracked beam, moment creates horizontal tensile bending stresses; shear stresses at this point are zero. When the magnitude of the tensile bending stresses reaches the modulus of rupture, i.e., the tensile bending strength, a vertical crack develops. As the crack moves up from the bottom surface, the shear stresses begin to increase. These stresses, combined with tensile bending stresses, cause the crack to bend and move upward at a reduced slope. Of course, as the crack propagates upward, the shear and the bending stresses on the cross section change in intensity. The crack will extend upward to the vicinity of the neutral axis of the cross section before stopping. After the initial crack forms, the cross section of the beam available to carry shear is reduced; therefore the shear stresses rise sharply on the uncracked section above the crack. At the neutral axis of the beam, bending stresses are zero, but shear stresses attain their maximum intensity. Now shear determines the behavior of the crack. The state of stress at this point can be studied by cutting out an infinitesimal square element of concrete at the level of the neutral axis (labeled 2 in Fig. 4.5). From well-known principles of strength of materials, shear stresses on vertical planes are known to induce shear stresses of equal intensity on horizontal planes at the same point. To maintain moment equilibrium, the pairs of stresses on the horizontal and vertical planes are directed so that they form two sets of equal and opposite couples (Fig. 4.6a). By the Mohr's circle method, a state of pure shear at a point in a stressed body can be shown to be equivalent to a state of pure tension and compression on diagonal planes oriented at 45° to the vertical (Fig. 4.6b). This analysis shows that the diagonal stresses are equal in magnitude to the shear stresses (Fig. 4.6c).

That a state of pure shear produces diagonal tension can also be seen by considering the deformations produced by shear stresses. In Fig. 4.6d the shear stresses cause point B to move upward and to the left to B′ and point D to displace downward and to the right to D′. As a result of these displacements, the diagonal from B to D must elongate, creating tension stresses in the direction of diagonal BD that produce diagonal cracks when the tensile strength of the concrete is exceeded. From the preceding analysis it is clear that the 45° slope of the diagonal cracks at middepth is due to diagonal tension created by shear.

As the diagonal crack extends into the compression zone with increasing load, the cross section above the crack continues to decrease and the shear stresses continue to increase. Even though moment produces large horizontal compression stresses that reduce the tension created by shear, the Mohr's circle analysis shows that tensile stresses will still develop on planes that slope upward and to the right, at a slight angle (Fig. 4.6g). The state of stress existing in this zone (Fig. 4.6e) is

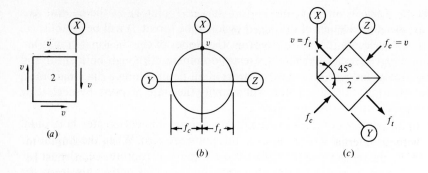

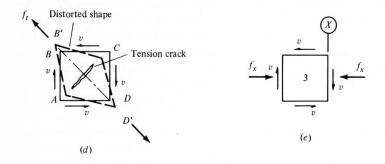

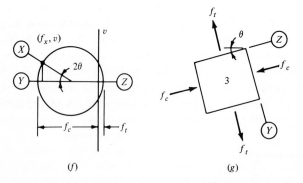

**Figure 4.6** Principal stresses on the cross section of a beam. State of stress at neutral axis: (*a*) pure shear, (*b*) Mohr's circle, (*c*) principal stresses. (*d*) Shear deformation of an element at the neutral axis. State of stress in the compression zone: (*e*) stresses, (*f*) Mohr's circle, (*g*) principal stresses.

established in Fig. 4.6*f* by the Mohr's circle analysis of the stresses on an infinitesimal square element of concrete at point 3 in Fig. 4.5.

Figure 4.7 shows the principal tension stresses and their orientation at the three points just investigated. It is evident that the shape of the diagonal crack is consistent with the orientation of these stresses.

Once the diagonal crack has opened, failure will occur unless the cracked cross section can transfer the applied loads to the supports. In Fig. 4.8 a free-body diagram of the portion of the beam to the left of a major diagonal crack is shown, together with the internal forces that develop in the member. If a shear failure is to be averted, $V_u$, the resultant of the external loads in the vertical direction, must be equilibrated by the forces $V_c$ and $V_D$ and the vertical component of $V_a$. $V_c$ represents the shear carried by the compression zone above the crack. $V_D$, the dowel force, stands for the shear carried by the flexural steel through dowel action. $V_a$ denotes the interface shear, or aggregate interlock, developed by friction along the irregular surface of the diagonal crack. If a beam can develop sufficient additional shear resistance from the above sources, it will not fail when the diagonal crack forms. Since it is impossible to estimate this additional shear capacity, which may decrease with time, the designer neglects it. Thus the shear force existing at the section where the diagonal crack opens is considered a measure of the ultimate shear strength of the section.

## Overload Behavior of a Beam Explained by Arch Action

After the formation of a major diagonal crack, the capacity of a beam to support additional load will depend on its ability to undergo an internal adjustment of stress and to develop a new structural form that has a greater capacity for load than the original system. Although no reserve of strength after diagonal cracking is used in design, one can develop further understanding of beam behavior after diagonal cracking by considering the beam as a tied arch. Such a study is useful in bringing the designer's attention to certain details that will increase the ultimate strength of heavily cracked beams. Figure 4.9 shows a beam that has developed large diagonal cracks. In addition to diagonal cracking, extensive flexural cracking has also occurred throughout the tension zone, wherever the moments created by load have exceeded the cracking moment of the cross section. The heavily cracked concrete in the tension zone contributes little to the load-carrying capacity of the beam. If the cracked concrete in the tension zone is disregarded, the beam can be seen to resemble a tied arch (Fig. 4.9*b*). The tension steel now acts

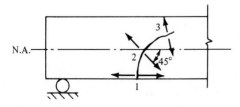

**Figure 4.7** Orientation of principal tension stresses along a diagonal crack.

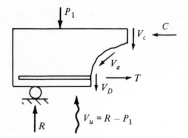

**Figure 4.8** Shear forces on a partially cracked section; $V_c$ = shear carried by compression zone above crack, $V_a$ = interface (or aggregate) interlock shear produced by friction that develops along surfaces of crack, $V_D$ = shear carried by reinforcing steel in dowel action.

as a tension tie to carry the horizontal component of the arch thrust. As an arch, the structure can be assumed to function primarily in direct stress. Figure 4.9c shows an idealized model of the arch. Points $ABCD$ represent the centerline of the concrete arch. The forces in the arch members can be established in terms of the applied forces and the geometry of the arch by considering equilibrium of forces at joints $A$ and $B$.

Several failure modes of the arch are possible. If the diagonal crack causes too great a reduction in the area of the compression zone at points $B$ or $C$, failure of the concrete by overstress in compression, termed a *shear-compression failure*, can occur. Failure can also occur if the compressive strength of member $AB$ is exceeded. A compression failure in this region is most likely in a member with a thin web.

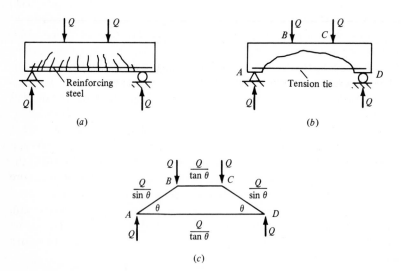

**Figure 4.9** Arch action in a heavily loaded beam: (*a*) cracked beam, (*b*) equivalent tied arch, (*c*) idealized model of tied arch.

When the member functions as a beam, the stress in the tension steel is a direct function of the moment. In Sec. 3.3 it was shown that the stress in the tension steel of a concrete beam at services loads can be computed by

$$f_s = \frac{M_s}{A_s \, jd} \tag{3.13}$$

At the end of a simply supported reinforced concrete beam, where the moment is zero, this means that the steel stress at that point will also reduce to zero and anchorage of the steel to the concrete will not be a problem. However, once the service loads have been exceeded and the beam becomes heavily cracked, significant arch action develops. Now the steel functions primarily as a tension tie in which the stress is constant throughout its length. If collapse is not to occur, the steel must be able to transmit the tie force into the base of the concrete arch at points $A$ and $D$. If the bond between the steel and the concrete is not adequate to transfer the tie reaction into the end of the arch, the tie will pull out and total failure of the arch will result. Recognizing the importance of preventing a failure of this type, ACI Code §12.12 contains a number of provisions to ensure that the flexure steel running into the end supports is properly anchored (Sec. 6.9).

## 4.3 DESIGN PROCEDURE FOR SHEAR

### Introduction

The design of reinforced concrete beams for shear is empirical and indirect for three reasons: (1) Because of the heterogeneous nature of reinforced concrete and its inability to carry tension without cracking, no equation is available to compute accurately the maximum value of shear stress on a particular cross section of a beam. (2) Shear failures do not occur on vertical planes in the direction of the shear force; instead, tensile stresses associated with shear or with shear and moment cause failure on diagonal planes. (3) The tensile strength of concrete is highly variable, and the ability of a concrete beam to carry the diagonal tension associated with shear cannot be predicted accurately.

The uncertainties involved in predicting accurate values of shear stress and in establishing the diagonal tension strength of concrete make a rational design procedure for shear using stresses on diagonal planes impossible; therefore the design procedure here and abroad is based on the assumption that a shear failure at a particular section occurs on a vertical plane when the shear force at that section due to factored service loads exceeds the concrete's fictitious vertical shear strength. Experimental studies are used to relate the fictitious vertical shear strength to the compressive strength $f'_c$ of the concrete and to certain properties of the beam's cross section.

By limiting stress intensities on cross sections and by calling the designer's attention to critical reinforcing details the ACI specifications ensure that all possible modes of brittle failure associated with shear will be eliminated.

### Computation of Shear Stresses; Nominal Stresses

The only equation available to relate shear stress to shear force is derived for a beam of constant cross section constructed of a homogeneous elastic material. Any standard strength-of-materials text gives the equation for shear stress $v$ as

$$v = \frac{VQ}{Ib} \tag{4.1}$$

where $v$ = shear stress at point on cross section
$V$ = shear force on cross section
$I$ = moment of inertia of gross section about centroidal axis
$b$ = thickness at level at which $v$ is computed
$Q$ = moment about centroidal axis of area lying between point at which the shear stress is being computed and outside surface

Unfortunately Eq. (4.1) cannot be applied to reinforced concrete beams for the following reasons:

1. Reinforced concrete, a combination of two distinct materials whose strength and stiffness differ significantly, is obviously not a homogeneous material.
2. Concrete is subject to creep and therefore is not elastic.
3. Cross sections may be cracked or uncracked. Since the extent of cracking at a specific location along the length of a beam is unpredictable, the actual cross-sectional properties on which to base computations of moment of inertia, moment of area, and so forth, cannot be determined.
4. Because of cracking, which varies unpredictably in length and location, the effective cross section of a reinforced concrete beam varies along the length of the member.

For these reasons a precise analytical evaluation of shear-stress intensity is not possible for a reinforced concrete beam. The ACI has therefore adopted a simple procedure for establishing the order of magnitude of the average shear stress on a cross section. The shear stress is computed by dividing the shear force by $b_w d$, the effective area of concrete

$$v = \frac{V}{b_w d} \tag{4.2}$$

where $v$ = nominal shear stress
$V$ = shear force at section
$b_w$ = width of beam web
$d$ = distance between compression surface and centroid of tension steel

Since Eq. (4.2) does not account for cracking, the shear stress $v$ may differ significantly from both the average and the maximum shear stress on the cross section. To emphasize that $v$ is not an actual stress but merely a measure of the shear-stress intensity, it is termed a *nominal* stress.

The nominal shear stress $v$ predicted by Eq. (4.2) can be expected to be significantly smaller than the actual value of maximum shear stress on the cross section. For example, if the beam is uncracked and behavior is assumed to be elastic, the shear stresses will vary parabolically from zero on the top and bottom surfaces to a maximum at the centroid. Figure 4.10 compares the maximum value of shear stress with the average value of shear stress for a rectangular and a T-shaped section. In this comparison $h$ is used instead of $d$ and the nominal stress in the T beam is based on the area of the web and not on the total area.

If vertical cracking caused by moment has occurred, the area available to carry shear may be significantly reduced (Fig. 4.11). Since studies indicate that cracks extend approximately to middepth of the cross section under service loads,[2] the maximum shear stress on a cracked section could be 2 or 3 times greater than the nominal stress predicted by Eq. (4.2); therefore, in establishing a design procedure, the shear strength must also be expressed in terms of nominal stresses so that valid comparisons can be made between level of shear stress and magnitude of shear strength.

## The Shear Strength of Concrete; Members with No Web Reinforcement

The shear strength of a concrete cross section reinforced for moment only can be established by loading a beam until a shear failure occurs. Either a complete rupture of the section or the formation of a major diagonal crack constitutes failure. The extent to which a diagonal crack must open to be considered a shear failure requires a subjective decision by the investigator conducting the test. The shear force at the section where the diagonal crack forms represents the shear strength $V_c$ of the cross section. Using Eq. (4.2), the shear strength can be expressed as a nominal stress by dividing $V_c$ by the effective area of the cross section, $b_w d$

$$v_c = \frac{V_c}{b_w d} \tag{4.3}$$

where $v_c$ = nominal shear strength per unit area of concrete
$\quad\quad V_c$ = shear force required to produce shear failure
$\quad\quad b_w$ = width of beam's web

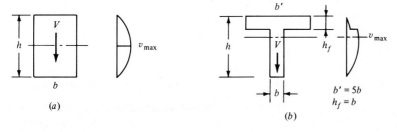

**Figure 4.10** Comparison of maximum shear stress with average shear stress assuming elastic behavior for (a) rectangular section, $v_{max}/v_{av} = 1.5$; (b) T beam, $v_{max}/v_{av} = 1.46$.

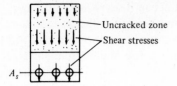

$A_s$

**Figure 4.11** Shear stresses acting on a cracked cross section.

From experimental studies early investigators concluded that the shear strength $v_c$ increases as the compressive strength $f'_c$ increases. Figure 4.12 shows a plot of a large number of experimental values of $v_c$ plotted against the compressive strength of the concrete used to construct the test beams. Most beams in these tests were loaded to failure by one or two concentrated loads. The wide scatter of test points in Fig. 4.12 indicates that a large variation in shear strength exists for beams constructed of concrete of the same strength and demonstrates that the shear strength of concrete does not correlate well with the compressive strength. Since shear failures are actually diagonal-tension failures, shear strength would be expected to correlate better with the concrete's tensile strength, a function of $\sqrt{f'_c}$. Moreover, the discussion in Sec. 4.2 also indicated that a shear failure is strongly influenced by the intensity of the moment that acts concurrently with the shear on the cross section. If the moment is small, the cross section is either un-cracked or lightly cracked. With most of the cross section available to resist shear, the shear capacity is relatively high. Tests indicate that a nominal shear strength $v_c$ of approximately $3.5\sqrt{f'_c}$ lb/in$^2$ $(0.29\sqrt{f'_c}$ MPa) can develop on a cross section before failure occurs.

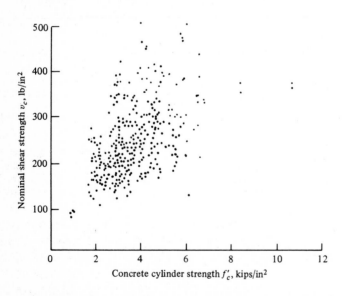

**Figure 4.12** Nominal shear strength $v_c$ as a function of compressive strength.[3]

If the moment acting on the cross section is large, extensive flexural cracking occurs and reduces the uncracked area of the cross section. Both the reduction in cross section and the high bending stresses created by moment can reduce the nominal shear strength $v_c$ to a value as low as $1.9\sqrt{f'_c}$ lb/in$^2$ ($0.16\sqrt{f'_c}$ MPa).

Investigators have also established that the shear strength of a concrete section is influenced by the percentage of longitudinal steel as well as the tensile strength of the concrete. For example, Rajagopalan and Ferguson[1] have recommended that the nominal shear strength of concrete be evaluated by

$$v_c = (0.8 + 100\rho)\sqrt{f'_c} \text{ lb/in}^2 \leq 2\sqrt{f'_c} \text{ lb/in}^2 \tag{4.4}$$

A large reinforcement ratio increases shear strength in several ways. First, the length and width of cracks is reduced as the area of steel increases. Reducing the length of cracks provides a greater area of uncracked concrete to carry shear. Minimizing the width of cracks results in close contact between the surfaces of the crack and increases the amount of aggregate interlock shear that can be developed by friction across the crack (Fig. 4.8). A large area of longitudinal steel also increases the dowel-shear capacity. Several investigators have estimated that aggregate interlock shear may carry between 33 and 50 percent of the total shear transferred across a partially cracked section.

A final factor that influences the shear strength of a cross section is the depth and slenderness of a member. For example, short deep beams that support loads on their top surface show a greater nominal shear strength than beams of moderate depth. The ACI Code classifies beams loaded on the compression surface as deep beams when the clear span-to-depth ratio $l_n/d$ is less than 5. Since the behavior of deep beams differs in many ways from that of moderate-depth beams, special provisions are required for the design and analysis of these members. The design of deep beams will not be covered in this section.

At present the ACI Code uses an empirical expression to predict the shear strength of a cross section that fails by the formation of a flexure-shear crack

$$v_c = \begin{cases} 1.9\sqrt{f'_c} + 2500\rho_w \dfrac{V_u d}{M_u} \leq 3.5\sqrt{f'_c} & \text{lb/in}^2 \quad \text{USCU} \\[3mm] 0.16\sqrt{f'_c} + 17.2\rho_w \dfrac{V_u d}{M_u} \leq 0.29\sqrt{f'_c} & \text{MPa} \quad \text{SI} \end{cases} \tag{4.5}$$

where $V_u d/M_u$ must not be taken greater than 1, $V_u$ and $M_u$ are the values of shear and moment due to factored loads that act on the cross section being investigated, and $\rho_w = A_s/b_w d$.

Equation (4.5) also gives a conservative estimate of the shear strength when failure occurs by web-shear cracking; therefore it can be used safely under all design conditions.

Notice that if $M_u$ is large in Eq. (4.5), the second term becomes small and $v_c$ approaches $1.9\sqrt{f'_c}$. On the other hand, if $M_u$ approaches zero, the second term becomes very large, the empirical equation no longer applies, and the upper limit

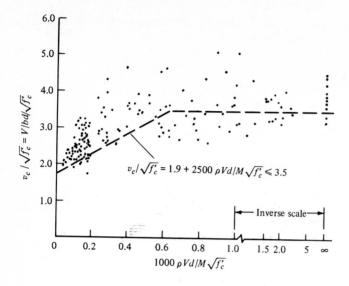

**Figure 4.13** Correlation of Eq. (4.5) with experimental data.[4]

of $v_c = 3.5\sqrt{f'_c}$ controls. In Fig. 4.13 Eq. (4.5) is plotted in nondimensional form together with the test results from which it was established. While considerable scatter still exists, the correlation between shear strength and the variables against which it is plotted is better than in Fig. 4.12.

As an alternative to Eq. (4.5), the ACI Code permits the shear strength to be more simply evaluated by Eq. (4.6), which is the value of the first term in Eq. (4.5) rounded off to the nearest whole number

$$v_c = \begin{cases} 2\sqrt{f'_c} & \text{lb/in}^2 \quad \text{USCU} \\ 0.17\sqrt{f'_c} & \text{MPa} \quad \text{SI} \end{cases} \qquad (4.6)$$

Use of Eq. (4.6) will speed design and will not make a significant difference in the total amount of the shear reinforcement required.

In the 1977 edition of the ACI Code, the design of beams for shear is carried out in terms of forces not nominal stresses. To establish the ACI equations for shear strength $V_c$ the nominal shear stress given by Eqs. (4.5) or (4.6) is multiplied by $b_w d$, the effective area of the beam web, to give

$$V_c = \begin{cases} \left(1.9\sqrt{f'_c} + 2500\rho_w \dfrac{V_u d}{M_u}\right) b_w d & \text{USCU} \\[3mm] \left(0.16\sqrt{f'_c} + 17.2\rho_w \dfrac{V_u d}{M_u}\right) b_w d & \text{SI} \end{cases} \qquad (4.7)$$

but $V_c$ must not exceed $3.5\sqrt{f_c'}\,b_w d(0.29\sqrt{f_c'}\,b_w d)$ and the quantity $V_u d/M_u$ must not be taken greater than 1; or

$$V_c = \begin{cases} 2\sqrt{f_c'}\,b_w d & \text{USCU} \\ 0.17\sqrt{f_c'}\,b_w d & \text{SI} \end{cases} \tag{4.8}$$

## The Influence of Axial Load on Shear Strength

When axial load exists simultaneously with shear on a cross section, the magnitude of the diagonal tension created by the shear is modified. Axial tension will use a portion of the tensile capacity of the concrete available for diagonal tension, thereby reducing the capacity of the beam to carry shear. Axial compression, on the other hand, will decrease the diagonal tension created by shear, thereby raising the concrete's shear capacity. The influence of tension and compression stresses on the diagonal tension stresses $f_t$ created by shear are shown by the Mohr's circles in Fig. 4.14

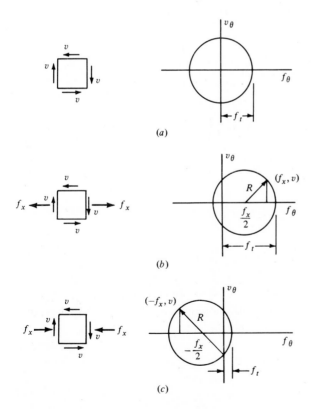

**Figure 4.14** Mohr's circles showing the influence of axial stress on diagonal tension: (*a*) pure shear, (*b*) shear and tension, (*c*) shear and compression.

The ACI equations for the nominal shear strength $V_c$ of a cross section that carries an axial compression force of $N_u$ are specified in ACI Code §11.3 and summarized below:

$$V_c = \left(1.9\sqrt{f'_c} + 2500\rho_w \frac{V_u d}{M_m}\right)b_w d \qquad \text{where } \rho_w = \frac{A_s}{b_w d} \qquad (4.9)$$

$$M_m = M_u - N_u \frac{4h - d}{8}$$

where $V_u$, $M_u$ = shear and moment due to factored loads acting on the section
where $V_c$ is calculated
$b_w$ = width of web
$d$ = effective depth of section
$h$ = overall depth
$N_u$ = axial force, lb
$f'_c$ = 28-day compressive strength of concrete
$A_s$ = area of flexural steel

In Eq. (4.9) the quantity $V_u d / M_m$ may be greater than 1, but $V_c$ given by Eq. (4.9) must not exceed

$$V_c = 3.5\sqrt{f'_c}b_w d\sqrt{1 + \frac{N_u}{500A_g}} \qquad (4.10)$$

where $A_g$ is the gross area in square inches.
As an alternative to Eq. (4.9) $V_c$ may also be computed by

$$V_c = 2\left(1 + \frac{N_u}{2000A_g}\right)\sqrt{f'_c}b_w d \qquad (4.11)$$

If $N_u$ is a tensile force, $V_c$ is evaluated by

$$V_c = 2\left(1 + \frac{N_u}{500A_g}\right)\sqrt{f'_c}b_w d \qquad (4.12)$$

where $N_u$ is to be taken negative. If a member carries significant axial tension, ACI Code § 11.3.1.3 requires that shear reinforcement be designed to carry the entire value of factored shear $V_u$.

**Lightweight concrete** When the shear capacity of a lightweight concrete member is to be evaluated, the term $f_{ct}/6.7$ should be substituted for $\sqrt{f'_c}$ as long as $f_{ct}/6.7$ is less than $\sqrt{f'_c}$. If $f_{ct}$, the split-cylinder tensile strength, is not available, expressions for $V_c$ using $\sqrt{f'_c}$ should be reduced by 0.75 for "all" lightweight aggregate concrete and by 0.85 for lightweight concrete made with sand and lightweight aggregate. Linear interpolation is permitted when partial sand replacement is used.

**Example 4.1:** Determine $V_c$, the nominal shear strength of the concrete, for the cross section shown in Fig. 4.15 for (a) $V_u = 40$ kips and $M_u = 30$ ft · kips; (b) the same as part (a) but with a tension force $N_u = -10$ kips; (c) the same as part (a) but with an axial compression force $N_u = 10$ kips; $f'_c = 3600$ lb/in$^2$ (24.82 MPa); $A_s = 3$ in$^2$.

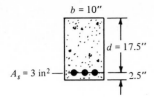

**Figure 4.15**

SOLUTION (a) Using Eq. (4.7) gives

$$V_c = \left(1.9\sqrt{f'_c} + 2500\rho_w \frac{V_u d}{M_u}\right) b_w d \leq 3.5\sqrt{f'_c} b_w d$$

where

$$\frac{V_u d}{M_u} = \frac{40(17.5)}{30(12)} = 1.94$$

but $V_u d/M_u$ must not exceed 1

$$\rho_w = \frac{A_s}{bd} = \frac{3}{10(17.5)} = 0.017$$

$$3.5\sqrt{f'_c} b_w d = \frac{3.5\sqrt{3600}\,10(17.5)}{1000} = 36.75 \text{ kips}$$

$$V_c = [1.9\sqrt{3600} + 2500(0.017)(1)]\frac{10(17.5)}{1000}$$

$$= 27.4 \text{ kips} < 36.75 \text{ kips} \quad \text{OK}$$

Alternately, $V_c$ may be evaluated by Eq. (4.8) as

$$V_c = 2\sqrt{f'_c} b_w d = \frac{2\sqrt{3600}(10)(17.5)}{1000} = 21 \text{ kips}$$

Obviously, since $V_u = 40$ kips exceeds $V_c$, the cross section must be reinforced for shear.

(b) Using Eq. (4.12)

$$V_c = 2\left(1 + \frac{N_u}{500A_g}\right)\sqrt{f'_c} b_w d$$

$$= 2\left[1 + \frac{-10,000}{500(10)(20)}\right]\frac{\sqrt{3600}}{1000}(10)(17.5) = 18.9 \text{ kips}$$

(c) Using Eq. (4.9)

$$V_c = \left(1.9\sqrt{f'_c} + 2500\rho_w \frac{V_u d}{M_m}\right) b_w d$$

but not to exceed

$$3.5\sqrt{f'_c} b_w d\sqrt{1 + \frac{N_u}{500A_g}} = \frac{3.5\sqrt{3600}}{1000}(10)(17.5)\sqrt{1 + \frac{10,000}{500(10)(20)}} = 38.54 \text{ kips}$$

where

$$M_n = M_u - N_u \frac{4h - d}{8} = 30(12) - \frac{10[4(20) - 17.5]}{8} = 281.9 \text{ in} \cdot \text{kips}$$

and

$$\frac{V_u d}{M_n} = \frac{40(17.5)}{281.9} = 2.48$$

$$V_c = [1.9\sqrt{3600} + 2500(0.017)(2.48)] \frac{10(17.5)}{1000} = 38.4 < 38.54 \text{ kips}$$

Alternatively $V_c$ can be evaluated by Eq. (4.11) as

$$V_c = 2\left(1 + \frac{N_u}{2000A_g}\right)\sqrt{f'_c}b_w d$$

$$= 2\left[1 + \frac{10,000}{2000(10)(20)}\right]\frac{\sqrt{3600}}{1000}(10)(17.5) = 21.53 \text{ kips}$$

## 4.4 DESIGN OF BEAMS WITH NO SHEAR REINFORCEMENT

Shear failures, which are actually diagonal-tension failures, can result in complete rupture of a beam. Such failures can be dangerous since they occur without warning and the falling concrete can damage elements that are hit. To ensure ductile failure of beams, the ACI Code establishes provisions that produce beams failing in bending by yielding of the tension steel at a load well below that required to produce a shear failure. Such beams may crack and sag excessively if overloaded, but they do not fall apart.

As a general requirement, the ACI Code requires all beams to be reinforced for shear. Shear reinforcement inhibits the growth of diagonal cracks, provides ductility, and reduces the likelihood of a complete rupture if a large diagonal crack opens. Shear reinforcement can be omitted from members only when the shear is low. Specifically, the ACI Code permits the omission of shear reinforcement from those sections of a concrete beam where the shear $V_u$ is less than $\phi V_c/2$ (half the available shear strength of the concrete). For shear a reduction factor $\phi = 0.85$ is specified by the Code. In addition, shear reinforcement is not required in shallow members such as slabs, footings, floor joists (that conform to the provisions of ACI Code §8.11), and wide shallow beams as long as the shear capacity of the member $\phi V_c$ is greater than the factored shear $V_u$. Tests verify that shear failures never occur before bending failures in shallow members. To be classified as a shallow beam, the depth of a beam must satisfy at least one of the following requirements:

1. Depth not greater than 10 in (254 mm)
2. Depth not greater than $2\frac{1}{2}$ times the flange thickness
3. Depth not greater than one-half the width of the web

**Example 4.2:** The cross section of the beam in Fig. 4.16 is reinforced for moment only. What is the maximum shear force $V_u$ permitted on the section by the ACI Code? $f'_c = 3$ kips/in$^2$ (20.68 MPa), $f_y = 60$ kips/in$^2$ (413.7 MPa).

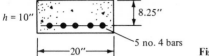

$h = 10''$  8.25"

5 no. 4 bars

|← 20" →|

**Figure 4.16**

SOLUTION Since $h = 10$ in and no shear reinforcement is used, the cross section can carry a maximum shear of $V_u = \phi V_c$. Using Eq. (4.8) for $V_c$ gives

$$V_u = \phi V_c = 0.85(2\sqrt{f'_c}b_w d) = 0.85(2\sqrt{3000})(20)(8.25) = 15{,}364 \text{ lb}$$

**Example 4.3:** The cross section shown in Fig. 4.17 is reinforced for moment only. What is the maximum shear force permitted on the section by the ACI Code? $f'_c = 3$ kips/in$^2$ (20.68 MPa), $f_y = 60$ kips/in$^2$ (413.7 MPa).

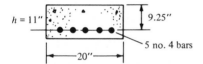

$h = 11''$  9.25"

5 no. 4 bars

|← 20" →|

**Figure 4.17**

SOLUTION Since no shear reinforcement is provided, ACI Code §11.5.5.1 requires $V_u = \frac{1}{2}\phi V_c$. The beam is not shallow.

$$V_u = 0.5(0.85)(2\sqrt{f'_c}b_w d)$$

$$= 0.5(0.85)(2\sqrt{3000})(20)(9.25) = 8613 \text{ lb}$$

Although the depths of the cross sections in the above examples differ by only 1 in, a strict application of the ACI Code provisions produces an 80 percent drop in shear capacity as the depth is increased 1 in. Such a sharp drop in capacity appears inconsistent with the small change in depth.

## Floor-Joist Construction

ACI Code §8.11.8 permits a shear of $1.1V_c$ to be carried by the ribs of joist construction, i.e., closely spaced T beams with tapered webs. To qualify for the 10 percent increase in shear capacity, the proportions of the joists must conform to the provisions of ACI Code §8.11, which specifies that ribs must be at least 4 in (102 mm) wide, have a depth not more than 3.5 times the minimum width of a rib, and have a clear spacing between faces of ribs not exceeding 30 in (762 mm).

**Example 4.4:** A one-way joist floor spans 20 ft (6.1 m) and carries a live load of 100 lb/ft$^2$ (4.79 kPa) (Fig. 4.18). The dead weight of the floor system is 68 lb/ft$^2$ (3.26 kPa). $f'_c = 3$ kips/in$^2$ (20.68 MPa), and $f_y = 60$ kips/in$^2$ (413.7 MPa). (a) What is the maximum shear

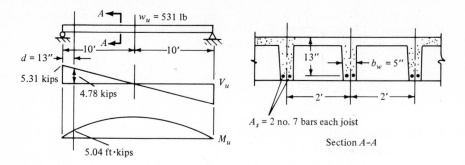

**Figure 4.18**

$V_u$ permitted on a rib if no shear reinforcement is used? (b) If the critical section for shear is located a distance $d$ out from the centerline of the support, are the ribs capable of carrying the shear created by the design loads?

SOLUTION (a) Design load supported by a typical 2-ft T-beam unit is

$$w_u = 1.4(68.0 \text{ lb/ft}^2)(2) + 1.7(100 \text{ lb/ft}^2)(2) = 531 \text{ lb/ft} = 0.531 \text{ kip/ft}$$

Compute $V_u$ a distance $d = 13$ in out from the support (ACI Code §11.1.3.1):

$$V_u = R - wd = 5.31 \text{ kips} - (0.531 \text{ kip/ft})(\tfrac{13}{12}) = 4.73 \text{ kips}$$

Compute $V_c$ using ACI Code §11.3

$$V_c = 2\sqrt{f_c'}\, b_w d = 2\sqrt{3000}(5)(13) = 7.12 \text{ kips}$$

The maximum shear $V_u$ permitted on a rib = $1.1V_c = 1.1(7.12) = 7.83$ kips

(b) 7.83 kips > 4.73 kips

Therefore, no shear reinforcement is required.

## 4.5 DESIGN OF SHEAR REINFORCEMENT

When the factored shear force $V_u$ at a cross section exceeds the shear capacity of the concrete, vertical reinforcement can be added to the member to increase the shear capacity. *Stirrups*, U-shaped bars that pass around the tension steel and are hooked at the top for anchorage in the compression zone, are most commonly used. If significant torsion is also present, a closed stirrup is required (Fig. 4.19).

Stirrups are normally fabricated from a single size of reinforcing bar. Since the factored shear $V_u$, a measure of the required shear strength, varies along the length of the member, the stirrup spacing may be varied at intervals, say several feet, to adjust the amount of reinforcement supplied. A conservative design can be produced by computing the required stirrup spacing at the section of maximum shear and then spacing stirrups at that interval until in the judgment of the designer

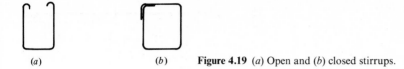

(a)        (b)       **Figure 4.19** (a) Open and (b) closed stirrups.

the shear reduces sufficiently to permit a larger spacing. At that point, the designer can compute the stirrup spacing for the next interval. Generally a particular spacing is used until the shear reduces enough to permit the distance between stirrups to be increased by several inches.

If possible, stirrups should be spaced not less than 3 in (76 mm) apart to facilitate the flow of concrete between reinforcing bars. If analysis indicates that a closer spacing is required for the bar size initially selected, the designer may increase the spacing either by increasing the diameter of the stirrup or by using a multileg stirrup (see Fig. 4.4f).

Before diagonal cracking occurs, the stirrups remain essentially unstressed. After diagonal cracking, the stress in the stirrups increases as they pick up a portion of the load formerly carried by the uncracked concrete. In addition to carrying part of the shear after the concrete cracks, stirrups also improve the behavior of beams by restricting the length of diagonal cracks, thereby increasing the concrete area available to carry shear. Stirrups also add ductility to the beam by holding the member together after a major diagonal crack forms.

## Spacing of Stirrups

The design of shear reinforcement is based on the assumption that the shear force $V_u$ produced by factored loads must not exceed the total shear capacity of the cross section. This relationship can be stated as

$$V_u \leq \phi V_n \tag{4.13}$$

where $V_n$ equals the nominal shear strength of the cross section and $\phi$, the reduction factor, equals 0.85. When shear reinforcement is used, $V_n$ can be written as

$$V_n = V_c + V_s \tag{4.14}$$

where $V_c$ is the nominal shear strength of the concrete as predicted by Eqs. (4.7) to (4.12) and $V_s$ is the shear capacity of the reinforcement, either stirrups or bent bars. If Eq. (4.14) is substituted into Eq. (4.13), we can write

$$V_u \leq \phi(V_c + V_s) \tag{4.15}$$

Equation (4.15), which forms the basis of stirrup design, can be expressed in terms of the properties of the cross section and the spacing of the stirrups by considering the equilibrium of a free body of the portion of a reinforced concrete beam to the left of a critical diagonal crack (Fig. 4.21). The free body is cut from the beam in Fig. 4.20 by a vertical section 1-1 that passes through the top of a diagonal crack.

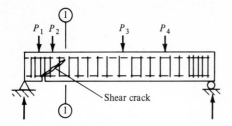

**Figure 4.20** An impending shear failure in a beam reinforced with stirrups.

The crack, which extends into the compression zone, does not split the beam but produces sufficient strain to yield the stirrup steel. Since the beam is in an impending state of shear failure, all sources of shear strength—aggregate interlock, dowel shear, and so forth—are fully mobilized.

The shear capacity $V_s$ of the reinforcement, which is assumed to be uniformly spaced across the diagonal crack can be expressed as

$$V_s = A_v f_y n \tag{4.16}$$

where $n$ is the number of stirrups crossing the diagonal crack and $A_v f_y$ is the force in each stirrup crossing the crack. The quantity $A_v$ stands for the total area of the stirrup legs crossing the crack. For a stirrup with two legs, this would represent two bar areas. Finally $f_y$ is the yield point of the stirrup steel. If the horizontal projection of the diagonal crack is assumed (conservatively) to be equal to the effective depth $d$ of the cross section, the number of stirrups crossing the crack can be computed as

$$n = \frac{d}{s} \tag{4.17}$$

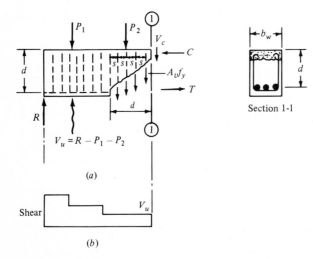

**Figure 4.21** Free body to the left of a critical diagonal crack: (a) internal and external forces, (b) shear curve.

where $s$ represents the spacing of the stirrups. If Eq. (4.17) is substituted into Eq. (4.16), we can write

$$V_s = \frac{A_v f_y d}{s} \tag{4.18}$$

Since the designer typically specifies $A_v$ and $f_y$, Eq. (4.18) can be solved for the required stirrup spacing $s$ by rearranging terms to give

$$s = \frac{A_v f_y d}{V_s} \tag{4.19}$$

In Eq. (4.19) the value of $V_s$ is computed from Eq. (4.15) by using only the equals sign, to give

$$V_s = \frac{V_u}{\phi} - V_c \tag{4.20}$$

## Code Provisions to Ensure Ductile Behavior

Equation (4.19), used to establish the required spacing of vertical stirrups, was derived by considering the equilibrium of the vertical forces on the free body of a beam to one side of a major diagonal crack as a shear failure impends. The crack is assumed to extend into the lower region of the compression zone but not to split the beam. The assumption is made that while the heavily cracked beam remains intact, the formation of the crack produces enough deformation to strain the stirrup steel to the yield point. To preclude the possibility of other brittle modes of failure developing when a diagonal crack opens, restrictions must also be placed on the maximum intensity of shear permitted on the cross section. These provisions, as well as details of maximum stirrup spacing and anchorage of stirrups, are discussed below.

**Compression failures produced by shear**  To prevent a *shear-compression* failure (crushing of the concrete by high compressive and shear stresses) in the region directly above the tip of a diagonal crack, ACI Code §11.5.6.8 requires that $V_s$, the shear carried by the web reinforcement, not exceed $8\sqrt{f_c'}\,b_w d$. By limiting the intensity of shear permitted on the cross section this provision also prevents crushing of the web by ensuring that the diagonal compression stresses produced by shear near middepth (Fig. 4.6c) will be below the compressive strength of the concrete (Fig. 4.22).

**Maximum stirrup spacing**  The derivation of (4.18) is based on the assumption that several stirrups cross each potential diagonal crack. To ensure that this requirement is satisfied, ACI Code §11.5.4.2 requires that stirrups not be spaced more than $d/2$ in apart when $V_s \le 4\sqrt{f_c'}\,b_w d$. This provision also ensures that a diagonal crack (assumed to extend horizontally approximately $d$ in) will not split

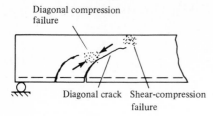

Diagonal compression
failure

Diagonal crack   Shear-compression
failure

**Figure 4.22** Brittle modes of failure produced by high shear stresses.

a beam into two sections between stirrups (Fig. 4.23). In regions of high shear where $V_s$ is greater than $4\sqrt{f_c'}\,b_w d$ but less than $8\sqrt{f_c'}\,b_w d$, the maximum permitted spacing is reduced to $d/4$ to ensure that each potential diagonal crack will be crossed by approximately three stirrups.

**Tension failure of flexural steel**   The maximum spacing requirements of ACI Code §11.5.4.2 also provide closely spaced stirrups to hold the longitudinal tension steel to the body of the beam. Closely spaced stirrups reduce the possibility of failure caused by the steel's tearing through the thin concrete cover or slipping relative to the concrete (Fig. 4.24).

**Minimum area of shear reinforcement**   The instant a diagonal crack forms the tension formerly carried by the concrete must be transferred to the stirrups if the beam is not to split into two sections. To ensure that the stirrups will have sufficient strength to absorb the diagonal tension in the concrete without snapping because the tensile strength of the steel is exceeded, ACI Code §11.5.5.3 requires all stirrups to have a minimum area $A_v$, given as

$$A_{v,\,\text{min}} = \begin{cases} \dfrac{50 b_w s}{f_y} & \text{USCU} \\[2ex] 0.34\dfrac{b_w s}{f_y} & \text{SI} \end{cases} \tag{4.21}$$

where $s$ = spacing of stirrups
  $b_w$ = width of web
  $f_y$ = yield point of steel, lb/in$^2$ or MPa

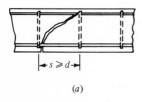

$s \geqslant d$

(a)

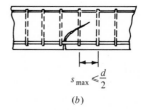

$s_{\text{max}} \leqslant \dfrac{d}{2}$

(b)

**Figure 4.23** Maximum stirrup spacing: (a) diagonal crack traverses the entire depth without crossing a stirrup, (b) maximum spacing of $d/2$ ensures that at least one stirrup with adequate anchorage will be available to hold the upper and lower sections together.

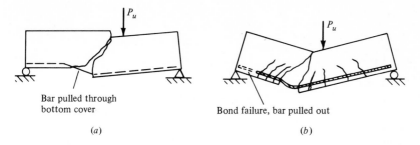

Bar pulled through
bottom cover

Bond failure, bar pulled out

(a)

(b)

**Figure 4.24** Failures of flexural steel induced by diagonal cracking: (a) vertical displacement of flexural steel cracks off concrete cover, (b) increase in tensile stress due to arch action produces a bond failure.

**Anchorage of stirrups**   To prevent failure of a beam by pull out of stirrups that are stressed to their yield point, ACI Code §12.14 details anchorage requirements. Running stirrups as high as possible into the compression zone and adding hooks is effective in anchoring the ends. Bending stirrups around the longitudinal steel reduces the high local bearing stresses under the hook and prevents crushing of the concrete. If the concrete were to crush, the stirrup would tear out (Fig. 4.25).

**Control of crack width**   The width of diagonal cracks is a direct function of the strain in the shear reinforcement. To limit the crack width so that the appearance of the beam is not marred and to ensure that the sides of the crack will remain in close contact so that aggregate interlock shear can develop, ACI Code §11.5.2 requires that the yield strength of shear reinforcement not exceed 60 kips/in$^2$. In addition, tests indicate that stirrups of higher-yield-point steels do not always undergo sufficient deformation as a result of diagonal cracking to strain the shear reinforcement above 60 kips/in$^2$.

   Limiting the yield point of the shear reinforcement also limits the force that must be anchored by the stirrup hooks and reduces the likelihood of a local bearing failure where the inside of the hook bears against the concrete (Fig. 4.25).

## Spacing of Stirrups Adjacent to a Support

ACI Code §11.1.3 permits the stirrups adjacent to a support to be sized for the shear that exists $d$ in out from the face of the support (Fig. 4.26). This provision recognizes that a crack adjacent to a support whose reaction induces compression

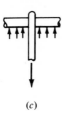

**Figure 4.25** Hooking stirrups around longitudinal steel to improve anchorage: (a) stirrups anchored in the compression zone (b) bearing stresses created by the curvature of a hook, (c) longitudinal steel acts as a baseplate to reduce compressive stresses under hook.

(a)

(b)

(c)

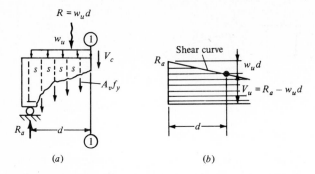

(a)                                    (b)

**Figure 4.26** Design of stirrups adjacent to a support that induces compression into the end of the beam: (a) Free body cut by potential diagonal crack nearest support. Vertical equilibrium indicates that the member must be designed with a shear strength of $V_u = R_a - w_u d$. (b) Shear curve showing the value of shear $V_u$ for which stirrups within a distance $d$ of support must be designed.

into a beam will have a horizontal projection of at least $d$ in; therefore the maximum shear force that must be transmitted across the potential failure plane closest to the support (section 1-1 in Fig. 4.26) will be equal to the reaction $R$ reduced by any external forces applied to the beam within a distance $d$ of the support.

If a support does not introduce compression into a member, the critical section for shear should be taken at the face of the support. For this situation the state of stress in the joint should also be studied to determine whether special reinforcement is required (Fig. 4.27).

### Spacing of Inclined Bars

If a series of equally spaced inclined bars is used as shear reinforcement (Fig. 4.28), the shear capacity of the reinforcement $V_s$ as given by Eq. (4.18) is modified by the inclusion of the factor $\sin \alpha + \cos \alpha$, to give

$$V_s = \frac{A_v f_y (\sin \alpha + \cos \alpha) d}{s} \qquad (4.22)$$

where $\alpha$ = angle between inclined bars and horizontal
$s$ = horizontal distance between bars
$A_v$ = area of shear reinforcement within distance $s$

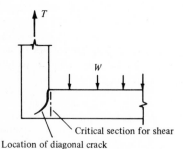

Location of diagonal crack

**Figure 4.27** Location of critical diagonal crack.

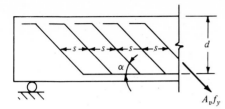

$A_v f_y$    **Figure 4.28** Inclined shear reinforcement.

If all the reinforcement is bent up at a single point, the ACI Code specifies that $V_s$ be computed as

$$V_s = A_v f_y \sin \alpha \qquad (4.23)$$

but $V_s$ must not be greater than $3\sqrt{f'_c}\,b_w d$ (USCU) or $0.25\sqrt{f'_c}\,b_w d$ (SI).

In addition, the Code considers that only the center three-fourths of the inclined bar is effective as shear reinforcement.

### Summary of the ACI Design Procedure for Shear

**Step 1** Draw the shear and moment curves using the factored service loads.

**Step 2** Establish the shear strength $V_c$ of the concrete. For members subject to shear and moment only use either

$$V_c = 2\sqrt{f'_c}\,b_w d \qquad (4.8)$$

or

$$V_c = \left(1.9\sqrt{f'_c} + 2500\rho_w \frac{V_u d}{M_u}\right) b_w d \qquad (4.7)$$

where $V_u d/M_u$ is not to exceed 1 and $V_c$ is not to exceed $3.5\sqrt{f'_c}\,b_w d$. [Equation (4.7) will predict a significantly greater value of $V_c$ than Eq. (4.8) only in regions where the moment $M_u$ is small.] If an axial force $N_u$ also acts, $V_c$ is established by Eqs. (4.9) to (4.12).

**Step 3** For slabs, footings, and shallow beams, i.e., members that are generally reinforced for moment only, verify that $V_u \leq \phi V_c$. If $V_u$ exceeds $\phi V_c$, the shear capacity of the cross section can be raised by increasing the depth of the cross section. Stirrups are not very effective in shallow members because the compression zone lacks the depth required to anchor the force in the stirrups.

**Step 4** For beams that are not shallow, reinforcement is required in regions where $V_u \geq \phi V_c/2$. The procedure to establish the stirrup spacing is a function of the relative magnitude between $V_u$ and $\phi V_c$.

*Case 1*

$$V_u > \frac{\phi V_c}{2} \quad \text{but} \quad V_u < \phi V_c$$

The shear capacity of the concrete exceeds the shear due to factored loads. Use minimum reinforcement; spacing is the smaller of

$$s = \frac{d}{2} \quad \text{and} \quad s = \frac{A_v f_y}{50 b_w}$$

*Case 2*

$$V_u \geq \phi V_c \quad \text{and} \quad V_s \leq 8\sqrt{f_c'} b_w d$$

Use Eq. (4.20) to establish $V_s$, the required shear capacity of the stirrups

$$V_s = \frac{V_u}{\phi} - V_c$$

After $V_s$ has been determined, use Eq. (4.19) to establish $s$

$$s = \frac{A_v f_y d}{V_s}$$

The stirrup spacing is also subject to the following restrictions:

$$s \leq \begin{cases} \dfrac{A_v f_y}{50 b_w} & \\[2mm] \dfrac{d}{2} & V_s \leq 4\sqrt{f_c'} b_w d \\[2mm] \dfrac{d}{4} & V_s > 4\sqrt{f_c'} b_w d \end{cases}$$

*Case 3* If $V_s$ exceeds $8\sqrt{f_c'} b_w d$, the cross section is too small. To prevent one of several possible brittle modes of failure associated with high shear stress, the cross-sectional area must be increased.

**Example 4.5:** Determine the required spacing of no. 3 stirrups, $A_v = 0.22$ in$^2$ (142 mm$^2$), in the T-beam section (Fig. 4.29) for the following values of factored shear $V_u$. Use Eq. (4.8) to evaluate $V_c$: (a) $V_u = 12$ kips (53.4 kN), (b) $V_u = 36$ kips (160.1 kN), (c) $V_u = 42$ kips (186.8 kN). $f_c' = 3$ kips/in$^2$ (20.68 MPa), and $f_y = 60$ kips/in$^2$ (413.7 MPa).

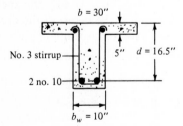

$b = 30''$

No. 3 stirrup

2 no. 10

$5''$  $d = 16.5''$

$b_w = 10''$

**Figure 4.29**

**SOLUTION** (*a*) Since the beam is not shallow, stirrups are required wherever $V_u$ exceeds $\phi V_c/2$

$$V_c = 2\sqrt{f_c'}b_w d = 2\sqrt{3000}(10)(16.5) = 18,100 \text{ lb} = 18.1 \text{ kips (80.5 kN)}$$

$$\frac{\phi V_c}{2} = 0.85\frac{18.1}{2} = 7.7 \text{ kips}$$

Since $V_u = 12$ kips exceeds $\phi V_c/2$, stirrups are required. Use minimum steel

$$s = \frac{d}{2} = \frac{16.5 \text{ in}}{2} = 8.25 \text{ in} \qquad \text{use 8 in (203 mm)}$$

$$A_{v,\min} = \frac{50b_w s}{f_y} = \frac{50(10)(8)}{60,000} = 0.07 \text{ in}^2$$

$$A_{v,\text{supp}} = 0.22 \text{ in}^2 > 0.07 \text{ in}^2 \qquad \text{OK}$$

(*b*) Since $V_u = 36$ kips exceeds $\phi V_c/2$, stirrups are required

$$V_s = \frac{V_u}{\phi} - V_c = \frac{36}{0.85} - 18.1 = 24.3 \text{ kips}$$

$$s = \frac{A_v f_y d}{V_s} = \frac{0.22(60)(16.5)}{24.3} = 8.96 \text{ in}$$

$$4\sqrt{f_c'}b_w d = 36.2 \text{ kips}$$

Since $V_s$ less than $4\sqrt{f_c'}b_w d$, $s = d/2 = 8.25$ in controls. Use 8 in (203 mm).

(*c*) $V_u = 42$ kips

$$V_s = \frac{V_u}{\phi} - V_c = \frac{42}{0.85} - 18.1 = 31.3 \text{ kips}$$

$$s = \frac{A_v f_y d}{V_s} = \frac{0.22(60)(16.5)}{31.3} = 6.96 \text{ in (177 mm)}$$

Since $V_s$ less than $4\sqrt{f_c'}b_w d$, $s$ should not exceed $d/2 = 8.25$ in

$$6.96 \text{ in controls} \qquad \text{use 6.5 in}$$

**Example 4.6:** The hollow box beam in Fig. 4.30 spans 30 ft (9.14 m). If the beam is reinforced with no. 3 stirrups, $A_v = 0.22 \text{ in}^2$ (142 mm$^2$), locate the region where (*a*) stirrups can be omitted and (*b*) stirrups can be spaced $d/2$ in apart. $f_c' = 3$ kips/in$^2$ (20.68 MPa), $f_y = 60$ kips/in$^2$ (413.7 MPa).

**SOLUTION** Shear capacity of cross section (use $b_w = 10$ in)

$$V_c = 2\sqrt{f_c'}b_w d = 2\sqrt{3000}(10)(17 \text{ in}) = 18,622 \text{ lb} = 18.6 \text{ kips (82.7 kN)}$$

No stirrups are required if

$$V_u \le \phi\frac{V_c}{2} = 0.85\frac{18.6 \text{ kips}}{2} = 7.9 \text{ kips (35.1 kN)}$$

With reference to the shear curve, no stirrups are required for a distance of 2.9 ft to right of the 30 kip load.

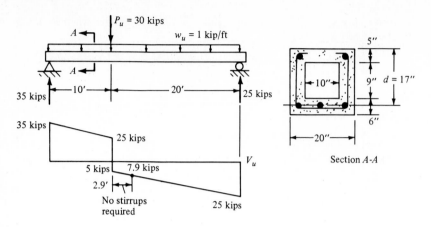

**Figure 4.30**

(b) Compute the shear capacity $V_u$ of the cross section with stirrups spaced $d/2$ in apart

$$V_u = \phi(V_c + V_s) \qquad \text{where } V_s = \frac{A_v f_y d}{s}$$

$$= 0.85\left[18.6 \text{ kips} + \frac{0.22(60 \text{ kips/in}^2)(17 \text{ in})}{8.5 \text{ in}}\right] = 38.25 \text{ kips (170.1 kN)}$$

Since the shear strength with stirrups spaced at $d/2$ exceeds the maximum shear in the beam, stirrups at $d/2$ can be used everywhere that $V_u > 7.9$ kips.

**Example 4.7: Design of stirrups** In Fig. 4.31 the simply supported beam spans 20 ft (6.1 m) and carries a uniformly distributed service load of 1.45 kips/ft (21.2 kN/m) dead load and 3.5 kips/ft (51.1 kN/m) live load. The weight of the beam is included in the dead load. If the beam is reinforced with a constant area of flexural steel $A_s = 6.06$ in$^2$ (3910 mm$^2$), design the stirrups using Eq. (4.7) to evaluate $V_c$; $f_c' = 2.5$ kips/in$^2$ (17.24 MPa), $f_y = 50$ kips/in$^2$ (344.7 MPa), $b = 16$ in (406.4 mm) and, $d = 22$ in (559 mm).

SOLUTION Compute the factored load

$$w_u = 1.45(1.4) + (3.5)(1.7) = 7.98 \qquad \text{use 8 kips/ft (116.8 kN/m)}$$

Compute the shear and moment on the cross section a distance $d = 22$ in out from the left support

$$V_u = 80 \text{ kips} - (8 \text{ kips/ft})(\tfrac{22}{12}) = 65.3 \text{ kips (290.45 kN)}$$

$$M_u = \frac{(80 + 65.3)}{2}\frac{22}{12} = 133.19 \text{ ft} \cdot \text{kips (180.6 kN} \cdot \text{m)}$$

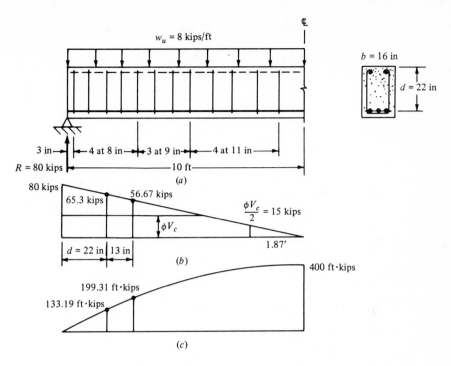

**Figure 4.31** Left half of beam: (a) spacing of stirrups, (b) shear curve, (c) moment curve.

Compute the nominal shear strength of the concrete using Eq. (4.7)

$$V_c = \left(1.9\sqrt{f_c'} + 2500\rho_w \frac{V_u d}{M_u} b_w d\right) \leq 3.5\sqrt{f_c'} b_w d$$

$$\frac{V_u d}{M_u} = \frac{(65.3 \text{ kips})(22 \text{ in})}{133.19(12)} = 0.9 < 1 \qquad \text{OK}$$

$$\rho_w = \frac{A_s}{b_w d} = \frac{6.06}{16(22)} = 0.0172$$

$$V_c = [1.9\sqrt{2500} + 2500(0.0172)(0.9)](16)(22) = 47{,}062 \text{ lb}$$

$$= 47.06 \text{ kips (209.3 kN)} < 3.5\sqrt{3500}(16)(\tfrac{22}{1000}) = 61.6 \text{ kips} \qquad \text{OK}$$

$$\frac{\phi V_c}{2} = \frac{0.85(47.06)}{2} = 20 \text{ kips}$$

Since $V_u = 65.3$ kips $> \phi V_c/2$, stirrups are required. The required stirrup spacing is

$$V_s = \frac{V_u}{\phi} - V_c = \frac{65.3}{0.85} - 47.06 = 29.76 \text{ kips (132.37 kN)}$$

$$s = \frac{A_v f_y d}{V_s} = \frac{0.22(50)(22)}{29.76} = 8.13 \text{ in (206.5 mm)} \qquad \text{use 8 in}$$

Check the maximum spacing. Since $V_s = 29.76$ kips $< 4\sqrt{f'_c}b_wd = 70.4$ kips,

$$s_{max} = \frac{d}{2} = 11 \text{ in} \qquad \text{use } s = 8 \text{ in}$$

Then

$$A_{v,min} = \frac{50b_ws}{f_y} = \frac{50(16)(8)}{50,000} = 0.128 \text{ in}^2 \ (82.6 \text{ mm}^2)$$

$$A_{v,supp} = 0.22 \text{ in}^2 > 0.128 \text{ in}^2 \qquad \text{OK}$$

The first stirrup will be spaced 3 in (76 mm) from the end of the beam. This provides over 2 in (51 mm) of cover to protect the steel. The next four stirrups will be spaced 8 in on center. This arbitrary arrangement establishes the shear reinforcement for the first 35 in of the beam. Another designer might choose to run the stirrups a little farther. At this point, the shear $V_u$ has dropped by more than 10 kips (44.48 kN): therefore a larger spacing between stirrups is possible. Again, the location at which a particular stirrup spacing is terminated is the decision of the designer.

Compute $V_u$ and $M_u$ on the cross section 35 in from the support

$$V_u = 80 \text{ kips} - \tfrac{35}{12}(8 \text{ kips/ft}) = 56.67 \text{ kips (252.07 kN)}$$

$$M_u = \frac{(80 + 56.67)}{2}\frac{35}{12} = 199.31 \text{ ft} \cdot \text{kips (270.26 kN} \cdot \text{m)}$$

Compute $V_c$ with Eq. (4.7)

$$V_c = \left[1.9\sqrt{2500} + 2500(0.0172)\frac{56.67(22)}{199.31(12)}\right](16)(22) = 41.33 \text{ kips}$$

Stirrup spacing is

$$V_s = \frac{V_u}{\phi} - V_c = \frac{56.67}{0.85} - 41.33 = 25.34 \text{ kips (112.71 kN)}$$

$$s = \frac{A_vf_yd}{V_s} = \frac{0.22(50)(22)}{25.34} = 9.55 \text{ in (242.57 mm)} \qquad \text{use 9 in}$$

Run three stirrups at 9 in. Computations (not shown) indicate that the balance of stirrups can be run at $d/2 = 11$ in (Fig. 4.31).

Determine the location where stirrups are not required, $V_u < \phi V_c/2$. Using Eq. (4.8), we get

$$V_c = 2\sqrt{f'_c}b_wd = \frac{2\sqrt{2500}}{1000}(16)(22) = 35.2 \text{ kips}$$

$$V_u = \frac{\phi V_c}{2} = \frac{0.85(35.2)}{2} = 14.96 \text{ kips} \qquad \text{say 15 kips}$$

With reference to the shear curve in Fig. 4.31, stirrups are required to within 1.875 ft of the midspan section.

**Example 4.8: Redesign stirrups using Eq. (4.8) to compute $V_c$**

SOLUTION As indicated at the end of Example 4.7, Eq. (4.8) gives $V_c = 35.2$ kips (156.6 kN). For stirrup spacing at $d = 22$ in (558.8 mm) out from the support $[V_u = 65.3$ kips (290.45 kN)]

$$V_s = \frac{V_u}{\phi} - V_c = \frac{65.3}{0.85} - 35.2 = 41.62 \text{ kips (185.1 kN)}$$

$$s = \frac{A_v f_y d}{V_s} = \frac{0.22(50)(22)}{41.62} = 5.8 \text{ in (147.3 mm)} \qquad \text{use 5 in}$$

Space one stirrup at 3 in from the end; then run seven at 5 in.
Compute required stirrup spacing 38 in from the support ($V_u = 54.7$ kips)

$$V_s = \frac{V_u}{\phi} - V_c = \frac{54.7}{0.85} - 35.2 = 29.15 \text{ kips}$$

$$s = \frac{A_v f_y d}{V_s} = \frac{0.22(50)(22)}{29.15} = 8.3 \text{ in} \qquad \text{use 8 in}$$

Space the next three stirrups 8 in apart.
Check required spacing 62 in from support ($V_u = 38.7$ kips)

$$V_s = \frac{V_u}{\phi} - V_c = \frac{38.7 \text{ kips}}{0.85} - 35.2 = 10.3 \text{ kips}$$

$$s = \frac{A_v f_y d}{V_s} = \frac{0.22(50)(22)}{10.3} = 23.56 \text{ in (596.9 mm)}$$

but $s \leq d/2 = 22/2 = 11$ in controls; space balance of stirrups at 11 in on center.

## Alternate Procedure for Spacing Stirrups

The designer can also establish a pattern of stirrups by selecting arbitrary values of stirrup spacing and evaluating the corresponding shear strength of the reinforced concrete cross section. A particular spacing can be used until the shear reduces to the value of the shear strength associated with the next larger interval of spacing. The designer might start the procedure with the maximum spacing permitted between stirrups, i.e., the distance $d/2$, and reduce the spacing by increments of 2 or 3 in (51 or 76 mm).

By plotting on the shear curve the shear strength of the cross section for each value of spacing the region in which a particular value of spacing is adequate to carry the shear can be established.

If the first trial produces stirrups that are reasonably spaced, the design is complete, but if the procedure results in closely spaced stirrups over a major portion of the member's length, the diameter of the stirrup steel should be increased and the procedure repeated. The locations at which a particular value of stirrup spacing changes may vary slightly with individual designers; however, the total number of stirrups used to reinforce a particular beam will be approximately the same.

**Example 4.9:** Repeat the design of stirrups for the beam in Example 4.7 using the alternate method described above. Since the shear curve is symmetrical, only half the beam need be investigated; $f'_c = 2.5$ kips/in², $f_y = 50$ kips/in.²

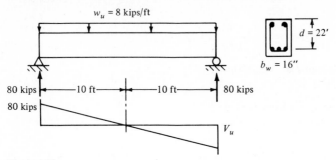

**Figure 4.32**

SOLUTION Compute the values of shear $V_u$ that can be carried by no. 3 stirrups spaced 11, 9, 7, and 5 in apart. Use Eq. (4.8) to evaluate $V_c$

$$V_c = 2\sqrt{f'_c}b_w d = \frac{2\sqrt{2500}(16)(22)}{1000} = 35.2 \text{ kips}$$

$$V_u = \phi\left(V_c + \frac{A_v f_y d}{s}\right) \tag{6}$$

where $A_v = 0.22$ in², $d = 22$ in, $f_y = 50$ kips/in², $\phi = 0.85$

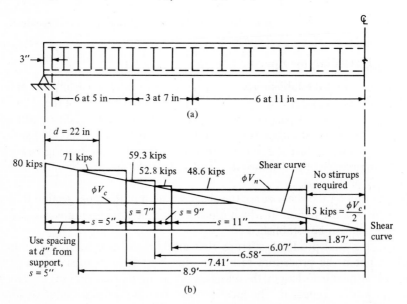

**Figure 4.33** Variation of shear capacity $\phi V_n$: (a) stirrup spacing, (b) shear capacity $\phi V_n$ as a function of stirrup spacing plotted on shear curve. Since 9-in spacing is permitted in such a limited region, 7-in spacing is carried through until 11-in spacing is permitted.

Substituting into Eq. (6) gives

$$V_u = \begin{cases} 48.6 \text{ kips} & \text{for } s = 11 \text{ in} \\ 52.8 \text{ kips} & s = 9 \text{ in} \\ 59.3 \text{ kips} & s = 7 \text{ in} \\ 71.1 \text{ kips} & s = 5 \text{ in} \end{cases}$$

Locate the values of shear strength $V_u$ for the various stirrup spacings on the shear curve (see Fig. 4.33). Use $\Delta V_u = \Delta X w_u$ to establish the position of the values of $V_u$. The distance $\Delta X$ is measured from midspan.

## 4.6 PUNCHING SHEAR

A heavy load applied to a slab or footing within a small area may cause a shear failure by pushing or punching out a truncated cone or pyramid of slab concrete (Fig. 4.34). Circular loaded areas produce conical failure surfaces, and rectangular areas produce pyramid-shaped failure surfaces.

The sloping failure plane is produced by the diagonal tension stresses created by the shear stresses in the slab adjacent to the column (Fig. 4.35). The stress condition in the slab below the column is very complex as a result of vertical compressive stresses from the applied load, cracking of the concrete on the tension side of the slab due to moment, and biaxial compression due to bending in two directions. Based on extensive experimental studies, the ACI Code specifies an empirical design procedure for evaluating the punching-shear strength. This procedure assumes that failure occurs on a fictitious vertical surface located a distance $d/2$ out from the face of the column (Fig. 4.36). The perimeter of the failure surface is similar in shape to that of the column or of the loaded area.

If no shear reinforcement is used, the maximum shear force $V_c$ that can be transmitted through the failure surface is specified in ACI Code §11.11.2 as

$$V_c = \left(2 + \frac{4}{\beta_c}\right)\sqrt{f'_c}\, b_0 d \qquad (4.24)$$

but

$$V_c \text{ must not exceed } 4\sqrt{f'_c}\, b_0 d \qquad (4.25)$$

where $V_c$ = punching shear strength
$b_0$ = perimeter of failure surface, located distance $d/2$ out from face of column
$\beta_c$ = ratio of long side to short side of column or loaded area

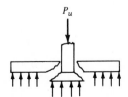

**Figure 4.34** Punching shear failure of a column footing.

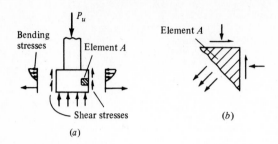

Bending stresses

$P_u$

Element A

Shear stresses

(a)

Element A

(b)

**Figure 4.35** State of stress in footing below column: (*a*) free-body diagram of column and footing, (*b*) orientation of diagonal-tension plane.

Equation (4.25) controls the punching-shear strength as long as the ratio of the long to short side of the loaded area does not exceed 2. To determine the available punching-shear strength, $V_c$ given by Eqs. (4.24) and (4.25) should be reduced by $\phi = 0.85$.

If shear reinforcement in the form of bars or wire cages is provided, the maximum shear $V_u$ on the critical section can be increased 50 percent to $\phi 6\sqrt{f'_c}\,b_0 d$; however, the shear force carried by the concrete then must not exceed $2\sqrt{f'_c}\,b_0 d$. Reinforcement must be sized to carry the balance of the applied shear.

If *shearheads*, welded structural sections, are used, the shear force $V_u$ transmitted through the critical section can be increased 75 percent to $\phi 7\sqrt{f'_c}\,b_0 d$. Details of shearhead design are covered in ACI Code §11.11.4.

**Example 4.10:** The axially loaded footing in Fig. 4.37 must be designed to support a factored column load, $P_u = 200$ kips (889.6 kN). If the footing is 16 in (406 mm) deep with an average $d = 12.5$ in (318 mm), is the punching-shear capacity of the concrete adequate to transfer the load from the column into the footing? $f'_c = 2.5$ kips/in$^2$ (17.24 MPa).

SOLUTION Compute the uniform soil pressure $q_u$ under the base of the footing

$$q_u = \frac{P_u}{A} = \frac{200 \text{ kips}}{5(5)} = 8 \text{ kips/ft}^2$$

The force that must be transmitted in shear through the failure surface equals the column load minus the force produced by the soil pressure $q_u$ acting on the area of the base within the failure plane (Fig. 4.37*b*), i.e.,

$$V_u = 200 \text{ kips} - \left[ \frac{26.5}{12} \frac{26.5}{12} (8 \text{ kips/ft}^2) \right] = 160.99 \text{ kips} \qquad \text{call 161 kips}$$

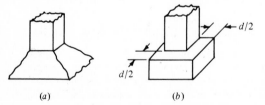

(a)

$d/2$

$d/2$

(b)

**Figure 4.36** (*a*) Actual and (*b*) assumed failure surface for punching shear.

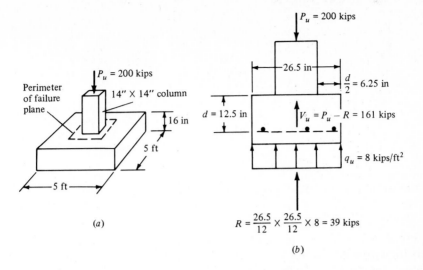

**Figure 4.37** Axially loaded footing: (*a*) details of footing, (*b*) free-body diagram of the column and footing within the failure surface.

Using Eq. (4.24) we compute the punching-shear capacity of the rectangular failure surface as

$$\phi V_c = \phi 4\sqrt{f'_c}b_0 d = 0.85 \frac{4\sqrt{2500}}{1000}(26.5)(4)(12.5)$$

$$= 225.25 \text{ kips} > V_u = 161 \text{ kips } (716.08 \text{ kN})$$

capacity adequate

## 4.7 SHEAR FRICTION

Exterior edges of members or narrow projecting elements acted upon by large concentrated loads are subject to a type of failure called *shear friction*. Large forces cause the vulnerable part of the member to shear off the main body of the member along a plane on which high shear stresses act. For a shear-friction failure to occur, the moment on the failure surface must be small; otherwise diagonal tension is the failure mode. Typical examples of shear-friction failures are shown in Fig. 4.38 by dashed lines.

For reinforcing against a potential shear-friction failure the design procedure is based on positioning steel reinforcement across the potential failure surface in order to produce a definite value of compression across the failure surface (Fig. 4.39).

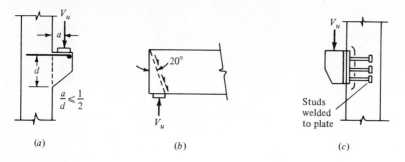

(a)    (b)    (c)

**Figure 4.38** Examples of shear-friction failures (potential failure surface shown by a dashed line): (a) narrow column bracket, (b) end support of a precast beam, (c) metal bracket.

Figure 4.39a shows a block of concrete anchored to a concrete surface by a steel dowel with an area $A_{vf}$. If a horizontal force $V_u$ is applied, tension will develop in the steel as it is elongated by the movement of the block to the right and upward. The upward displacement is caused by the block riding up on the surface irregularities. Equilibrium in the vertical direction requires that a compression force $N$ normal to the slip surface develop to balance the tension in the steel (Fig. 4.39b). $N$ represents the resultant of the compressive stresses that develop along the failure surface. From vertical equilibrium of the block we can write

$$\Sigma F_y = 0 \quad \text{and} \quad N = F_s = A_{vf} f_y$$

The normal force $N$ then permits a frictional force $F_r$ to develop

$$F_r = \mu N = \mu A_{vf} f_y$$

where $\mu$, the coefficient of friction, is given in Table 4.1 for various surfaces. Equating $V_u$ to $F_r$ and reducing the nominal frictional strength by a reduction factor $\phi = 0.85$ gives the ACI design equation

$$V_u = \phi F_r = \phi(A_{vf} f_y \mu) \qquad (4.26)$$

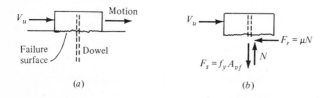

(a)    (b)

**Figure 4.39** Frictional resistance created by shear-friction reinforcement: (a) failure impends, (b) forces on block. $A_{vf}$ = area of shear friction steel.

## Table 4.1 Value of the coefficient of friction as a function of the failure surface

| Characteristic of failure surface | Coefficient of friction $\mu$ |
|---|---|
| Monolithic | 1.4 |
| On hardened roughened concrete | 1.0 |
| On unpainted steel | 0.7 |
| On hardened smooth concrete | 0.7† |
| To steel plate | 1.0† |

† Values from Ref. 6.

To determine the required area of dowel steel Eq. (4.26) is solved for $A_{vf}$, giving

$$A_{vf} = \frac{V_u}{\phi \mu f_y} \tag{4.27}$$

For a satisfactory design the Code also specifies that the following requirements be satisfied:

1. $V_n$, the nominal shear-friction strength must not exceed either $0.2f'_c A_c$ or $(800 \text{ lb/in}^2)A_c$, where $A_c$ represents the effective area of the concrete failure plane.
2. The tension steel must be well distributed over the failure plane and fully anchored on both sides of the failure surface. If space is limited, bars can be anchored by welding to plates or angles embedded in the concrete.
3. If shear friction is to be developed along a surface that is cast in two stages, the first surface should be roughened with grooves at least $\frac{1}{4}$ in (6.4 mm) deep.
4. $f_y$ of the steel must not exceed 60 kips/in$^2$ (413.7 MPa) to ensure that the strain required to stress the steel to failure is not excessive.

## QUESTIONS

**4.1** What does *nominal shear stress* mean? Why is it used?

**4.2** Why is the shear strength $v_c$ always expressed as a function of $\sqrt{f'_c}$?

**4.3** Describe the two modes of shear failure.

**4.4** How does axial force influence the shear capacity of a cross section?

**4.5** Explain how the magnitude of bending moment on a cross section influences shear strength.

**4.6** Why does the ACI Code specify that $V_s < 8\sqrt{f'_c}b_w d$?

**4.7** As specified by the ACI Code, the shear strength of concrete varies between $2\sqrt{f'_c}$ and $3.5\sqrt{f'_c}$. Since a shear failure is actually a diagonal-tension failure, how do you explain the fact that the shear strength is so much lower than the tensile strength predicted by the modulus of rupture or the split-cylinder tensile strength?

## PROBLEMS

**4.1** An ultimate moment $M_u = 15$ ft · kips acts on the cross section in Fig. P4.1. If no shear reinforcement is provided, what is the maximum value of shear force $V_u$ permitted on the cross section? $f'_c = 3$ kips/in$^2$.

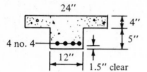

24"

4 no. 4

4"

5"

12"

1.5" clear      **Figure P4.1**

**4.2** A shear force of 13.4 kips acts on the cross section in Fig. P4.2. Are stirrups required? If so, what spacing should be used for no. 3 stirrups? $f'_c = 3$ kips/in$^2$, and $f_y = 50$ kips/in$^2$.

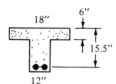

18"      6"

15.5"

12"      **Figure P4.2**

**4.3** A shear force of 28 kips acts on the cross section in Fig. P4.3. Are stirrups required? $f'_c = 3.6$ kips/in$^2$.

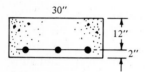

30"

12"

2"      **Figure P4.3**

**4.4** In Fig. P4.4 locate the region of the beam where stirrups can be spaced $d/2$ in apart. Specify the stirrup spacing required at section x-x. $f'_c = 2.5$ kips/in$^2$, and $f_y = 50$ kips/in$^2$.

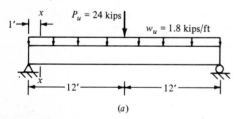
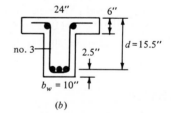

$x$      $P_u = 24$ kips

1'→|

$w_u = 1.8$ kips/ft

$x$

12'————|———— 12'————|

24"      6"

no. 3

2.5"      $d = 15.5$"

$b_w = 10$"

(a)      (b)

**Figure P4.4**

**4.5** If the beam in Fig. P4.5 is reinforced for shear with the maximum stirrup steel permitted by the ACI Code, what is the value of the total shear force $V_u$ allowed? If no. 3 stirrups are used, what spacing is required? Can no. 4 stirrups be used if they are spaced by Eq. (4.19)? Explain. $f'_c = 3$ kips/in$^2$, and $f_y = 60$ kips/in$^2$.

17.5"

12"      **Figure P4.5**

**4.6** The box beam in Fig. P4.6 is reinforced with no. 4 stirrups in each web; $f_c = 2.5$ kips/in$^2$, and $f_y = 50$ kips/in$^2$.

(*a*) What is the maximum value of shear force $V_u$ permitted on the cross section by the ACI Code?

(*b*) What is $V_{u, \max}$ if no. 3 stirrups are used at 6 in on center?

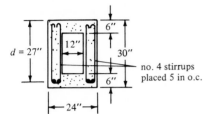

**Figure P4.6**

**4.7** In Fig. P4.7, what stirrup spacing is required (*a*) at section *A* and (*b*) at section *B*? Indicate the region of the beam where stirrups are not required. Use no. 3 stirrups. $f'_c = 2.5$ kips/in$^2$, and $f_y = 60$ kips/in$^2$.

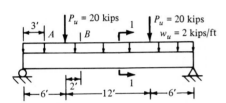

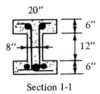

Section 1-1

**Figure P4.7**

**4.8** In Fig. P4.8 what is the maximum allowable spacing of no. 3 stirrups permitted by the ACI Code at sections 1 and 2? $f'_c = 3$ kips/in$^2$, and $f_y = 60$ kips/in$^2$. Use both the short and long forms of the equation for $V_c$.

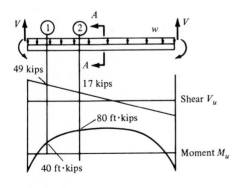

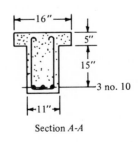

Section *A-A*

**Figure P4.8**

**4.9** Select the stirrup spacing for the beam in Fig. P4.9. Loads shown on the beam are service loads. $f'_c = 3$ kips/in$^2$, and $f_y = 60$ kips/in$^2$. Use no. 3 stirrups. Show your results on a scaled sketch. Use Eq. (4.8) to evaluate $V_c$.

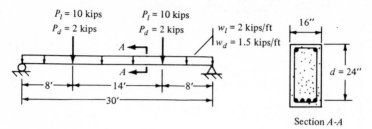

**Figure P4.9**

**4.10** For the frame in Fig. P4.10:

(a) At what point in member $AB$ would you expect a shear failure to occur if no shear reinforcement is used? Explain.

(b) If a shear crack opens at point $F$, sketch its shape.

(c) What stirrup spacing is required in the column at point $E$, located 4 ft below the centerline of the girder? Use no. 3 stirrups; $f'_c = 3$ kips/in$^2$, and $f_y = 50$ kips/in$^2$.

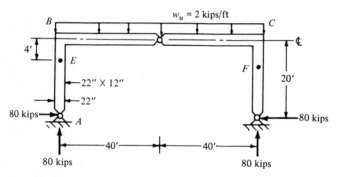

**Figure P4.10**

**4.11** In Fig. P4.11, a one-way slab is reinforced with no. 4 bars at 5 in on center. Considering both flexure and shear strength, determine the maximum value of factored uniform load $w_u$ the slab can support. $f'_c = 3$ kips/in$^2$, and $f_y = 50$ kips/in$^2$,

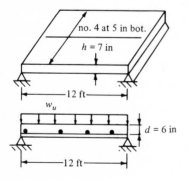

**Figure P4.11**

# REFERENCES

1. ASCE Committee 426: The Shear Strength of Reinforced Concrete Members, *Proc. ASCE J. Struct. Div.*, vol. 99, no. ST6, June 1973, pp. 1091–1187.
2. J. G. MacGregor and J. M. Hanson: Proposed Changes in Shear Provisions for Reinforced and Prestressed Concrete Beams, *J. ACI*, vol. 66, pp. 276–288, April 1969.
3. ACI-ASCE Committee 326: Shear and Diagonal Tension, p. 1, *J. ACI*, vol. 59, no. 1, p. 29, January 1962.
4. ACI-ASCE Committee 326: Shear and Diagonal Tension, p. 2, *J. ACI*, vol. 59, no. 2, February 1962.
5. Commentary on Building Code Requirements for Reinforced Concrete (ACI 318-77), American Concrete Institute, Detroit, 1977.
6. "PCI Design Handbook," Prestressed Concrete Institute, Chicago, 1971.

Dallas City Hall, Dallas, Texas; I. M. Pei architect. (*Photograph by I. M. Pei.*)

# FIVE

## TORSION

## 5.1 INTRODUCTION

Shear stresses due to torsion create diagonal-tension stresses that produce diagonal cracking. If a member is not properly reinforced for torsion with closed stirrups and longitudinal steel, a sudden brittle fracture can occur. The addition of steel reinforcement reduces crack width, raises the torsional strength, and imparts ductility. Since shear and moment usually develop simultaneously with torsion, a rational design should logically account for the interaction of these three forces; however, cracking of unpredictable extent, the inelastic behavior of concrete, and the complex state of stress produced by the interaction of shear, moment, and torsion make an exact analysis almost impossible. The current design procedure for torsion, which first appeared in the 1971 ACI Code, is based on the following simplifying assumptions:[1]

1. No interaction exists between flexure and torsion. Reinforcement for each force is designed independently.
2. When shear and torsion act simultaneously, the strength of the concrete can be apportioned between the two forces by a circular interaction curve.
3. Lateral reinforcement for shear and for torsion can be designed independently and then combined.

Experimental studies verify that these assumptions produce a conservative design.[1] Torsional reinforcement (the longitudinal component) increases flexural strength, and flexural steel and the compression produced by bending increase torsional capacity.

When sizing interior floor beams that are components of monolithic floor systems, the designer can normally neglect the small amounts of torsion that develop without creating problems (excessive rotation, overstress, or excessive crack width) because (1) the shear and the moment capacities of a beam are not reduced by small amounts of torque and (2) the stressing of adjacent members as the beam twists permits a redistribution of forces to these members and reduces the torque that must be supported by the beam. However, reinforcement for torsion is often required for spandrel beams (exterior building members), curved girders, and girders in which the resultant of the applied loads acts with a large eccentricity from the centroid of the cross section.

The design procedure developed in this section applies to nonprestressed beams with either a rectangular cross section or a section composed of rectangular areas.

## 5.2 DISTRIBUTION OF SHEAR STRESSES CREATED BY TORSION ON A RECTANGULAR CROSS SECTION

The twisting of a rectangular concrete member by a torque creates shear stresses on cross sections normal to the longitudinal axis. Before cracking occurs these stresses vary from zero at the center of the section to a maximum value at the midpoints of the long sides. As shown on the face of the member in Fig. 5.1*a*, torsional shear stresses always act in opposite directions on opposite sides of a cross section but are consistent with the direction (clockwise or counterclockwise) of the torque. For an elastic material, the direction and the variation of shear stresses are shown in Fig. 5.1*b* along several radial lines extending outward from the center of the cross section. Diagonal tension and compression stresses, induced by the shear stresses, develop at all points throughout the member: the diagonal stresses are equal in magnitude to the shear stresses. Figure 5.1*a* shows the diagonal stresses on a typical surface element at middepth of the vertical face. On the opposite face the

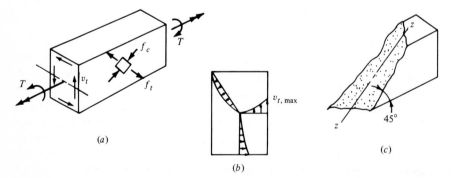

*(a)*

*(b)*

*(c)*

**Figure 5.1** Stresses produced in a rectangular bar by a torque: (*a*) principal stresses created by torsional shear stresses, (*b*) distribution of shear stresses on a rectangular section if behavior is elastic, (*c*) skewed failure plane produced by torsion in a brittle material.

direction of the tensile and compressive stresses reverses because the shear stresses change in direction.

When the diagonal-tension stresses created by the shear exceed the tensile strength of the concrete, a diagonal crack at a slope of approximately 45° opens near middepth on one of the long sides and spirals around the perimeter. If the concrete is not reinforced for torsion, development of the crack produces a sudden failure on a skewed plane (Fig. 5.1c). In a brittle material the failure mechanism resembles a bending failure about the principal axis of the failure plane (axis z-z in Fig. 5.1c), which is parallel to the longer side.

Although the addition of reinforcement does not change the value of torque at which diagonal cracking occurs, it does prevent the member from tearing apart and enables the cracked member to transmit a substantial torsional moment.

## 5.3 PREDICTION OF SHEAR STRESSES DUE TO TORQUE ON A RECTANGULAR CROSS SECTION

Theoretical studies indicate that the maximum stress produced on a rectangular cross section by torsion occurs at the midpoints of the long sides and can be predicted by

$$v_t = \frac{kT}{x^2 y} \tag{5.1}$$

where $v_t$ = maximum torsional shear stress at midpoint of long side
$x$ = length of short side
$y$ = length of long side
$T$ = torque
$k$ = a constant that is a function of the $y/x$ ratio

For elastic behavior $k$ varies from 3 for large $y/x$ ratios to 4.8 for $y/x = 1$.

As the basis for the ACI design equations that relate torsional shear stress to torque, a value of $k = 3$ is used in Eq. (5.1). Making that substitution and adding subscripts to $v_t$ and $T$ gives

$$v_{tu} = \frac{3T_u}{x^2 y} \tag{5.2}$$

Since Eq. (5.2) neglects both the influence of the cross section's proportions and the reduction of the cross section produced by flexural cracking, $v_{tu}$, the nominal torsional shear stress, must be considered only an approximate measure of the intensity of the diagonal-tension stress. [If a torsional failure is treated as an inclined bending failure, an expression identical to Eq. (5.2) results for the maximum diagonal tension at failure.] Equation (5.2) can be extended to members with a cross section composed of rectangular areas by substituting $\Sigma x^2 y$ for $x^2 y$ to give

$$v_{tu} = \frac{3T_u}{\Sigma x^2 y} \tag{5.3}$$

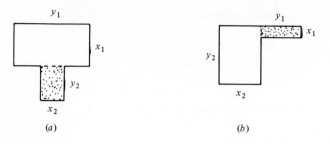

**Figure 5.2** Subdivision of a cross section into rectangular areas to evaluate $\Sigma x^2 y$: (a) T beam, (b) L-shaped section.

To evaluate $\Sigma x^2 y$, a cross section is subdivided into rectangular areas that maximize the value of $\Sigma x^2 y$. To achieve this result, an area should be subdivided so that the rectangle with the greatest width has the greatest length (see Fig. 5.2).

The flanges on either side of a beam web contribute to the torsional strength of a section; however, studies show that only a length of flange extending out from the face of the stem a distance equal to 3 times the flange thickness is effective in resisting torque (see ACI Code §11.6.1.1).

Since the shear stresses are low near the center of a cross section and act with a small arm from the center of rotation, a hollow box beam with a small opening at its center has about the same torsional strength as a solid beam with the same outside dimensions. Therefore, ACI Code §11.6.1.2 permits hollow beams whose wall thickness $h$ is at least $x/4$ to be treated as a solid beam when evaluating $\Sigma x^2 y$. If the wall thickness of a box beam falls between $x/4$ and $x/10$, it can still be analyzed as a solid section for torsion except that $\Sigma x^2 y$ must be multiplied by $4h/x$ to compensate for the reduction in strength produced by the opening. When beam walls are less than $x/10$ thick, a brittle mode of torsional failure is possible and the design of the cross section must consider the stiffness of the walls to prevent buckling.

To reduce the likelihood that closed stirrups will tear out the interior corner of a box beam, ACI Code §11.6.1.2 requires fillets to be provided at the interior corners of all box sections. These fillets also reduce stress concentrations at the corners and the possibility of honeycombing by permitting concrete to flow more easily into the bottom flange. The ACI Commentary recommends that fillets have a minimum leg size of $x/6$ when the longitudinal torsional reinforcement consists of less than eight bars. If eight or more bars are used, the leg of the fillet may be reduced to $x/12$ but need not be larger than 4 in. In the computation of $\Sigma x^2 y$, the area of the fillets can be neglected.

**Example 5.1:** Evaluate the quantity $\Sigma x^2 y$ for the box beam in Fig. 5.3 for (a) $h = 6$ in (152 mm) and (b) $h = 4$ in (102 mm).

SOLUTION (a) 
$$x = 18 \text{ in}, \quad y = 24 \text{ in}$$

$$\frac{x}{4} = \frac{18}{4} = 4.5 \qquad h > \frac{x}{4}$$

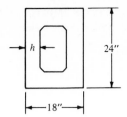

**Figure 5.3** Hollow box beam.

Treat like a solid section.

$$\Sigma x^2 y = 18^2(24) = 7776 \text{ in}^3 \ (127.43 \times 10^6 \text{ mm}^3)$$

(b)
$$\frac{x}{4} = \frac{18}{4} = 4.5 \qquad \frac{x}{10} < h < \frac{x}{4}$$

Use the properties of a solid section but reduce by $4h/x$.

$$\Sigma x^2 y = \frac{4h}{x} 18^2(24) = \frac{4(4)}{18} 18^2(24) = 6912 \text{ in}^3 \ (113.27 \times 10^6 \text{ mm}^3)$$

**Example 5.2:** Evaluate the quantity $\Sigma x^2 y$ for the T beam in Fig. 5.4. Include a 15-in length of flange equal to 3 times the thickness $h_f$.

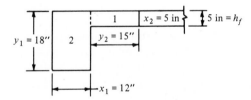

**Figure 5.4**

SOLUTION
$$\Sigma x^2 y = 12^2(18) + 5^2(15) = 2967 \text{ in}^3$$

## Minimum Torque

ACI Code §11.6.1 permits torsion to be neglected whenever the torque $T_u$ due to factored loads is less than

$$T_u = \phi(0.5\sqrt{f'_c}\Sigma x^2 y) \tag{5.4}$$

This equation is established by substituting a value of nominal stress $v_{tu}$ equal to $1.5\sqrt{f'_c}$ into Eq. (5.3), adding a reduction factor, $\phi = 0.85$, and solving for $T_u$. A value of $v_{tu} = 1.5\sqrt{f'_c}$ represents approximately 25 percent the torsional stress required to produce diagonal cracking of a rectangular member without torsion reinforcement. Studies indicate that a torque of the magnitude given by Eq. (5.4) will not reduce the strength of a section for either vertical shear or bending moment.

**Example 5.3:** The cantilever beam in Fig. 5.5 carries an 8-kip (35.58-kN) load applied 5 in (127 mm) off center. Should torsion be considered in the design of the member?

$$f'_c = 3000 \, \text{lb/in}^2 \, (20.68 \, \text{MPa}).$$

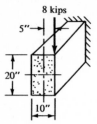

8 kips

5"

20"

10"

**Figure 5.5**

SOLUTION Use Eq. (5.4); the limiting torque is

$$T_u = \phi(0.5\sqrt{f'_c}\Sigma x^2 y) = 0.85[0.5\sqrt{3000}\,(10^2)(20)] = 46.6 \, \text{in} \cdot \text{kips}$$

The applied torque is

$$8 \, \text{kips} \, (5 \, \text{in}) = 40 \, \text{in} \cdot \text{kips} < 46.6 \, \text{in} \cdot \text{kips} \, (5.15 \, \text{kN} \cdot \text{m})$$

Neglect torsion.

## 5.4 REINFORCING FOR TORSION

Tests show that both longitudinal bars and closed stirrups must be provided if beams are to be effectively reinforced for torsion. Both are required to intercept the large number of diagonal tension cracks that form on all surfaces of a beam. To be effective the longitudinal steel, no smaller than no. 3 bars, must be distributed uniformly around the perimeter of the stirrups at a spacing not to exceed 12 in (305 mm). In addition at least one longitudinal bar must be positioned in each corner of the closed stirrup to anchor the sitrrup legs so they can develop their full tensile strength. Should the concrete inside the radius of the bend crush, the stirrup would slip and allow torsion cracks to widen. These requirements from ACI Code §11.6.8.2 are summarized in Fig. 5.6.

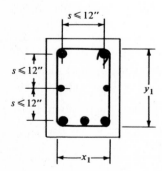

$s \leqslant 12''$

$s \leqslant 12''$

$s \leqslant 12''$

$y_1$

$x_1$

**Figure 5.6** Torsion reinforcement: $x_1$, the shorter side of stirrup, and $y_1$, the longer side of stirrup, are measured between centerlines of reinforcement.

Substituting $V_c = 2\sqrt{f'_c}b_w d$ and $T_c = 0.8\sqrt{f'_c}\Sigma x^2 y$ into the right side of Eq. (5.17), using the square of Eq. (5.15) to express $T^2_{cv}/V^2_{ct}$ in terms of $T_u$ and $V_u$, and replacing the $V_{ct}$ on the left side of the equation by $V_c$ gives

$$V_c = \frac{2\sqrt{f'_c}b_w d}{\sqrt{1 + (2.5C_t\,T_u/V_u)^2}} \qquad \text{where } C_t = \frac{b_w d}{\Sigma x^2 y} \qquad (5.18)$$

which specifies the shear strength of concrete in the presence of torsion. If $T_u = 0$, the denominator in Eq. (5.18) equals 1 and the equation reduces to Eq. (4.8), the basic equation for shear strength when no torsion acts.

Similarly, if each term in Eq. (5.16) is multiplied by $T^2_c$ and the resulting expression solved for $T_{cv}$, when $T_c$ replaces $T_{cv}$ on the left side, the result is

$$T_c = \frac{0.8\sqrt{f'_c}\Sigma x^2 y}{\sqrt{1 + (0.4V_u/C_t\,T_u)^2}} \qquad (5.19)$$

for the concrete's nominal torsional strength in the presence of shear.

## 5.7 SUMMARY OF DESIGN PROCEDURE FOR A MEMBER STRESSED BY TORSION, SHEAR, AND MOMENT

**Step 1**  Select $b$ and $d$ based on $M_u$. These dimensions may have to be revised if shear stresses are excessive. Square sections are best for torsion.

**Step 2**  If $T_u \le \phi(0.5\sqrt{f'_c}\Sigma x^2 y)$, neglect torsion. If

$$T_u > \frac{\phi 4\sqrt{f'_c}\Sigma x^2 y}{\sqrt{1 + (0.4V_u/C_t\,T_u)^2}}$$

increase dimensions. The critical section for shear and torsion is located a distance $d$ from the face of the support.

**Step 3**  Compute the nominal shear and torsional capacity of the concrete

$$T_c = \frac{0.8\sqrt{f'_c}\Sigma x^2 y}{\sqrt{1 + (0.4V_u/C_t\,T_u)^2}}$$

Where

$$C_t = \frac{b_w d}{\Sigma x^2 y}$$

If significant axial tension exists, see ACI Code §11.6.6.2,

$$V_c = \frac{2\sqrt{f'_c}b_w d}{\sqrt{1 + (2.5C_t(T_u/V_u))^2}}$$

**Step 4**  Size stirrups for shear

$$V_s = \frac{V_u}{\phi} - V_c \qquad \frac{A_v}{s} = \frac{V_s}{f_y d} \qquad f_y \leq 60 \text{ kips/in}^2$$

If $V_s > 8\sqrt{f'_c} b_w d$, increase the size of the cross section.

**Step 5**  Size stirrups for torsion

$$T_s = \frac{T_u}{\phi} - T_c \qquad \frac{A_t}{s} = \frac{T_s}{\alpha_t x_1 y_1 f_y}$$

where

$$\alpha_t = 0.66 + 0.33 \frac{y_1}{x_1} \leq 1.5$$

**Step 6**  Combine areas of shear and torsion steel. Express $A_v/s$ in terms of $A_t/s$ or vice versa and solve for the required spacing of closed stirrups. By dividing $A_v$ and $A_t$ by $s$, we place them on a unit basis, i.e., area/1 in, so they can be combined.

**Step 7**  Check minimum steel

$$A_v + 2A_t \geq \frac{50 b_w s}{f_y}$$

**Step 8**  Compute longitudinal area of steel (see Fig. 5.12)

$$A_l = \frac{2A_t(x_1 + y_1)}{s}$$

but not less than

$$A_l = \left( \frac{400 x s}{f_y} \frac{T_u}{T_u + V_u/3C_t} - 2A_t \right) \frac{x_1 + y_1}{s} \tag{5.20}$$

where $x$ = length of short side of cross section
$s$ = spacing of closed stirrups from Step 6

$A_l$ computed from this equation need not exceed that obtained by substituting $50 b_w d/f_y$ for $2A_t$.

$y_1 > x_1$     **Figure 5.12**

**Step 9**   Maximum spacing of closed stirrups for torsion not to exceed $x_1 + y_1/4$ or 12 in.

**Step 10**   Torsion reinforcement must be extended at least a distance $d + b$ beyond the section where $T_u = \phi(0.5\sqrt{f_c'}\Sigma x^2 y)$.

**Step 11**   Maximum spacing of stirrups based on shear must also be satisfied

$$\frac{d}{2} \quad \text{if } V_s \le 4\sqrt{f_c'}b_w d$$

$$\frac{d}{4} \quad \text{if } V_s > 4\sqrt{f_c'}b_w d$$

**Example 5.4: Design of a beam for torsion, shear, and moment**  Design the cantilever beam in Fig. 5.13 using no. 3 closed stirrups. The uniform load is applied 10 in (254 mm) off the centerline of the cross section. $f_c' = 3$ kips/in$^2$ (20.68 MPa), $f_y = 60$ kips/in$^2$ (413.7 MPa), $w_u = 1.5$ kips/ft (21.89 kN/m), and $l = 12$ ft (3.66 m).

SOLUTION 1. Find the minimum depth to control flexural deflections (see Table 3.1). Since

$$h_{\min} = \frac{l}{8} = \frac{(12 \text{ ft})(12)}{8} = 18 \text{ in (457 mm)}$$

estimate $d = 15.5$ in (394 mm).
 2. Design for moment. Try $b = 10$; using Table 3.4, read $F = 0.2$. Then compute $K_u$

$$K_u = \frac{M_u}{F} = \frac{108}{0.2} = 540$$

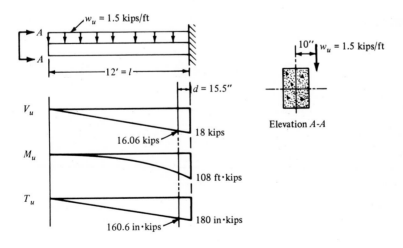

**Figure 5.13**

Interpolating from Table 3.3 gives $\rho = 0.0116$.

$$A_s = \rho bd = 1.8 \text{ in}^2 \text{ (1161 mm}^2\text{)}$$

3. Determine whether torsion can be neglected

$$T_u < \phi(0.5\sqrt{f_c'}\Sigma x^2 y) = 0.85 \frac{0.5\sqrt{3000} \times 10^2(18)}{1000} = 41.9 \text{ in} \cdot \text{kips (4.73 kN} \cdot \text{m)}$$

Since $160.7 \text{ in} \cdot \text{kips} > 41.9 \text{ in} \cdot \text{kips}$, consider torsion.

4. Check the maximum torsional capacity to ensure that the cross section will be adequate.

$$T_{u,\max} = \frac{\phi 4\sqrt{f_c'}\Sigma x^2 y}{\sqrt{1 + \left(\dfrac{0.4V_u}{C_t T_u}\right)^2}} \qquad C_t = \frac{b_w d}{x^2 y} = \frac{10(15.5)}{10^2 \times 18} = 0.086$$

where $\qquad V_u = 16.06 \text{ kips} \qquad T_u = 160.6 \text{ in} \cdot \text{kips} \qquad \phi = 0.85$
$\qquad\qquad T_{u,\max} = 303.9 \text{ in} \cdot \text{kips}$

At $d$ in $T_u = 160.7 \text{ in} \cdot \text{kips} < 303.9 \text{ in} \cdot \text{kips}$; section OK.

5. Compute $T_c$ and $V_c$ at $d$ in out from the support. Since the ratio of $V_u/T_u$ is constant along the length, $T_c$ and $V_c$ will be constant at all sections along the beam

$$T_c = \frac{0.8\sqrt{f_c'}\Sigma x^2 y}{\sqrt{1 + (0.4V_u/C_t T_u)^2}} = \frac{0.8(\sqrt{3000})(10^2)(18)}{\sqrt{1 + [0.4(16.06)/0.086(160.6)]^2}}$$

$$= 71,520 \text{ in} \cdot \text{lb} = 71.52 \text{ in} \cdot \text{kips (8.08 kN} \cdot \text{m)}$$

$$V_c = \frac{2\sqrt{f_c'}b_w d}{\sqrt{1 + (2.5C_t T_u/V_u)^2}} = \frac{2\sqrt{3000}(10)(15.5)}{\sqrt{1 + [2.5(0.086)(160.6)/16.07)]^2}} = 7160 \text{ lb} = 7.16 \text{ kips}$$

$$\text{(31.85 kN)}$$

6. Stirrups required for vertical shear

$$V_s = \frac{V_u}{\phi} - V_c = \frac{16.06}{0.85} - 7.16 = 11.73 \text{ kips}$$

$$V_s < 8\sqrt{f_c'}b_w d = 78.87 \text{ kips}$$

$$\frac{A_v}{s} = \frac{V_s}{f_y d} = \frac{11.73}{60(15.5)} = 0.0126 \text{ in}^2/\text{in}$$

Or in terms of $A_t/s = A_v/2s = 0.0063 \text{ in}^2/\text{in}$

7. Stirrups required for torsion (see Fig. 5.14)

$$T_s = \frac{T_u}{\phi} - T_c = \frac{160.6}{0.85} - 71.52 = 117.42 \text{ in} \cdot \text{kips (13.27 kN} \cdot \text{m)}$$

$$\frac{A_t}{s} = \frac{T_s}{\alpha_t x_1 y_1 f_y} = \frac{117.42}{1.4(6.6)(14.6)(60)} = 0.0145 \text{ in}^2/\text{in}$$

where $\alpha_t = 0.66 + 0.33 y_1/x_1 = 1.39$; use 1.4.

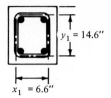

$y_1 = 14.6''$

$x_1 = 6.6''$

**Figure 5.14**

8. Combine shear and torsion steel

$$\frac{A_t}{s} = 0.0063 + 0.0145 = 0.0208 \text{ in}^2/\text{in}$$

For a no. 3 bar, $A_t = 0.11 \text{ in}^2$

$$s = \frac{A_t}{0.0208} = 5.29 \text{ in} \qquad \text{use 5 in}$$

$$s_{max} = \frac{x_1 + y_1}{4} = \frac{6.6 + 14.6}{4} = 5.3 \text{ in} \qquad s_{max} = 12 \text{ in}$$

Use 5 in. Since the maximum permitted spacing is 5.3 in, Use 5 in throughout.

9. Minimum steel

$$\frac{A_v}{s} + \frac{2A_t}{s} \geq \frac{50b_w}{f_y}$$

$$0.0126 + 2(0.0145) > 0.0083 \qquad \text{OK}$$

10. Longitudinal steel for torsion. Use the larger of the values from Eqs. (5.12) and (5.20)

$$A_l = \frac{2A_t(x_1 + y_1)}{s} = 2(0.0145)(6.6 + 14.6) = 0.61 \text{ in}^2$$

$$A_l = \left(\frac{400xs}{f_y} \frac{T_u}{T_u + V_u/3C_t} - 2A_t\right)\frac{x_1 + y_1}{s} \qquad (5.20)$$

$$= \left[\frac{400(10)(5)}{60,000} \frac{160.6}{160.6 + 16.06/[3(0.086)]} - 2(0.11)\right]\frac{6.6 + 14.6}{5}$$

$$= 0.086 \text{ in}^2 \ (55.48 \text{ mm}^2)$$

If $50b_w s/f_y$ is substituted for $2A_t$, the upper limit of $A_l$ for Eq. 5.20 is 0.84 in². Therefore, use $A_l = 0.61 \text{ in}^2 \ (394 \text{ mm}^2)$.

11. Combine flexure and torsion reinforcement (longitudinal steel). Divide $A_l$ into three equal areas for the top, bottom, and middepth of section.

$$\frac{A_l}{3} = \frac{0.61}{3} = 0.203 \text{ in}^2$$

Use two no. 3 ($A_s = 0.22$) at middepth and at the bottom. At top, add 0.203 to the flexural steel, $A_s = 1.8 \text{ in}^2$ (see step 2). $A_s = 2.0 \text{ in}^2$; use two no 9; then the area of steel supplied is 2.0 in² (see Fig. 5.15).

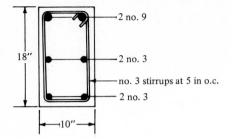

**Figure 5.15** Reinforcing details.

12. Find the point at which stirrups can be discontinued for torsion. When $T_u = \phi(0.5\sqrt{f_c'}\Sigma x^2 y) = 41.9$ in · kips, torsion need not be considered in the design. Then extend reinforcement a distance of $d + b$. From similar triangles (Fig. 5.16)

$$\frac{x}{41.9} = \frac{12}{180}$$

$$x = 2.79 \text{ ft}$$

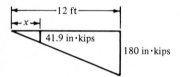

**Figure 5.16** Termination of torsional stirrups.

Distance from tip where stirrups can be terminated equals

$$x - (d + b) = 2.79 - \frac{(15.5 + 10)}{12} = 0.66 \text{ ft } (0.2 \text{ m})$$

Stirrups are required to within 0.66 ft from the tip, but extend the full distance to be conservative.

13. For sketch of reinforcement see Fig. 5.17.

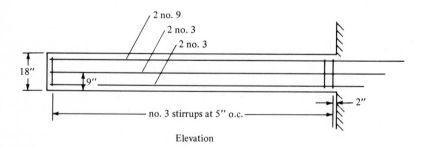

Elevation

**Figure 5.17** Elevation showing reinforcement.

## 5.8 EQUILIBRIUM AND COMPATIBILITY TORSION

In a statically determinate structure, only one path exists for the transmission of loads into the supports. The magnitude of the internal torque, termed the *equilibrium torque*, on any cross section can be determined by first cutting a free body perpendicular to the longitudinal axis and then summing moments of all forces about the longitudinal axis through the centroid of the section. For structures of this type (Fig. 5.18) ACI Code §11.6.2 specifies that all sections must be designed for a torsional capacity that is equal to or greater than that of the torque produced by factored loads.

In the design of members of a statically indeterminate structural system, ACI Code §11.6.3 allows the designer two options: (1) members can be designed for the torque established by an elastic analysis; or (2) if a redistribution of internal forces can occur, a member can be designed at the most critical section for a maximum torque of $\phi(4\sqrt{f'_c}\Sigma x^2 y/3)$. With this value of torque established, the structure is then analyzed elastically to establish the design forces in the balance of the members. The latter option, termed *compatibility torsion*, is allowable because a member reinforced for torsion develops many diagonal cracks and undergoes a large rotational deformation at a torque of approximately $\phi(4\sqrt{f'_c}\Sigma x^2 y/3)$ (Fig. 5.19). This deformation, which occurs at a constant value of torque, permits a redistribution of forces to adjacent sections of the member. Although extensive cracking develops as the rotation occurs, the crack width can be limited to acceptable values by the use of closely spaced torsion reinforcement.

The designer may wish to take advantage of the redistribution of forces permitted by compatibility torsion to reduce the torque for which a particular member such as a spandrel beam must be designed. Reduced forces often permit the use of smaller cross sections and allow greater flexibility in adapting the structural requirements to the architectural constraints.

Example 5.5 illustrates the use of a compatibility torsion analysis to reduce the torque for which a spandrel beam must be designed. As permitted by ACI Code §8.3.3 (see Sec. 9.8), the assumption is made that an exterior negative moment of $w_n l_n^2/24$ per foot of slab develops. To simplify the analysis, the moment between the slab and the spandrel beam is treated as if it acted at the centerline of the spandrel beam rather than at the face of the beam, as specified in the code.

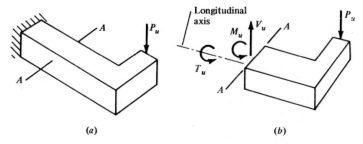

(a)                                                      (b)

**Figure 5.18** Equilibrium torque: (a) determinate member, (b) free body cut by section A-A.

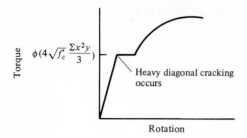

Figure 5.19 Torque-rotation curve.

For the analysis of the slab by moment distribution, a 1-ft width of slab is assumed to act as a continuous beam on simple supports, and the center-to-center distances between supports are used in the computation of fixed-end moments.

**Example 5.5: Use of a compatibility torsion analysis to reduce forces for which a spandrel beam must be designed** The one-way slab system in Fig. 5.20 carries a uniform factored load $w_u = 0.3$ kip/ft$^2$ (14.36 kPa). The spandrel beams along lines 1 and 3 have a depth of 18 in (457 mm), width $b = 12$ in (305 mm), and $d = 15$ in (381 mm); $f_c' = 3$ kips/in$^2$ (20.68 MPa). Determine the moments in a typical 1-ft (0.3-m) strip of slab and the maximum torque and the uniform load for which the spandrel beams along lines 1 and 3 must be designed when (a) the restraint of the spandrel beams along columns 1 and 3 applies to the slab a negative design moment of

$$\frac{w_u l_n^2}{24} = \frac{0.3(14^2)}{24} = 2.45 \text{ ft} \cdot \text{kips/ft} \quad (10.89 \text{ kN} \cdot \text{m/m})$$

(the above value of moment is recommended by ACI Code §8.3.3) and (b) a compatibility torsion analysis limits the torsion on the most critical section to $\phi(4\sqrt{f_c'}\Sigma x^2 y/3)$.

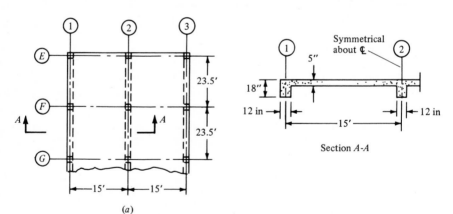

(a)

**Figure 5.20** One-way slab-and-beam system supported on 12-in square columns: (a) floor plan.

SOLUTION (a) If a moment of 2.45 ft · kips/ft develops between the slab and the spandrel beam (Fig. 5.21a), the torque $T_u$ on the most critical section located a distance $d = 15$ in from the face of the column equals

$$T_u = 2.45(10 \text{ ft}) = 24.5 \text{ ft} \cdot \text{kips}$$

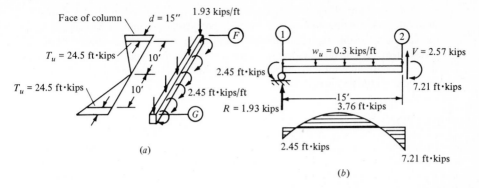

(a)

(b)

**Figure 5.21** Design forces to spandrel beam and slab: (a) spandrel beam, (b) typical strip of slab 1-ft wide. Reaction R represents the load that must be supported by the spandrel beam.

For a uniform factored load of 0.3 kip/ft and an end moment of 2.45 ft · kips/ft, the end moments in a 1-ft width of slab analyzed as a continuous beam can be determined by moment distribution. The results of this analysis are shown in Fig. 5.21b.

(b) A compatibility torsion analysis permits the torque on the critical section of the spandrel beam to be taken as

$$T_u = \phi\left(4\sqrt{f_c'}\,\frac{\Sigma x^2 y}{3}\right) = 0.85\left(4\sqrt{3000}\,\frac{2967}{3}\right) = 184{,}177 \text{ in} \cdot \text{lb} = 15.4 \text{ ft} \cdot \text{kips}$$

where $\Sigma x^2 y/3$ is computed in Example 5.2. Assuming that the slab applies a uniformly distributed moment to the spandrel beam, the moment per foot, as shown in Fig. 5.22a, is

$$\frac{15.4 \text{ ft} \cdot \text{kips}}{10 \text{ ft}} = 1.54 \text{ ft} \cdot \text{kips/ft}$$

The analysis of the slab for the total factored load and the end moment of 1.54 ft · kips/ft gives the moment curve shown in Fig. 5.22b.

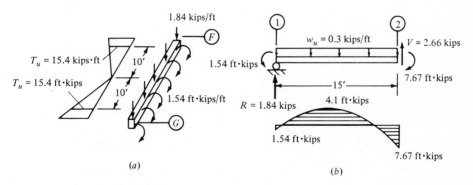

(a)

(b)

**Figure 5.22** Design forces to slab and beam using a compatibility torsion analysis: (a) spandrel beam, (b) typical strip of slab 1-ft wide. Reaction R represents the load that must be supported by the spandrel beam.

A comparison of the two methods of analysis as illustrated in Figs. 5.21 and 5.22 shows that the torsion and vertical load in the spandrel beam can be reduced by the compatibility-torsion analysis. This reduction in forces permits a smaller beam. The reduction in forces in the spandrel beam is accompanied by a small increase in the positive and the negative interior slab moments. The small increase in slab moments has no significant effect on the slab proportions.

## QUESTIONS

**5.1** Which of the three cross sections in Fig. Q5.1 can transmit a pure torque most effectively? Explain your choice.

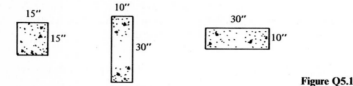

**Figure Q5.1**

**5.2** Use the torsion-failure mechanism to explain the function of longitudinal steel.

## PROBLEMS

**5.1** Evaluate $\Sigma x^2 y$ for the hollow box beam. All walls are 4 in thick.

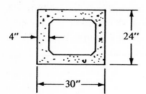

**Figure P5.1**

**5.2** A torque of 120 in · kips acts on the cross section. Must torsion be considered in design? $f'_c = 4$ kips/in$^2$.

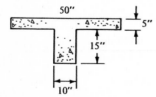

**Figure P5.2**

**5.3** If the cross section is properly reinforced, what is the maximum value of torsion for which the cross section can be designed? $f'_c = 3$ kips/in$^2$, and $V_u/T_u = \frac{1}{6}$.

**Figure P5.3**

# BOND, ANCHORAGE, AND REINFORCING DETAILS

## 6.1 SCOPE OF CHAPTER

In addition to creating longitudinal stresses in the reinforcement and the concrete, flexural deformations of a beam also create stresses between the reinforcement and the concrete. If the intensity of these stresses, called *bond stresses*, is not limited, they may produce crushing or splitting of the concrete surrounding the reinforcement. Failure of the concrete permits the reinforcement to slip. As slipping occurs, the stress in the reinforcement reduces to zero and the composite action between the steel and the concrete is lost. Once the reinforcement is unbonded, the beam, which behaves as if it were unreinforced, is subject to instant failure as soon as the concrete cracks. Initially, we will study the mechanics of bond strength and develop design equations that ensure the reinforcement is solidly anchored to the concrete.

In the concluding sections of this chapter, the following topics related to the termination and the detailing of reinforcement are considered:

1. Using standard hooks to improve anchorage
2. Checking local bond stresses (at simple supports and points of inflection) when the standard anchorage provisions are not fully applicable
3. Terminating a portion of the flexural reinforcement in regions where the area of reinforcement exceeds that required for moment
4. Detailing reinforcement to produce tough ductile members
5. Splicing reinforcement

Careful attention to the details discussed in this chapter will reduce the likelihood of local bond failures and assure that the full bending strength of the beam will develop at all sections.

## 6.2 BOND STRESSES

The design of beams is based on the underlying assumption that no slippage will occur between the reinforcement and the concrete when a load is applied. In other words, it is assumed that the steel will undergo the same tensile and compressive deformations as the concrete to which it is bonded (Fig. 6.1b). For the steel to deform in conjunction with the concrete, the concrete must exert shear-type stresses, termed *bond stresses*, on the surface of the reinforcement (Fig. 6.1c). The conventional representation of these bond stresses as a longitudinal shear stress is a simplification of the actual stresses exerted on the reinforcement by the concrete and neglects the high radial stresses, created by the bearing of the bar deformations against the concrete, that also act around the circumference of the bar. The action of these radial stresses (see Fig. 6.5), which produce longitudinal cracks in the concrete surrounding the reinforcement, will be discussed in detail in Sec. 6.4. Since the radial stresses do not influence the longitudinal equilibrium of the bar, they are omitted in the sketches of Fig. 6.1. As shown in Fig. 6.1d, the bond stresses $u(x)$ are oriented to balance the tension force $T$ in the reinforcement. If such bond stresses did not develop, the steel would remain unstressed and would not contribute to the bending strength of the cross section. A bending failure

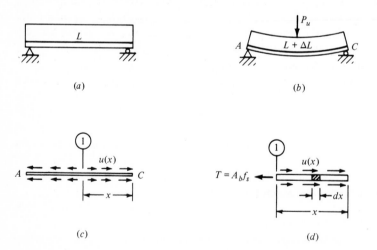

**Figure 6.1** Stressing of reinforcement by bending deformations: (a) unloaded beam, flexural steel unstressed, (b) steel elongated and stressed by bending deformations, (c) free-body diagram of reinforcement showing bond stresses $u(x)$, (d) free-body diagram of the right half of the reinforcement; tension force $T$ is balanced by bond stresses acting over the surface of reinforcing bar; $T = \int_0^x u(x)\Sigma_0\, dx$, where $\Sigma_0$ is the circumference of the bar.

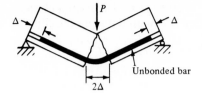

Unbonded bar  **Figure 6.2** Failure due to lack of bond between concrete
and steel; $\Delta$ = distance bar slips at each end.

would then occur in a sudden brittle manner when the tensile stresses in the concrete reached the modulus of rupture, the concrete behaving as if no steel were
present.

The behavior of a beam in which bond stresses do not develop between the
reinforcement and the concrete is illustrated in Fig. 6.2 by a hypothetical case: a
beam, reinforced with a smooth-surfaced, greased bar, is loaded to failure. Because
of the greased surface, the concrete cannot bond to the bar. When the beam is
deformed by load, the unstressed bar slips in relation to the stressed concrete and
pulls out an amount equal to the width of the crack $2\Delta$ as the beam splits into two
sections with the development of the first tension crack.

## 6.3 DISTRIBUTION AND VARIATION OF BOND STRESSES

Bond stresses, which vary in magnitude along the length of a reinforcing bar, are a
function of the rate at which the tension force in the reinforcement changes from
section to section. Both a change in bending moment and the development of
tension cracks cause the tension force in the reinforcement to vary and to induce
variable bond stresses on the surface of the reinforcement. In regions where the
bending moment changes rapidly and the shear is consequently high, the tensile
force in the reinforcement necessarily undergoes significant change in a short
distance $\Delta x$; therefore, large bond stresses, called *flexural bond stresses*, are required to equilibrate the difference in tension $\Delta T$ between the two sections
(Fig. 6.3).

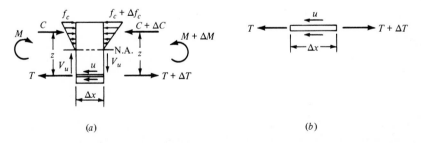

**Figure 6.3** Flexural bond stresses created by a moment gradient: (*a*) small beam segment showing
internal forces, (*b*) detail of the reinforcing bar showing bond stresses required to balance $\Delta T$. Horizontal
equilibrium gives $\Delta T = u\Sigma_0 \Delta x$.

Before 1971 the bond provisions in the ACI Code were based on the assumption that bond stresses were created only by a change in moment. Based on the equilibrium requirements for the free bodies shown in Fig. 6.3, expressions for bond stresses were related to the shear, a measure of the rate at which the moment changes. The computed bond stresses would then be compared with experimentally determined values of nominal bond strength to check against overstress.

However, once it was recognized that large local variations in bond stress are created by cracks, it became self-evident that if a crack was located at the point at which bond stresses were calculated, the actual bond stresses could deviate significantly from the computed bond stress. As a result, an alternate approach was developed in which the actual length of *anchorage*, the distance between the point of maximum stress and the near end of the bar, is compared with the minimum length required for assured anchorage.

The influence of cracking on bond stress[1] is shown qualitatively in Fig. 6.4, which shows a free body of a short length of beam that contains a crack. To simplify the discussion, it is further assumed that the element is cut from a region of constant moment so that the bond stresses are not created by a change in moment. At the cross section directly through the crack (section 2 in Fig. 6.4b), the total longitudinal tension force $T_s'$ is carried by the steel alone. At the cross sections that remain intact on either side of the crack (sections 1 and 3) the tension is carried by both the steel and the uncracked concrete. Therefore, between sections 1 and 2, the tension in the steel is reduced by an amount equal to that carried by the concrete. To balance

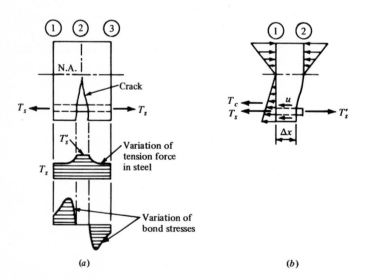

(a)                                    (b)

**Figure 6.4** Influence of flexural cracking on bond stress: (a) a small segment of beam with a flexural crack in a region of constant moment, (b) segment of beam between a flexural crack at section 2 and the uncracked section at 1. At the crack, the entire tension $T_s'$ is carried by reinforcement; on the uncracked section force $T_s$ is carried by steel and force $T_c$ by concrete. For horizontal equilibrium, average bond stresses created by the difference in steel forces are given as $u = (T_s' - T_s)/(\Delta x \, \Sigma_0)$.

the difference in tensions $(T'_s - T_s)$ between the cracked and the uncracked sections, bond stresses must develop (Fig. 6.4b). It is self-evident that the bond stresses are zero over the width of the crack since no contact exists between the exposed surface of the reinforcement and the concrete (Fig. 6.4a).

## 6.4 MECHANICS OF BOND STRENGTH

Factors that contribute to bond strength are chemical adhesion, friction, and bearing of the bar deformations (also referred to as ribs or lugs) against the concrete. The contribution of each of these components of bond resistance varies with the level of stress in the reinforcement. When members are lightly stressed, bond resistance is due primarily to chemical adhesion. Since the bond resistance that can develop from this source is limited [on the order of 200 to 300 lb/in² (1.38 to 2.07 MPa)], it is not a significant component of bond strength. Moreover, once the reinforcement slips relative to the concrete, even this limited bond resistance is lost (Fig. 6.5a).

After the adhesion is broken and some slight movement between the reinforcement and the concrete occurs, bond strength is supplied both by friction and by the bar deformations bearing against the concrete. Of these two sources of resistance to further slipping, the bearing stresses are the more significant.

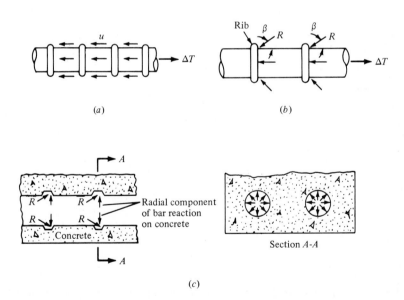

**Figure 6.5** Sources of bond strength: (a) bond stresses due to friction and chemical adhesion between concrete and steel, (b) free-body diagram of reinforcing bar showing reaction of concrete on ribs, (c) free-body diagram of concrete showing forces exerted by ribs on concrete; radial component of bar reactions acts to produce cracking and spalling of concrete surrounding the bar.[3]

Experimental studies[2] of internal crack patterns in the concrete surrounding the reinforcement indicate that the bearing stresses, whose direction is influenced by the slope of the rib face, are inclined to the longitudinal axis of the bar at an angle $\beta$ that varies from 45 to 80° (Fig. 6.5b); therefore, the component of the bearing stresses normal to the longitudinal axis of the bar is equal to or larger than the longitudinal component.

For most beams constructed with the flexural reinforcement positioned 2 to 3 in (51 to 76 mm) from the exterior surface, experimental studies indicate that bond failures are produced by the radial component of the bearing pressure (section *A-A* in Fig. 6.5). By creating circumferential tensile stresses in a *hypothetical* cylinder of concrete surrounding each reinforcing bar, the radial component of the bearing pressure creates a state of stress that tends to produce longitudinal splitting of the concrete. Figure 6.6 shows qualitatively the state of tensile stress that develops on typical diameters of the cylindrical area. Several investigators have suggested that the state of stress in the concrete surrounding each reinforcing bar is similar to that produced in a thick-walled pipe by an internal pressure.

The critical plane on which splitting is most likely to take place will pass through the center of the reinforcing bars in the direction in which the cylindrical wall is thinnest and correspondingly, the tensile stress the greatest.[4] The controlling thickness of the cylindrical wall is taken as the smallest of (1) the depth of clear cover $c_b$ or (2) half the clear spacing between adjacent bars $c_s$ or (3) the side cover $c_s$ (Fig. 6.7b). For example, if the bars are positioned near the bottom surface of a beam, the section of minimum wall thickness is oriented in a downward, vertical

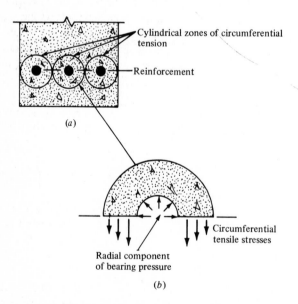

(a)

Cylindrical zones of circumferential tension

Reinforcement

Circumferential tensile stresses

Radial component of bearing pressure

(b)

**Figure 6.6** Tensile stresses created by the radial component of the bearing pressure: (a) zones of circumferential tension surrounding reinforcement, (b) approximate stress distribution in the concrete surrounding a reinforcing bar.

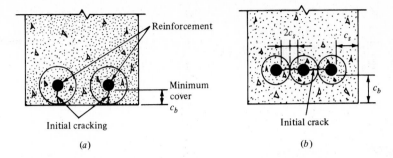

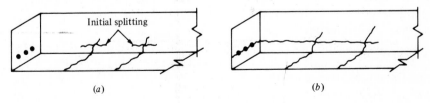

Figure 6.7 Influence of the depth of cover and the bar spacing on the location of initial splitting: (*a*) small depth of bottom cover; large spacing between bars; (*b*) closely spaced bars, large depth of cover.

direction; a splitting failure will be initiated by the development of the vertical cracks shown by the wavy line in Fig. 6.7*a*. If (Fig. 6.7*b*) the cover on the bars is sufficiently large (2.5 bar diameters or better) but the bars are closely spaced, the minimum wall thickness of the cylinders is equal to half the spacing between bars. In this latter situation splitting initiates along a horizontal plane extending through the row of reinforcement.

The longitudinal cracking that leads to a bond failure develops in stages. Initially, cracks of limited extent open where the local bond stresses are highest adjacent to diagonal tension or flexural cracks (Fig. 6.4). (The orientation of the plane of cracking is determined by the thickness of the cover or by the spacing between bars.) With an increase in load, the initial cracks lengthen until they join together to form a continuous crack that extends to the end of the beam (Fig. 6.8). Once the continuous crack forms, the bond between the reinforcement and the concrete is destroyed. As the reinforcement slips, total collapse of the beam takes place.

If both the cover over the bars and the spacing between bars are large, bond failures occur by pullout of the bar rather than by splitting of the concrete. Experimental studies indicate that pullout failures develop when the edge cover is greater than 2.5 bar diameters and the clear spacing between bars exceeds 5 bar diameters.[4] For a bar to pull out, the concrete keys between ribs must shear off (see Fig. 6.9), or the concrete in front of the ribs must crush. The likelihood of a pullout failure is increased if the concrete is weak or porous.

Figure 6.8 Development of a side-split failure in a region of high bond stress: (*a*) initial splitting adjacent to diagonal-tension cracks, (*b*) loss of cover by fully developed cracks producing bond failure.

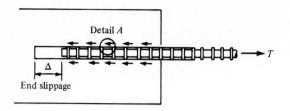

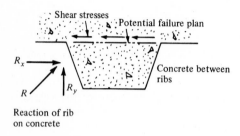

Reaction of rib
on concrete

**Figure 6.9** Pullout failure: (*a*) shear failure of concrete between ribs permits bar to pull out, (*b*) shear stresses in concrete between ribs produced by horizontal component of the bearing reaction.

Tests show that the bond strength of top bars, i.e., bars positioned above a 12-in (305-mm) or greater depth of concrete, is significantly less than that of bottom bars. The reduced bond strength is caused by air and water that rise from the freshly poured concrete beneath the bars and collect on the bottom surface of the top bar. By increasing the water content of the gel, by washing out particles of cement from the concrete, and by creating voids beneath the bar, the air and water reduce the bond strength of top bars by approximately 40 percent.

Experimental studies also indicate that the addition of transverse reinforcement along the anchorage length of a bar will provide a moderate increase in bond strength by reducing the tendency of the concrete to split. To be most effective, the transverse steel must be positioned to cross the potential failure plane produced by splitting.

## 6.5 NOMINAL BOND STRENGTH

The nominal bond strength of a reinforcing bar can be established experimentally by measuring the force needed to produce excessive slippage or pullout of a bar embedded in concrete. Although a wide variety of test specimens have been developed to study bond, the use of beam specimens permits the influence on bond strength of flexural cracks, thickness of concrete cover, shear, and the proximity of other bars to be taken into account.

The current expression for nominal bond strength is based primarily on beam tests carried out at the National Bureau of Standards (NBS) and at the University of Texas. The standard beam used by the NBS is shown in Fig. 6.10. As shown in the plan view, T-shaped ends are constructed so that the support reaction does not apply direct compression to the test bar. In this beam anchorage of the reinforcement at each end takes place in the distance *l* between points 1 and 2. In this

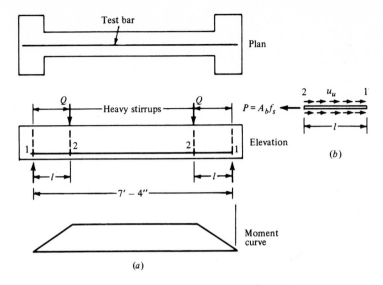

**Figure 6.10** NBS test beam: (a) details of beam, (b) free-body diagram of reinforcement between points 1 and 2.

distance the force in the bar reduces from a maximum value of $f_s A_b$ at point 2 to zero at point 1. If shear-type bond stresses $u_u$ of constant magnitude are assumed to act uniformly on the bar's surface over the anchorage length $l$, horizontal equilibrium of the bar between points 1 and 2 (Fig. 6.10b) gives

$$u_u \Sigma_0 l = P$$

$$u_u = \frac{P}{\Sigma_0 l} \tag{6.1}$$

where $u_u$ = nominal bond strength in units of stress
$P$ = bar force to be anchored
$\Sigma_0$ = circumference of bar
$l$ = length of embedment

Substituting experimental values of $P$ into the right side of Eq. (6.1) gives values of $u_u$ that can be approximated by the empirical expression

$$u_u = \frac{9.5\sqrt{f_c'}}{d_b} \leq 800 \text{ lb/in}^2 \tag{6.2}$$

where $d_b$ is the bar diameter. Equation (6.2) applies to bottom bars embedded in stone or gravel concrete. If the bars are positioned in the top of a form over a depth of concrete that is 12 in (305 mm) or greater, or if the concrete is made of lightweight aggregates, the bond strength will be smaller than that predicted by Eq. (6.2). Modifications to the bond strength for these conditions are discussed in Sec. 6.6. Since it is recognized that bond failures are produced by tensile cracking, bond strength in Eq. (6.2) is correlated with $\sqrt{f_c'}$ rather than $f_c'$.

Because Eq. (6.2) does not account for the influence of depth of cover, for the spacing between longitudinal reinforcement, or for the presence of transverse reinforcement (variables that are known to influence bond strength), experimental values of bond strength may deviate significantly from predicted values. Moreover, for beams that contain rows of closely spaced bars or bars with small cover, a significant percentage of test results are well below the nominal bond strength predicted by Eq. (6.2). The unrealistically high nominal bond strengths result from the use of test beams in which favorable conditions for bond existed. These beams were either heavily reinforced with stirrups along the anchorage length or were relatively wide and contained one test bar.

As an alternative to expressing bond strength in terms of stress, the nominal bond force per inch of length $U_u$ is often used. $U_u$ is related to $u_u$, the nominal bond strength in terms of stress, by

$$U_u = u_u \Sigma_0 \; (1 \text{ in}) \tag{6.3}$$

Using Eq. (6.2) to evaluate $u_u$ in Eq. (6.3), expressing $\Sigma_0$ as $\pi d_b$, and rounding the coefficient give

$$U_u = \frac{9.5\sqrt{f_c'}}{d_b} d_b \pi$$

$$U_u = 30\sqrt{f_c'} \tag{6.4}$$

Note that $f_c'$ is expressed in pounds per square inch and $U_u$ in pounds per inch.

## 6.6 DEVELOPMENT LENGTH OF TENSION STEEL

As indicated graphically in Fig. 6.4a, the magnitude and the direction of the bond stresses undergo sharp fluctuations in the vicinity of a crack. Therefore, if a crack opens at a point where the bond stresses are computed, the calculations are meaningless. Because the position at which a crack opens is unpredictable, a valid comparison between bond stress and bond strength (to determine the likelihood of a bond failure) is not possible.

Since the use of bond-stress calculations to ensure proper anchorage is merely a theoretical exercise whenever a crack occurs, the bond provisions in the current ACI Code are based on the length of the bar necessary for anchorage rather than the intensity of bond stress at a point. The Code requires that the actual length of bar available for anchorage (the distance is measured from the point of peak stress to the near end of the bar) be equal to or exceed the *development length*, $l_d$, defined as the minimum embedment length required to anchor a bar that is stressed to the yield point. The minimum anchorage length results when the maximum nominal bond strength $U_u$ is mobilized.

To establish the expression for development length in terms of the properties of a reinforcing bar and the strength of the concrete, the bar force $A_b f_y$ is equated to the anchorage resistance $U_u l_d$ to give

$$A_b f_y = l_d U_u$$

$$l_d = \frac{f_y A_b}{U_u} \tag{6.5}$$

The basic ACI equation for the development length of reinforcing bars (no. 11 and smaller) embedded in stone concrete is derived from Eq. (6.5) by multiplying the yield-point stress by 1.25 to provide a factor of safety, substituting $U_u = 30\sqrt{f_c'}$ [see Eq. (6.4)], and rounding off the coefficient.

$$l_d = \frac{0.04 f_y A_b}{\sqrt{f_c'}} \text{ USCU or } \frac{0.019 A_b f_y}{\sqrt{f_c'}} \text{ SI} \tag{6.6}$$

But $l_d$ is not to be taken less than

$$0.0004 d_b f_y \text{ USCU or } (0.058 d_b f_y) \text{ SI} \tag{6.7}$$

or 12 in (305 mm). Equation (6.7) usually governs for small-diameter bars.

Since a capacity reduction factor of 0.9 has already been applied when the reinforcement was sized for moment, and since the bar force to be anchored is based on $1.25 f_y$, no additional $\phi$ factor is applied to Eq. (6.6).

## Table 6.1 Modifiers for development length

|  | Modifier |
|---|---|
| Top reinforcement (reinforcement with 12 in or more of concrete under the bar; this increase accounts for the reduction in bond strength produced by water and air bubbles that rise from the concrete below the bar and collect on the underside of the reinforcement) | 1.4 |
| Bar with $f_y > 60,000$ lb/in$^2$ | $2 - \dfrac{60,000}{f_y}$ |
| For SI units | $2 - \dfrac{414}{f_y}$ |
| All lightweight concrete | 1.33 |
| Sand lightweight concrete | 1.18 |
| Bar spaced at least 6 in on center and at least 3 in from the side face of the member | 0.8 |
| Area of steel supplied > area of steel required | $\dfrac{A_{s,\,req}}{A_{s,\,sup}}$ |
| Bars enclosed within a spiral $\geq \frac{1}{4}$ in diameter with pitch $\leq 4$ in | 0.75 |

*Note*: For bundled bars the development length of each bar is to be increased by 20 percent for a three-bar bundle and 33 percent for a four-bar bundle.

Because tests show that the bond strength of no. 14 and no. 18 bars is less than that of the smaller bars, the ACI Code gives special equations for their development length.

$$l_d = \begin{cases} \dfrac{0.085 f_y}{\sqrt{f'_c}} \text{ USCU or } \dfrac{26 f_y}{\sqrt{f'_c}} \text{ SI} & \text{for no. 14 bars} & (6.8) \\[2.5em] \dfrac{0.11 f_y}{\sqrt{f'_c}} \text{ USCU or } \dfrac{34 f_y}{\sqrt{f'_c}} \text{ SI} & \text{for no. 18 bars} & (6.9) \\[2.5em] \dfrac{0.03 f_y}{\sqrt{f'_c}} \text{ USCU or } \dfrac{0.36 d_b f_y}{\sqrt{f'_c}} \text{ SI} & \text{for deformed wire} & (6.10) \end{cases}$$

where $f'_c$ in pounds per square inch or megapascals. To account for other variables that influence bond strength, Eqs. (6.6) to (6.10), the basic equations for development length, are to be multiplied by all the applicable factors listed in Table 6.1.

**Example 6.1:** Estimate the force required to pull out a no. 5 and a no. 9 bar embedded 5 in in a block of concrete; $f'_c = 3{,}600$ lb/in.$^2$, and $f_y = 60$ kips/in$^2$; see Fig. 6.11.

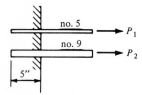

no. 5 → $P_1$

no. 9 → $P_2$

5″

**Figure 6.11**

SOLUTION Each bar can develop a nominal bond strength of $U_u = 30\sqrt{f'_c}$ for each 1 in of embedment

$$P_1 = P_2 = 30\sqrt{f'_c}\,(5 \text{ in}) = 9000 \text{ lb}$$

**Example 6.2:** Determine the minimum length of embedment required by the ACI Code to anchor a no. 5 ($A_b = 200$ mm$^2$) and a no. 9 ($A_b = 645$ mm$^2$) bar if each is stressed to its yield point. $f'_c = 24.82$ MPa, $f_y = 413.7$ MPa.

SOLUTION From the larger value given by Eq. (6.6) or (6.7)

$$l_{min} = l_d = \frac{0.019 A_b f_y}{\sqrt{f'_c}}$$

but not less than $0.058 d_b f_y$ or 305 mm.

$$l_d = \begin{cases} \dfrac{0.019(200)413.7}{\sqrt{24.82}} = 315.6 \text{ mm} & & \text{for no. 5} \\[2em] \text{or } 0.058(15.88)413.7 = 381 \text{ mm} & \text{controls} & \\[1.5em] \dfrac{0.019(645)413.7}{\sqrt{24.82}} = 1017.6 \text{ mm} & \text{controls} & \text{for no. 9} \\[2em] \text{or } 0.058(28.65)413.7 = 687.4 \text{ mm} & & \end{cases}$$

**Example 6.3:** An area of steel equal to 2.4 in² (1548.5 mm²) is required to carry the negative moment in the beam at support $A$. If two no. 10 bars ($A_s = 2.53$ in²) are used as reinforcement, (*a*) is sufficient embedment length available in the cantilever to anchor the bars properly? (*b*) If the bars are not properly anchored, determine the largest size bar that can be used as reinforcement $f'_c = 3$ kips/in², and $f_y = 60$ kips/in² (see Fig. 6.12).

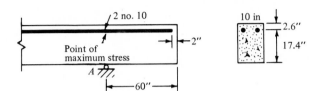

Figure 6.12

SOLUTION (*a*) Compute

$$l_d = \frac{0.04 f_y A_b}{\sqrt{f'_c}} (1.4) \frac{A_{s,\,req}}{A_{s,\,sup}}$$

$$= \frac{0.04(60,000)(1.27)}{\sqrt{3000}} (1.4) \frac{2.4}{2.53} = 73.9 \text{ in (1.88 m)} \qquad \text{controls}$$

but not less than

$$l_d = 0.0004 \, d_b f_y (1.4) \frac{2.4}{2.53} = 0.0004(1.27)(60,000)(1.4) \frac{2.4}{2.53} = 40.48 \text{ in}$$

Providing 2 in of end cover, distance in cantilever available for anchorage $= 60 - 2 = 58$ in. Since 73.9 in $> 58$ in, no. 10 bars are not properly anchored.

 (*b*) To compute the maximum bar size that can be properly anchored, set $l_d = 58$ in in the expression for $l_d$ and solve for $A_b$ (neglect the difference between $A_{s,\,req}$ and $A_{s,\,sup}$

$$58 \text{ in} = \frac{0.04(60,000 A_b)(1.4)}{\sqrt{3000}}$$

$$A_b = 0.94 \text{ in}^2$$

Reinforcing bars must be equal to or less than no. 8.

**Example 6.4:** Verify that the reinforcement in the one-way footing in Fig. 6.13 is properly anchored. Bars extend to within 3 in (76 mm) of the edge of the footing. The peak stress in the transverse footing steel is assumed to occur on a section at the face of the wall. $f'_c = 3$ kips/in² (20.68 MPa), $f_y = 60$ kips/in² (413.7 MPa).

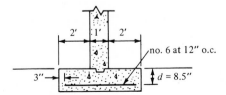

Figure 6.13 Cross section of wall footing.

SOLUTION Length $\Delta x$ available to anchor reinforcement from the face of wall

$$24 \text{ in} - 3 \text{ in} = 21 \text{ in (533 mm)}$$

Compute the development length. [Since the bar spacing exceeds 6 in (152 mm), a reduction of 0.8 is permitted.]

$$l_d = \frac{0.04 A_b f_y (0.8)}{\sqrt{f_c'}} = \frac{0.04(0.44)(60,000)(0.8)}{\sqrt{3000}} = 15.42 \text{ in (392 mm)} \qquad \text{controls}$$

Also

$$l_d \geq (0.0004 d_b f_y)(0.8) = 0.0004(0.75)(60,000)(0.8) = 14.4 \text{ in (366 mm)}$$

$$l_d \geq 12 \text{ in}$$

Since 21 in > 15.42 in, the anchorage is OK.


# 6.7 DEVELOPMENT LENGTH OF COMPRESSION STEEL

Smaller development lengths are required for reinforcement stressed in compression than for reinforcement stressed in tension. Both the lack of cracking and the resistance supplied by end bearing of the bar against the concrete reduce the length required to anchor compression steel. As specified in ACI Code §12.3, the development length in compression equals

$$l_d = \frac{0.02 f_y d_b}{\sqrt{f_c'}} \text{ USCU or } \frac{0.24 f_y d_b}{\sqrt{f_c'}} \text{ SI} \qquad (6.11)$$

but not less than

$$0.0003 f_y d_b \text{ USCU or } 0.044 f_y d_b \text{ SI} \qquad (6.12)$$

or 8 in (203 mm).

If excess steel area is supplied, $l_d$ in Eqs. (6.11) and (6.12) can be further reduced by the ratio $A_{s,\text{req}}/A_{s,\text{sup}}$.

If the reinforcement is encased in a spiral [the diameter must be at least $\frac{1}{4}$ in (6.4 mm) and the pitch must not exceed 4 in (102 mm)], the code permits a 25 percent reduction in the development length by Eq. (6.11). This modification recognizes the improvement in bond strength produced by the lateral pressures exerted by the closely spaced turns of the spiral.

**Example 6.5:** Four no. 8 dowels stressed to $f_y$ are required to transfer the axial compression force in a pier into the footing in Fig. 6.14. Determine the minimum extension of the dowels into the footing. $f_c' = 3.5 \text{ kips/in}^2$ (24.13 MPa), and $f_y = 60 \text{ kips/in}^2$ (4137 MPa). Although dowels are typically bent so that they can be positioned securely by wiring to the footing reinforcement, the bend does not contribute any resistance to compression.

SOLUTION Compute $l_d$ using Eqs. (6.11) and (6.12)

$$l_d = \frac{0.02 f_y d_b}{\sqrt{f_c'}} = \frac{0.02(60,000)(1)}{\sqrt{3500}} = 20.28 \text{ in (515 mm)} \qquad \text{controls}$$

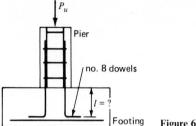

Figure 6.14

but not less than

$$0.0003 f_y d_b = 0.0003(60,000)(1) = 18 \text{ in } (457 \text{ mm})$$

or 8 in (203 mm). The 20.28 in extension requires that the footing be approximately 24 in deep. If less depth is available in the footing, a larger number of smaller dowels may be used to reduce the development length required to transfer the force from the dowels. For example, assume that two no. 6 dowels are substituted for each no. 8 dowel. Determine the minimum required extension into the footing. $A_s$ of two no. 6 dowels equals 0.88 in$^2$.

$$l_d = \frac{0.02 f_y d_b}{\sqrt{f'_c}} \frac{A_{s,\text{req}}}{A_{s,\text{sup}}} = \frac{0.02(60,000)(0.75)}{\sqrt{3500}} \frac{0.79}{0.88} = 13.65 \text{ in } (347 \text{ mm}) \qquad \text{controls}$$

but not less than

$$0.0003 f_y d_b = 0.0003(60,000)(0.75) \frac{0.79}{0.88} = 12.12 \text{ in } (308 \text{ mm})$$

or 8 in (203 mm).

## 6.8 STANDARD HOOKS

If a beam's dimensions are limited so that a tension bar cannot be extended in a straight line a distance equal to the required development length, a hook may be added to provide additional anchorage capacity. Depending on the particular configuration of the structural member, either a 90 or a 180° bend may be used to form a hook (see Fig. 6.15). The 180° hook is most suited to shallow members; the 90° hook is often used when reinforcement in one member is to be made continuous with the reinforcement in a second member.

If the force applied to a hook exceeds its nominal strength, several modes of failure are possible. For hooks located near the surface of a member, the radial component of the bearing stresses produced by the ribs on the concrete can produce failure by longitudinal splitting of the concrete cover. In addition, hooks can fail if the bearing stresses within the radius of the bend exceed the concrete's compressive strength and crush it.

Computations show that the length of bar used to fabricate a standard hook has approximately the same anchorage capacity when it is run straight. In other words, the curvature of the hook does not create significant additional anchorage

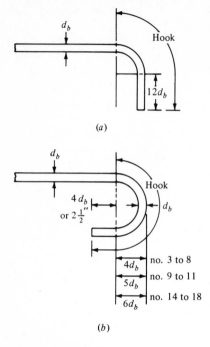

(a)

(b)

**Figure 6.15** Dimensions of standard hooks: (a) 90° hook, (b) 180° hook. (For no. 3 to no. 11 bars with 180° hooks that are fabricated from steel with $f_y = 40$ kips/in$^2$, the minimum required radius, measured to the outside of the bar, is $3.5d_b$.

resistance (see Example 6.6). Because concrete within the radius of a standard hook is vulnerable to crushing, adding additional length of bar to the end of a standard hook does not increase the force the hook can anchor.

A standard hook is considered to develop a tensile stress $f_h = \xi \sqrt{f'_c}$. Values of $\xi$ are given in Table 6.2 as a function of bar size. To compute the reduced development length of a bar with a standard hook, Eq. (6.6) is applied using the

**Table 6.2 $\xi$ values†**

| Bar no. | $f_y = 60$ kips/in$^2$ (413.7 MPa) | | | | $f_y = 40$ kips/in$^2$ (275.8 MPa) | |
|---|---|---|---|---|---|---|
| | Top bars | | Other bars | | All bars | |
| | USCU | SI | USCU | SI | USCU | SI |
| 3–5 | 540 | 45 | 540 | 45 | 360 | 30 |
| 6 | 450 | 37 | 540 | 45 | 360 | 30 |
| 7–9 | 360 | 30 | 540 | 45 | 360 | 30 |
| 10 | 360 | 30 | 480 | 40 | 360 | 30 |
| 11 | 360 | 30 | 420 | 35 | 360 | 30 |
| 14 | 330 | 27 | 330 | 27 | 330 | 27 |
| 18 | 220 | 18 | 220 | 18 | 220 | 18 |

† Based ACI Code 318-77, Table 12.5.1.

resultant stress to be anchored, $f_y - f_h$, in place of $f_y$. Hooks are not assumed to be effective in anchoring bars stressed in compression.

If lateral restraint (called *enclosure* in the Code) perpendicular to the plane of the hook is provided, the $\xi$ values in Table 6.2 can be increased 30 percent. Lateral restraint may be supplied by closed ties, stirrups, or spirals surrounding the hook or by external concrete. The term *external concrete* denotes sufficient thickness of concrete on each side of the hook to provide lateral confinement.

**Example 6.6: The strength of a standard hook** (*a*) Compute the force the standard hook on the no. 7 bar in Fig. 6.16 can anchor; determine (*b*) the length of bar required to form a standard hook for a no. 7 bar and (*c*) the length of a straight bar required to anchor the same force carried by the hook.

$$f'_c = 3.6 \text{ kips/in}^2 \text{ (24.82 MPa)}, \text{ and } f_y = 60 \text{ kips/in}^2 \text{ (413.7 MPa)}.$$

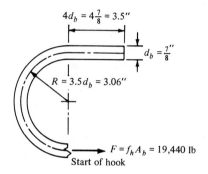

$4d_b = 4\frac{7}{8} = 3.5''$

$d_b = \frac{7}{8}''$

$R = 3.5d_b = 3.06''$

$F = f_h A_b = 19,440 \text{ lb}$

Start of hook            **Figure 6.16**

SOLUTION (*a*) For stress anchored by a standard hook use Table 6.2

$$f_h = \xi\sqrt{f'_c} = 540\sqrt{3600} = 32,400 \text{ lb/in}^2$$

$$F = f_h A_b = (32.4 \text{ kips/in}^2)(0.6 \text{ in}^2) = 19,440 \text{ lb}$$

(*b*) The length of a standard hook is

$$l = \pi R + 3.5 \text{ in}$$

$$= 9.61 + 3.5$$

$$= 13.11 \text{ in (333 mm)}$$

(*c*) The length of a straight bar to develop $f_h = 32.4 \text{ kips/in}^2$ is computed by replacing $f_y$ in Eq. (6.6) with $f_h$ to give

$$l = \frac{0.04 f_h A_b}{\sqrt{f'_c}} = \frac{0.04(32,400)(0.6)}{\sqrt{3600}} = 12.96 \text{ in (329 mm)}$$

The length of the hook is approximately equal to the length of straight bar required to anchor the force carried by the hook.

**Example 6.7:** In Fig. 6.17 the top steel in the beam has been designed for a flexural stress of $f_y$ at the face of the column, the section of maximum negative moment. If the

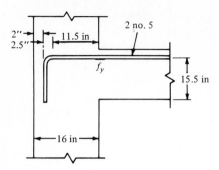

**Figure 6.17**

no. 5 bars terminate with a standard 90° hook, is sufficient anchorage length available in the column to develop the tensile strength of the steel fully? $f'_c = 3$ kips/in² (20.68 MPa), and $f_y = 60$ kips/in² (413.7 MPa).

SOLUTION  Required development length of a no. 5 top bar using Eqs. (6.6) and (6.7)

$$l_d = \frac{0.04A_b f_y(1.4)}{\sqrt{f'_c}} = \frac{0.04(0.31)(60,000)(1.4)}{\sqrt{3000}} = 19.02 \text{ in (483 mm)}$$

but not less than

$$0.0004d_b f_y(1.4) = 0.0004(\tfrac{5}{8})(60,000)(1.4) = 21 \text{ in (533 mm)} \qquad 21 \text{ in controls}$$

Length of straight bar available for anchorage

$$l = 16 \text{ in} - (2 \text{ in cover} + 4d_b) = 11.5 \text{ in} \qquad \text{where } d_b = \tfrac{5}{8} \text{ in}$$

The stress that can be anchored by 11.5 in of straight bar is

$$\frac{21 \text{ in}}{60,000 \text{ lb/in}^2} = \frac{11.5 \text{ in}}{f_s} \qquad \text{and} \qquad f_s = 32,857 \text{ lb/in}^2 \text{ (226.5 MPa)}$$

Stress that can be anchored by a standard hook (see Table 6.2) is

$$f_h = \xi\sqrt{f'_c} = 540\sqrt{3000} = 29,577 \text{ lb/in}^2 \text{ (205.6 MPa)}$$

Total stress that can be anchored by hook and straight section

$$f_s + f_h = 32,857 + 29,577 = 62,434 \text{ lb/in}^2 > 60,000 \text{ lb/in}^2 \qquad \text{anchorage OK}$$

## 6.9 ANCHORAGE OF STEEL AT SIMPLE SUPPORTS AND POINTS OF INFLECTION

Complete collapse of a beam will occur suddenly if the ends of the positive steel extending into a simple support or into a point of inflection are not properly anchored and slip out. Although the moment is zero at these points and the stress in the steel is low, the bond stresses, a function of the rate at which the moment is changing rather than a function of the absolute value of moment, may be large. The rate at which the moment changes is related to the shear, which is maximum at a simple support and often high at a point of inflection. To prevent a local bond failure, ACI Code §12.12.3 requires the diameter of the positive reinforcement to be limited so that

$$l_d \leq \begin{cases} \dfrac{M_n}{V_u} + l_a & \text{at point of inflection} & (6.13) \\[3ex] \dfrac{1.3M_n}{V_u} + l_a & \text{at simple support} & (6.14) \end{cases}$$

where $V_u$ = shear at simple support or point of inflection

$M_n$ = nominal moment capacity of section based on area of positive reinforcement extending into support or into point of inflection

$l_a$ = embedment length beyond centerline of support or beyond point of inflection; $l_a$ not to exceed $12d_b$ or $d$ at point of inflection. If a hook is used, it must be converted to an equivalent length of straight bar with the same anchorage capacity.

The need for these equations, which account for the influence of the shape of the moment curve on the bond stresses in uniformly loaded beams, can be understood by considering the uniformly loaded beam in Fig. 6.18. This beam is reinforced with a bar whose length from the centerline of support $A$ to point $C$ at midspan is just equal to the bar's development length $l_d$. (The idea of using a bar whose length between midspan and the support equals $l_d$ is based on Ferguson's work.[6]) It is further assumed that the bar has been sized so the steel at midspan, the point of maximum moment, is stressed to $f_y$ under the factored design loads. If the arm between the steel and the resultant of the compressive bending stresses is assumed to have a constant value at all sections, the tension force in the bar at any section will be proportional to the ordinate of the parabolic moment curve at that same section. Although the length of the reinforcing bar between the support where the stress is zero and midspan, where the stress equals $f_y$, is equal to the development length, *the bar is not properly anchored*. This fact becomes evident when the force in the bar at any other point is compared with the length of bar available for anchorage between that point and the support. For example, at the quarter point $B$ of the span the force in the bar has reduced to $\frac{3}{4}f_y A_b$ because the ordinate of the parabolic moment curve decreases by 25 percent between midspan and the quarter point. Since the force to be anchored at the quarter point is three-fourths of the midspan force, a minimum length of embedment of at least $\frac{3}{4}l_d$ is required; however, between point $B$ and the support only a length of $\frac{1}{2}l_d$ is available.

Figure 6.18*b* further clarifies the variation of bond stresses along the length of the reinforcing bar in Fig. 6.18*a*. From equilibrium of the two segments of reinforcing bar it can be seen that the average bond stresses are 3 times larger between the support and the quarter point than between the quarter point and midspan. Therefore, although the average bond stresses over the entire length may not exceed the bond strength, the high local bond stresses near the support may result in a local bond failure and slipping of the steel that will lead to the collapse of the beam.

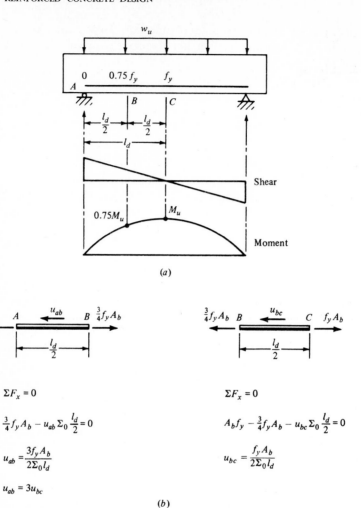

**Figure 6.18** Anchorage of reinforcement at a simple support in a uniformly loaded beam: (*a*) uniformly loaded beam reinforced with a bar whose length from midspan to the support equals $l_d$, (*b*) free-body diagrams of segments $AB$ and $BC$.

Since bond specifications in the current ACI Code are expressed in terms of development length only (equivalent to considering average bond stresses), Eqs. (6.13) and (6.14) were added to provide protection against a local bond failure. The basis for these equations can be developed by considering the uniformly loaded beam in Fig. 6.19. If the moment is assumed to change at a constant rate given by $V_u$, the shear at the support, a linear moment curve (dashed) results. It implies that bond stresses are constant and equal to the maximum value occurring at the support. As the stress in the steel increases from zero to $f_y$ in the distance $\Delta x$, the bending moment in the beam changes from zero to $M_n$. If the change in

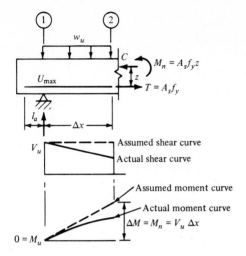

Figure **6.19** Anchorage of steel at a simple support of a uniformly loaded beam.

moment $M_n$ is equated to the area under the assumed shear curve $V_u \Delta x$, $\Delta x$ can be expressed as

$$\Delta x = \frac{M_n}{V_u} \tag{6.15}$$

If the distance $\Delta x$ given by Eq. (6.15) is greater than or equal to the development length, the bar is safely anchored. This criterion can be stated as

$$l_d \leq \Delta x \tag{6.16}$$

substituting the value of $\Delta x$ in Eq. (6.15) into Eq. (6.16) gives

$$l_d \leq \frac{M_n}{V_u} \tag{6.17}$$

If the reinforcement extends past the support or point of inflection a distance $l_a$, the total length available to anchor the positive steel is $\Delta x + l_a$. Under these circumstances, Eq. (6.17) can be written

$$l_d \leq \frac{M_n}{V_u} + l_a \tag{6.13}$$

To account for the improvement in bond strength created by the confining effect of the compressive stresses from the reaction at a simple support, ACI Code §12.12.3 permits the first term on the right-hand side of Eq. (6.13) to be increased by 30 percent. With this modification Eq. (6.13) becomes

$$l_d \leq \frac{1.3M_n}{V_u} + l_a \tag{6.14}$$

If the steel extending into the point of inflection or into a simple support does not satisfy the criterion given by Eq. (6.13) or (6.14) respectively, smaller-diameter bars may be used, the steel may be hooked, or the area of steel running into the support may be increased.

**Example 6.8:** In Fig. 6.20 determine whether the flexural reinforcement running into the end support satisfies the requirements of ACI Code §12.12.3. Bars extend 4 in past the centerline of the supports. $f'_c = 3$ kips/in², and $f_y = 60$ kips/in².

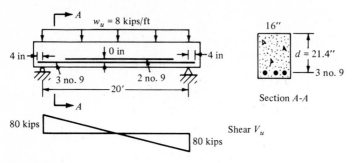

**Figure 6.20**

SOLUTION Verify that $l_d \le 1.3 M_n / V_u + l_a$.

$$l_d = \frac{0.04 A_b f_y}{\sqrt{f'_c}} = \frac{0.04(1)(60,000)}{\sqrt{3000}} = 43.8 \text{ in} \qquad \text{controls}$$

but not less than

$$l_d = 0.0004 d_b f_y = 0.0004(1.128)(60,000) = 27.1 \text{ in}$$

Using Eq. (3.24),

$$a = \frac{A_s f_y}{0.85 f'_c b} = \frac{3(60)}{0.85(3)16} = 4.4 \text{ in}$$

$$M_n = T\left(d - \frac{a}{2}\right) = 3(60)(21.4 - 2.2)(\tfrac{1}{12}) = 288 \text{ ft} \cdot \text{kips}$$

$$\frac{1.3 M_n}{V_u} + l_a = \frac{1.3(288)(12)}{80} + 4 \text{ in} = 60.2 \text{ in}$$

Since 43.8 in < 60.2 in, reinforcement is properly anchored at the supports.

## 6.10 CUTOFF POINTS

In the typical continuous beam of constant depth, the cross-sectional dimensions are initially established at the point of absolute maximum moment. Then areas of flexural steel are computed at all points of maximum positive and maximum negative moment. At these locations, where a state of impending failure exists under

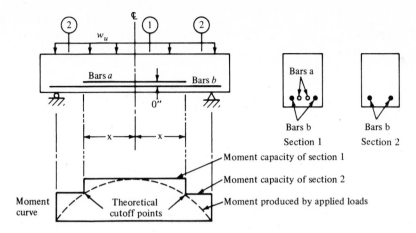

**Figure 6.21** Establishing the location of the theoretical cutoff points.

the factored loads, the steel is assumed to be stressed to its yield point. If the areas of steel required at the points of maximum moment are run continuously throughout each region of positive or negative moment, the beam will be overdesigned at all points except those where the moment is maximum. Although using a constant area of reinforcement throughout each region of positive or negative moment is certainly safe, it is often desirable, particularly in heavily reinforced members, to terminate a portion of the steel when the moment decreases significantly. Reducing the area of reinforcement in regions where the moment is low reduces the volume of reinforcement, the expensive component, and lowers the cost of the member. In addition, for heavily reinforced members with many bars, the reduction in congestion produced by terminating excessive bars facilitates the flow of concrete into all regions of the forms and permits improved compaction.

One or more bars may be terminated at the *theoretical cutoff point*, at which the moment produced by the factored loads reduces to the moment capacity of the cross section reinforced with the continuing bars. Normally, the designer terminates 50 to 60 percent of the steel area sized at the point of maximum moment. Figure 6.21 illustrates how the theoretical cutoff point can be located by using the moment curve and the moment capacity $\phi M_n$ of the cross section containing the continuing bars. To account for the possibility of higher than anticipated moment at the cutoff point due to possible variations in the position of the live load, settlements of support, or other causes, ACI Code §12.11.3 requires that reinforcement be extended beyond the theoretical cutoff point a distance equal to $12d_b$ or the effective depth $d$, whichever is larger (see Example 6.9 for the computations required to locate a theoretical cutoff point to establish the length of the terminated bars).

**Example 6.9:** To carry the midspan moment of 229.9 ft · kips, an area of steel of 2.72 in$^2$ is required in the simply supported beam shown in Fig. 6.22. Two no. 7 and two no. 8 bars, which provide an $A_s = 2.77$ in$^2$, are selected; $f'_c = 3$ kips/in$^2$, and $f_y = 60$ kips/in$^2$.

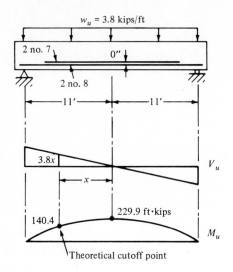

**Figure 6.22**

(*a*) Establish the theoretical cutoff point at which the no. 7 bars can be terminated. (*b*) Considering the extension beyond the cutoff points required by ACI Code §12.11.3, determine the minimum distance the no. 7 bars should extend on each side of the center-line. (*c*) Verify that the no. 7 bars are properly anchored.

SOLUTION (*a*) To locate the theoretical cutoff point of the no. 7 bars, establish the moment capacity of the cross section reinforced with two no. 8 bars ($A_s = 1.57$ in$^2$)

$$T = A_s f_y = 1.57(60 \text{ kips/in}^2) = 94.2 \text{ kips}$$

$$C = T \qquad ab(0.85f'_c) = 94.2 \text{ kips}$$

Therefore

$$a = 3.08 \text{ in}$$

$$M_n = T\left(d - \frac{a}{2}\right) = 94.2 \text{ kips}\left(21.4 - \frac{3.08}{2}\right)\frac{1}{12} = 156 \text{ ft} \cdot \text{kips}$$

$$M_u = \phi M_n = 0.9(156) = 140.4 \text{ ft} \cdot \text{kips (190.4 kN} \cdot \text{m)}$$

Locate the point on the moment curve at which the moment equals 140.4 ft · kips. The change in moment from midspan to the theoretical cutoff point equals $229.9 - 140.4 = 89.5$ ft · kips. Equating the area under the shear curve to $\Delta M$, solve for the distance $x$ between midspan and the cutoff point

$$229.9 - 140.4 = \frac{(x)(3.8x)}{2}$$

$$x = 6.86 \text{ ft (2.09 m)}$$

(b) Total length of the no. 7 bars on each side of the center line equals 6.86 ft plus an extension equal to $12d_b$ or the effective depth, whichever is larger

$$12d_b = 12(\tfrac{7}{8}) = 10.5 \text{ in}$$

$$d = 21.4 \text{ in} \qquad \text{controls}$$

$$\text{Length of no. 7} = 6.86 \text{ ft} + \frac{21.4}{12} = 8.47 \text{ ft} \qquad \text{use 8 ft 9 in}$$

(c) Check the anchorage of no. 7 bars

$$l_d = \frac{0.04 A_b f_y}{\sqrt{f_c'}} = \frac{0.04(0.6)(60,000)}{\sqrt{3000}} = 26.3 \text{ in} = 2.2 \text{ ft (0.67 m)}$$

but not less than $0.0004d_b\, f_y = 0.0004(\tfrac{7}{8})60,000 = 21$ in. Since 8.75 ft > 2.2 ft, the anchorage is OK.

To ensure that the bending strength of a beam will not be reduced excessively, in regions of small moment the ACI Code limits the area of reinforcement that can be terminated. ACI Code §12.12.1 requires that at least one-third of the reinforcement from a section of maximum positive moment be extended (along the face of the member) into a simple support at least 6 in. If the member is continuous, at least one-quarter of the area of reinforcement from the section of maximum positive moment must extend (along the face of the member) into the support at least 6 in (Fig. 6.23).

ACI Code §12.13.3 limits the area of negative steel that can be terminated by specifying that at least one-third of the negative reinforcement from the section of maximum moment be extended beyond the point of inflection a distance not less than (1) the effective depth $d$, (2) $12d_b$, or (3) one-sixteenth of the clear span, whichever is largest. These requirements for terminating negative reinforcement are illustrated in Fig. 6.23.

## Shear Requirements at Cutoff Points in Tension Zones

At the point in a tension zone where flexural steel is terminated, the stress in the continuing steel must increase sharply since a smaller area of steel in the continuing bars must carry the same tension force (Fig. 6.24). For the stress in the continuing bars to increase, the beam must undergo a large local increase in strain at the cutoff point. If this increase in strain results in the formation of large tension cracks, the cross-sectional area available to carry shear will be sharply reduced and the possibility of a diagonal-tension failure increased. To reduce the likelihood of a diagonal-tension failure developing at a cutoff point, ACI Code §12.11.5 requires that at least one of the following conditions be satisfied:

1. The shear $V_u$ due to factored loads must not exceed two-thirds of the available shear strength of the cross section; i.e.,

$$V_u \leq \tfrac{2}{3}\phi(V_c + V_s)$$

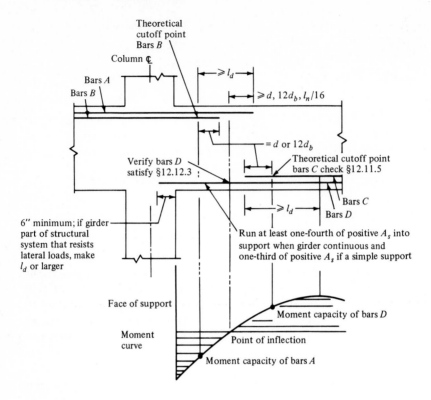

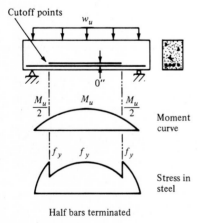

**Figure 6.23** ACI Code requirements for terminating and anchoring reinforcement.

**Figure 6.24** Variation of tensile stress in the flexural steel in a beam with terminated bars; 50 percent of steel terminated when midspan moment reduces by one-half.

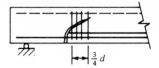

**Figure 6.25** Position of extra stirrups at a cutoff point.

2. For no. 11 and smaller bars, the area of steel supplied by the continuing bars must be twice that required for moment at the cutoff point and the shear due to factored loads must not exceed three-fourths of the shear capacity of the cross section.

$$A_{s,\,\mathrm{sup}} \geq 2A_{s,\,\mathrm{req}}$$

$$V_u \leq \tfrac{3}{4}\phi(V_c + V_s)$$

3. Extra stirrups, in addition to those required for shear and torsion, are supplied. They are to be positioned along each terminated bar for a distance of three-fourths the effective depth of the member from the cutoff point. The area of the additional stirrups $A_v$ is not to be less than $60\,b_w s/f_y$ or $0.41\,b_w s/f_y$. In addition the spacing $s$ of the additional stirrups is not to exceed $d/8\beta_b$, where $\beta_b$ is defined as the ratio of the area of steel terminated to the total area of steel at the cutoff point (Fig. 6.25).

## 6.11 SPLICING REINFORCEMENT

At times reinforcement must be fabricated in the shop in convenient lengths to facilitate placement in the forms. Later these lengths must be connected in the field to produce continuity of the reinforcement in accordance with the design requirements. These splices occur most frequently in column construction or in complicated structural configurations.

Reinforcing bars can be spliced, i.e., made continuous, by welding, by using mechanical connectors, or by lap splices. A lap splice is formed by extending bars past each other far enough to permit the force in one bar to be transferred by bond stress through the concrete and into the second bar. Although bars joined by a lap splice are usually wired together with their sides in direct contact, forces can also be transferred effectively between bars whose transverse spacing does not exceed one-fifth of the lap length $l_s$ or 6 in (152 mm), whichever is smaller (Fig. 6.26).

Each method of splicing bars has its advantages and disadvantages. To restore continuity of the reinforcement, butt welding is superior to all other methods. When properly made, the strength of a welded joint is normally greater than that of the original cross section because the area of the joint can be made larger than that of the original cross section. In addition welding rods, made of high-quality alloy steel, produce weld material whose strength and ductility are superior to that of the reinforcing bars being connected.

A variety of mechanical connectors are available to join reinforcing bars. Most connectors consist of a sleeve into which the ends of the bars to be joined are

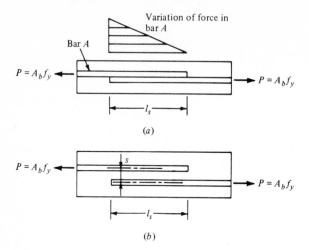

(a)

(b)

**Figure 6.26** Lap splice: (a) bars wired together, (b) bars separated wired together, (b) bars separated, and $s \leq l_s/5$ or 6 in, whichever is smaller.

inserted. Wedges, some type of mechanical grips, or molten metal are then used to join the ends of the bars together securely. Although mechanical connectors can be used with any grade of steel, excessive transverse cracking in the concrete may develop at the splice if the connector permits slippage of the bars to occur as the strength of the joint is mobilized.

If the reinforcement is heavily stressed at a splice point (the area of reinforcement is less than twice that required to carry the design force) ACI Code §12.15.3 requires the strength of splices made by butt welding or by mechanical connectors to exceed by at least 25 percent the strength of the reinforcing bar stressed to its yield point. Overdesigning the splice ensures that a ductile failure will occur in the unspliced region of the bar by yielding of the steel rather than by rupture of the splice. At splice locations where the stresses are lower, i.e., the area of reinforcement is at least twice that required by analysis, the strength of welds or connectors may be reduced. At these locations ACI Code §12.16.4 permits splices to be designed for twice the tension force determined by analysis but not less than the force produced by the area of steel stressed to 20 kips/in$^2$ (137.9 MPa). In addition splices must be staggered at least 24 in (610 mm).

Although lap splices are the simplest and most economical method of joining bars, particularly when the diameter of the bars is small, they also have a number of drawbacks. In the region of the splice where bars overlap, the congestion of the reinforcement is increased. Also at the ends of each splice, where the area of steel decreases, transverse cracks develop due to stress concentrations. By increasing the local bond stresses, transverse cracks also tend to produce longitudinal splitting of concrete, which reduces the ductility of the member. Since few experimental studies have been carried out using lap splices made with no. 14 or no. 18 bars, their behavior is not well established; therefore, the current edition of the ACI Code permits lap splices only of no. 11 and smaller bars.

Although experimental studies[5] indicate that the current design procedure for lap splices uses only a limited portion of the available bond strength of the

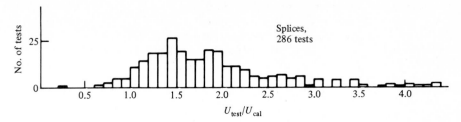

**Figure 6.27** Comparison between average bond stresses measured in splice tests with ACI Code values of permitted bond stress.[5]

concrete and produces splice designs that are very conservative (Fig. 6.27), it is still considered good practice to locate splices at sections where stresses are low and to stagger the location of lap splices for individual bars. If a splice is defective due to poor workmanship, the strength and ductility of a member will be reduced.

## Tension Splices

Since the strength of a lap splice is based on the transfer of force between bars by bond stress, the minimum required lap length of a splice is expressed in terms of development length. Currently the ACI Code defines three classes of splices, class A, class B and class C, with minimum lap lengths of $l_d$, $1.3l_d$, and $1.7l_d$, respectively, but not less than 12 in. The class of splice to be used at a particular section depends on the level of stress in the reinforcement and the percentage of steel to be spliced. The longest lap length is required when the reinforcement is highly stressed and all bars are spliced at the same location. Table 6.3 summarizes the conditions which control the type of splice to be specified.

When bundled bars are spliced, the lap lengths must be increased to account for the reduced surface of contact between the bars and concrete. To splice a bar in a three-bar bundle, ACI Code §12.15.2.2 requires the lap length to be made 20 percent longer than that required for an individual bar in the bundle. If a bundle contains four bars, a 33 percent increase in lap length is specified. Moreover, no overlap of individual bar splices is permitted for bundled bars.

**Table 6.3 Conditions determining the splice requirements†**

| $\dfrac{A_{s,\,sup}}{A_{s,\,req}}$ | Maximum percent of $A_s$ spliced within required lap length | | |
|---|---|---|---|
| | 50 | 75 | 100 |
| $\geq 2$ | Class A | Class A | Class B |
| $\leq 2$ | Class B | Class C | Class C |

† Based on table 12.16 of the ACI Code.

**Example 6.10: Design of a tension splice** To facilitate construction of a retaining wall, the vertical wall steel in Fig. 6.28 is to be spliced to dowels from the foundation. If the flexural steel is fully stressed to its yield point at the bottom of the wall, what splice length is required? $f'_c = 4$ kips/in$^2$ (27.58 MPa), and $f_y = 60$ kips/in$^2$ (413.7 MPa).

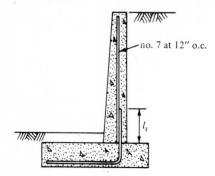

no. 7 at 12" o.c.

$l_s$

**Figure 6.28**

SOLUTION  Since 100 percent of all bars are to be spliced in a region where the steel is fully stressed $A_{s,\,\text{sup}}/A_{s,\,\text{req}} = 1$, a class C splice is required. Because the bars are spaced more than 6 in (152 mm) on center, a reduction of 0.8 is applied to $l_d$

$$l_d = \frac{0.04 A_b f_y(0.8)}{\sqrt{f'_c}} = \frac{0.04(0.6)(60{,}000)(0.8)}{\sqrt{4000}} = 18.2 \text{ in } (462 \text{ mm}) \qquad \text{controls}$$

but not less than

$$l_d = 0.0004 d_b f_y(0.8) = 0.0004(0.875)(60{,}000)(0.8) = 16.8 \text{ in } (427 \text{ mm})$$

$$l_s = 1.7 l_d = 1.7(18.2) = 30.94 \text{ in } (787 \text{ mm}) \text{ use } 32 \text{ in}$$

## Compression Splices

The minimum lap length required for a compression splice is equal to the development length for compression bars [see Eq. (6.11)] but is not to be less than $0.0005 f_y d_b$ or 12 in $(0.073 f_y d_b$ or 305 mm). If $f_y$ exceeds 60 kips/in$^2$ (413 MPa), the limit of $0.0005 f_y d_b$ is increased to $(0.0009 f_y - 24) d_b$ $[(0.13 f_y - 24) d_b]$. An additional 33 percent increase in lap length is required when $f'_c$ is less than 3 kips/in$^2$ (20.68 MPa).

If the reinforcement in a compression member is confined by ties with an effective area in each direction equal to or greater than $0.0015 hs$ (where $h$ is the dimension of the column perpendicular to the direction of the tie legs and $s$ is the spacing of the column ties), the length of lap splice can be reduced by 0.83. For reinforcement confined within a spiral, the splice length can be reduced by 0.75 because of the increase in strength produced by the confinement of the concrete. For both tied and spiral columns, the lap length must be at least 12 in (305 mm).

When bars are always in compression, the compressive force in a bar may be transmitted to a second bar by direct end bearing. For bars to transmit compression by end bearing, a suitable device must be supplied to hold the ends of the bars in alignment, closed ties or spirals must be used to confine the reinforcement being spliced, and the ends of the bars must be cut accurately at a right angle to the longitudinal axis; i.e., the plane of the bottom surface must not deviate more than 1.5° from a true right angle for either bar. If an end-bearing splice were to be used in a column, additional reinforcement in each face of the column would be required by ACI Code §12.18.3 to transmit a tensile force equal to 25 percent of the yield strength of the reinforcement in that face.

**Example 6.11:** The reinforcement in the column of Fig. 6.29 is spliced just above the floor level. If no. 7 bars in the upper section of column are spliced to no. 8 bars in the lower section, determine the required length of splice. $f'_c = 3.6$ kips/in², and $f_y = 60$ kips/in².

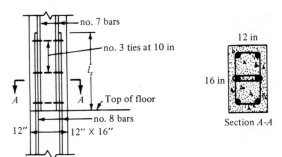

Figure 6.29

SOLUTION Since the lap must be long enough to transfer the force out of the no. 7 bars, the splice length is based on the properties of the no. 7 bars. Compute

$$l_d = \frac{0.02 d_b f_y}{\sqrt{f'_c}} = \frac{0.02(\frac{7}{8})(60,000)}{\sqrt{3600}} = 17.5 \text{ in}$$

but not less than 12 in or

$$0.0005 f_y d_b = 0.0005(60,000)(\tfrac{7}{8}) = 26.3 \text{ in} \qquad \text{controls}$$

Determine whether the splice can be reduced by 0.83. The area of ties in each direction must not be less than $0.0015 hs$

In the long direction, $h = 12$ in, 2 bars:

$$A_s = 2(0.11) = 0.22 \text{ in}^2$$

$$0.0015 hs = 0.0015(12)(10) = 0.18 \text{ in}^2 < 0.22 \text{ in}^2; \qquad \text{OK}$$

In the short direction, $h = 16$ in; 4 bars:

$$A_s = 4(0.11) = 0.44 \text{ in}^2$$

$$0.0015 hs = 0.0015(16)(10) = 0.24 \text{ in}^2 < 0.44 \text{ in}^2; \qquad \text{OK}$$

The required splice length is

$$l_s = 0.83 l_d = 0.83(26.3 \text{ in}) = 21.8 \text{ in} \qquad \text{use 22 in}$$

## 6.12 COMPREHENSIVE DESIGN EXAMPLE

Chapter 6 concludes with a comprehensive example to illustrate the details of terminating and anchoring reinforcement in accordance with ACI Code specifications.

**Example 6.12: Comprehensive design problem** Design the rectangular beam in Fig. 6.30 in accordance with ACI Code 318-77. Use a width of 13 in. Locate the theoretical cutoff points of reinforcement, establish the length of all bars, and verify that all reinforcement is properly anchored; $f'_c = 3$ kips/in$^2$, and $f_y = 50$ kips/in$^2$.

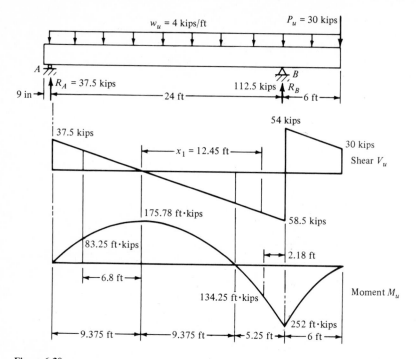

**Figure 6.30**

SOLUTION 1. Using Eq. (3.46), establish dimensions of cross section based on $M_{u,\,max}$ of 252 ft · kips. Try $\rho = 0.018 < \frac{3}{4}\rho_b = 0.0204$

$$M_u = \phi\rho f_y bd^2\left(1 - \frac{\rho f_y}{1.7f'_c}\right) \qquad (3.46)$$

$$252(12) = 0.9(0.018)(50)(13d^2)\left[1 - \frac{0.018(50)}{1.7(3)}\right]$$

$$d = 18.67 \text{ in} \qquad \text{use } 18.7 \text{ in}$$

The required depth is

$$h = d + 2.6 \text{ in} = 21.3 \text{ in} \qquad \text{use } 22 \text{ in}$$

Recompute $d = 22 - 2.6 = 19.4$ in. With $d = 19.4$ in and $b = 13$ in use Eq. (3.46) to compute $\rho = 0.0163$

$$A_{s,\,req} = \rho bd = 0.0163(13)(19.4) = 4.11 \text{ in}^2$$

Use two no. 9 and two no. 10

$$A_{s,\,sup} = 4.53 \text{ in}^2 > 4.11 \text{ in}^2$$

2. With $d = 19.4$ in, $b = 13$ in, and $M_u = 175.78$ ft · kips solve Eq. (3.46) to give $\rho = 0.00107$; then $A_s = 2.70 \text{ in}^2$. Use two no. 7 and two no. 8.

$$A_{s,\,sup} = 2.77 \text{ in}^2$$

3. Preliminary sketch of steel; the location and the approximate lengths of flexural reinforcement are shown in Fig. 6.31. Additional reinforcement to reduce creep and to position stirrups is also required.

4. Compute stirrup spacing at left end. Critical section for shear at $d = 19.4$ in

$$V_u = 37.5 - 4\frac{19.4}{12} = 31.03 \text{ kips}$$

$$V_c = 2\sqrt{f_c'}b_wd = 2\sqrt{3000}(13)(19.4)(\tfrac{1}{1000}) = 27.63 \text{ kips}$$

Since $V_u > \phi V_c/2$, stirrups are required. Use no. 3 stirrups, $A_v = 0.22 \text{ in}^2$

$$s = \frac{A_v f_y d}{(V_u/0.85) - V_c} = \frac{0.22(50)(19.4)}{(31.03/0.85) - 27.63} = 24.0 \text{ in}$$

$$s_{max} = \frac{d}{2} = \frac{19.4}{2} = 9.7 \text{ in}$$

Space stirrups at 9 in on center.

5. The value of $V_u$ beyond which stirrups are not required (ACI Code §11.5.5.1) is

$$V_u = \frac{\phi V_c}{2} = \frac{0.85(27.63)}{2} = 11.74 \text{ kips}$$

6. To satisfy the requirement that at least one-third of the positive steel from the section of maximum moment extend a minimum of 6 in (152 mm) into the support at $A$, run the no. 7 bars into the left support. Also extend no. 7 bars along the bottom face into the cantilever to position stirrups and to reduce creep deflections (Fig. 6.31).

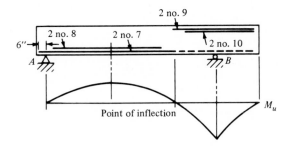

Figure 6.31

7. Anchorage of no 7 bars; to ensure anchorage of bottom steel at simple supports and points of inflection, ACI Code §12.12.3 requires

$$l_d \leq \begin{cases} \dfrac{1.3M_n}{V_u} + l_a & \text{at simple support} & \text{(6.14)} \\[2em] \dfrac{M_n}{V_u} + l_a & \text{at point of inflection} & \text{(6.13)} \end{cases}$$

where

$$a = \frac{A_s f_y}{0.85 f'_c b} = \frac{1.2(50)}{0.85(3)(13)} = 1.8 \text{ in}$$

$$M_n = A_s f_y \left( d - \frac{a}{2} \right)$$

$$M_n = 1.2(50)(19.4 - 0.9) = 1110 \text{ in} \cdot \text{kips}$$

$$l_d = \frac{0.04 A_b f_y}{\sqrt{f'_c}} = \frac{0.04(0.6)(50,000)}{\sqrt{3000}} = 21.9 \text{ in controls}$$

but not less than 12 in or

$$l_d = 0.0004 d_b f_y = 0.0004(\tfrac{7}{8})(50,000) = 17.5 \text{ in}$$

At support $A$

$$\frac{1.3(1110)}{37.5} + 6 \text{ in} = 44.48 \text{ in} > l_d = 21.9 \text{ in} \qquad \text{anchorage OK}$$

At point of inflection

$$\frac{M_n}{V_u} + l_a \geq l_d$$

where $l_a$ equals larger of $d$ or $12d_b$.

$$l_a = d = 19.4 \text{ in controls} \qquad 12d_b = 12(\tfrac{7}{8}) = 10.5 \text{ in}$$

$$\frac{1110}{37.5} + 19.4 \text{ in} = 49.0 \text{ in} > 21.9 \text{ in} \qquad \text{anchorage OK}$$

8. Cutoff points for no. 8 bars. When the magnitude of the positive moment reduces to a value that can be carried by the cross section reinforced with two no. 7 bars, the no. 8s can be cut off. Moment capacity of cross section reinforced with two no. 7 (see step 7 for $M_n$) equals

$$M_u = \phi M_n = 0.9(1110 \text{ in} \cdot \text{kips}) = 999 \text{ in} \cdot \text{kips} = 83.25 \text{ ft} \cdot \text{kips}$$

Locate section at which the positive $M_u = 83.25$ ft · kips (see Fig. 6.32)

$$\Delta M = \text{area under shear curve}$$

$$175.78 - 83.25 = \tfrac{1}{2}(4x)x$$

$$x = 6.8 \text{ ft}$$

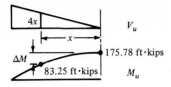

**Figure 6.32** Cutoff point of no. 8 bars.

For the extension to no. 8 bar beyond the cutoff point, use the greater of $d$ or $12d_b$:

$$d = 19.4 \text{ in} \qquad 12d_b = 12(1) = 12 \text{ in}$$

The length of no. 8 bars on each side of the point of maximum moment is

$$6.8 \text{ ft} + \frac{19.4 \text{ in}}{12} = 8.42 \text{ ft}$$

Verify that the length of the no. 8 bar on each side of the point of maximum moment is equal to or greater than $l_d$

$$l_d = \frac{0.04A_b f_y}{\sqrt{f_c'}} = \frac{0.04(0.79)(50,000)}{\sqrt{3000}} = 28.8 \text{ in} = 2.4 \text{ ft controls}$$

$$l_d = 0.0004d_b f_y = 0.0004(1)(50,000) = 20 \text{ in}$$

Since 8.42 ft > 2.4 ft, the anchorage is OK.

9. Since the no. 8 bars are cut off in a tension zone, one of the three conditions of ACI Code §12.11.5 must be satisfied. Verify that at the cutoff point $V_u \leq \frac{2}{3}\phi V_n$

$$V_u = (6.8 \text{ ft})(4 \text{ kips/ft}) = 27.2 \text{ kips}$$

Compute $\phi V_n$ (no. 3 stirrups at 9 in on center)

$$\phi V_n = \phi(V_c + V_s)$$

$$\phi V_n = \phi\left(2\sqrt{f_c'}b_w d + \frac{A_v f_y d}{s}\right)$$

$$= 0.85\left[\frac{2\sqrt{3000}}{1000}(13)(19.4) + \frac{0.22(50)(19.4)}{9}\right] = 43.64 \text{ kips}$$

$$27.2 \text{ kips} < \tfrac{2}{3}(43.64 \text{ kips}) = 29.1 \text{ kips OK}$$

10. Check anchorage of the no. 10 bars in the cantilever

$$l_d = \frac{0.04A_b f_y}{\sqrt{f_c'}} = \frac{0.04(1.27)(50,000)}{\sqrt{3000}} = 46.37 \text{ in} \qquad \text{controls}$$

or

$$l_d = 0.0004d_b f_y = 0.0004(\tfrac{10}{8})(50,000) = 25 \text{ in}$$

Modify $l_d$

$$l_d = 46.37(1.4)\frac{A_{s,\text{req}}}{A_{s,\text{sup}}}\frac{4.11}{4.53} = 58.9 \text{ in}$$

Length available to anchor (assume 2 in cover at end of bar)

$$(6 \text{ ft})(12) - 2 \text{ in} = 70 \text{ in} > 58.9 \text{ in} \qquad \text{OK}$$

Similarly the smaller no. 9 bars extending into cantilever are properly anchored.

11. Locate the section at which the no. 10 bars can be cut off to the left of support $B$. Compute the capacity of the cross section reinforced with two no. 9 bars

$$M_u = \phi A_s f_y \left( d - \frac{a}{2} \right) \qquad \text{where } a = \frac{2(50)}{0.85(3)(13)} = 3 \text{ in}$$

$$= 0.9(2)(50)(19.4 - 1.5)(\tfrac{1}{12}) = 134.25 \text{ ft} \cdot \text{kips}$$

Locate the point at which $M_u = -134.25$ ft · kips. Between the point of maximum positive moment and $-134.25$ equate $\Delta M$ to the area under the shear curve (see the $V_u$ and $M_u$ curves in Fig. 6.30)

$$\Delta M = 175.78 + 134.25 = 310.03 = \frac{x_1(4x_1)}{2}$$

$$x_1 = 12.45 \text{ ft}$$

The distance of the cutoff point from support $B$ is

$$24 \text{ ft} - (9.375 + 12.45) = 2.18 \text{ ft}$$

For the total length of the no. 10 bar to the left of support $B$, add the larger of $d$ or $12d_b$ to 2.18 ft. Also verify that the bar extends from the support a distance equal to or greater than its development length

$$\text{Length of bar with extension} = 2.18 \text{ ft} + \frac{19.4 \text{ in}}{12 \text{ in}} = 3.8 \text{ ft or } 45.6 \text{ in}$$

$$\text{Development length} = 58.9 \text{ in} \qquad \text{controls}$$

Extend no. 10 bars 5 ft from support. Since the no. 10 bars are cut off in a tension zone, the requirements of ACI Code §12.11.5 must be checked to ensure adequate shear capacity; these computations are not shown here.

12. Cutoff point for no. 9 bars to left of support $B$. Bars should extend beyond the point of inflection a distance of $d$, $12d_b$, or $l_n/16$, whichever is greater

$$(5.25 \text{ ft})(12) + 19.4 \text{ in} = 82.4 \text{ in}$$

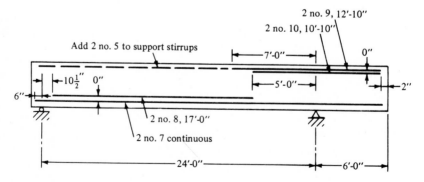

**Figure 6.33** Details of flexural steel (stirrups required for shear omitted).

The no. 9 bars must also extend at least $l_d$ beyond the theoretical cutoff point of the no. 10 bars.

$$l_d = \frac{0.04A_b\, f_y\, 1.4}{\sqrt{f_c'}} = \frac{0.04(1)(50,000)(1.4)}{\sqrt{3000}} = 51 \text{ in}$$

The minimum required extension to the left of support $B$ is

$$2.18(12) + 51 = 77.2 \text{ in}$$

Since 82.4 in controls, run no. 9 bars 84 in (7 ft) beyond support $B$. Cutoff points of flexural reinforcement are shown in Fig. 6.33.

## QUESTIONS

**6.1** What factor makes it impossible to predict bond stresses with certainty?

**6.2** What determines whether a bond failure occurs by splitting of the concrete or pullout of the bar?

**6.3** What factors determine the magnitude of bond strength?

**6.4** Why is it more logical to express bond strength as a function of $\sqrt{f_c'}$ than $f_c'$?

**6.5** When a bond failure occurs by splitting of the concrete, what determines the orientation of the failure plane?

**6.6** How is bond strength established?

**6.7** Define the term *development length*.

**6.8** How is development length used to ensure proper anchorage of a bar?

**6.9** Why do top bars require a longer development length for full anchorage than bottom bars?

**6.10** Why is the development length smaller for bars in compression than for bars in tension?

**6.11** Why can't the anchorage capacity of a standard hook be increased by adding an extension to the end of the hook?

**6.12** The ACI Code lists special conditions that must be satisfied when bars are cut off in a tension zone. What type of failure do these conditions try to prevent?

**6.13** List three methods for splicing bars and give the advantages and the disadvantages of each method.

## PROBLEMS

**6.1** The beam in Fig. P6.1 is reinforced with two rows of no. 11 bars. Are the bars in row 1 long enough to be properly anchored with regard to ACI Code bond requirements? Given $A_{s,\,\text{sup}} = A_{s,\,\text{req}}$; $f_c' = 3.6 \text{ kips/in}^2$, and $f_y = 60 \text{ kips/in}^2$.

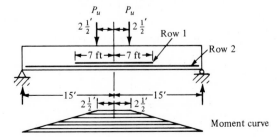

Moment curve    **Figure P6.1**

**6.2** Computations indicate that 3.06 in² of top steel is required in the beam at support *B* (Fig. P6.2).

(*a*) If two no. 11 bars are selected, will ACI bond requirements be satisfied? $f'_c = 3.6$ kips/in², and $f_y = 60$ kips/in².

(*b*) If bond requirements are not satisfied in span *BC*, what is the maximum diameter of the bar permitted if no hooks are used?

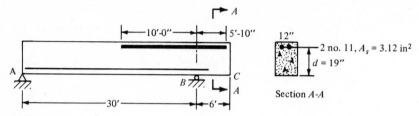

**Figure P6.2**

**6.3** In the beam in Fig. P6.3 the area of steel supplied at the point of maximum moment is 5 percent greater than that required. All reinforcement is no. 10 bars. Are the bars properly anchored by ACI Code requirements? $f'_c = 2.5$ kips/in², and $f_y = 60$ kips/in².

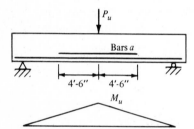

**Figure P6.3**

**6.4** Can a no. 5 bar (Fig. P6.4) stressed to 90 percent of its yield point at the face of the column be safely anchored if the bar ends in a standard hook? $f'_c = 3.75$ kips/in², and $f_y = 60$ kips/in².

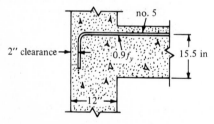

**Figure P6.4**

**6.5** If a no. 11 bar with a standard hook (Fig. P6.5) is stressed to $0.95f_y$ over the support where the moment is maximum, what is the minimum required anchorage length in the cantilever? $f'_c = 3.6$ kips/in², and $f_y = 50$ kips/in².

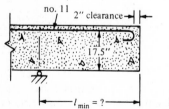

**Figure P6.5**

**6.6** If no. 9 bars in a column are to be spliced to no. 10 bars, and if the bars are confined by a closely spaced spiral (Fig. P6.6), what is the minimum required lap length for the splice? $f'_c = 4$ kips/in$^2$, and $f_y = 60$ kips/in$^2$ (spiral not shown).

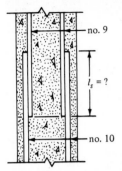

no. 9

$l_s = ?$

no. 10

**Figure P6.6**

**6.7** In Fig. P6.7, $d = 25.4$ in, $f'_c = 3$ kips/in$^2$, and $f_y = 60$ kips/in$^2$.

  (a) Draw the shear and the moment curves for the beam and locate the points of maximum moment and the points of inflection.

  (b) Determine the area of flexural steel required at support $B$.

  (c) If four no. 10 bars are used as reinforcement at support $B$, locate the theoretical cutoff point where two of the no. 10 bars can be discontinued.

  (d) If the two no. 10 bars in row 1 extend 6 ft on either side of the centerline at support $B$, are they properly anchored?

  (e) Determine the area of steel required to carry the maximum positive moment in span $AB$.

  (f) Three no. 8 bars are placed in the bottom of the beam to carry the positive moment. Two of these bars are extended 2 in past the centerline of the supports at $A$ and at $C$. Are they properly anchored according to ACI specifications?

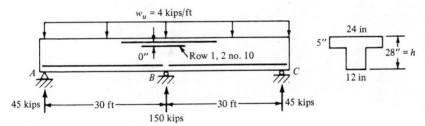

$w_u = 4$ kips/ft

24 in

5″

28″ = h

12 in

0″  Row 1, 2 no. 10

$A$

$B$

$C$

45 kips

30 ft

30 ft

45 kips

150 kips

**Figure P6.7**

# REFERENCES

1. R. M. Mains: Measurement of the Distribution of Tensile and Bond Stresses along Reinforcing Bars, *J. ACI*, vol. 48, p. 225, November 1951.
2. Y. Goto: Cracks formed in Concrete around Deformed Tensioned Bars, *J. ACI*, vol. 68, p. 244, April 1971.
3. L. Lutz and P. Gergely: Mechanics of Bond and Slip of Deformed Bars in Concrete, *J. ACI*, vol. 64, p. 711, November 1967.
4. C. Orangon, J. O. Jirsa, and J. E. Breen: A Reevaluation of Test Data on Development Length and Splices, *J. ACI*, vol. 74, p. 114, March 1977.
5. J. O. Jirsa, L. A. Lutz, and P. Gergely: Rationale for Suggested Development, Splice and Standard Hook Provisions for Deformed Bars in Tension, *Concrete Int.*, vol. 1, no. 7, pp. 47–61, July 1979.
6. P. M. Ferguson: "Reinforced Concrete Fundamentals," 3rd ed., Wiley, New York, 1973.

Val Restal Viaduct, Trento, Italy, designed by Ufficio Tecnico Provinciale of Trento. Extremely long box column supports a curved segmental bridge. (*Photograph by Arvid Grant.*)

# DESIGN OF COLUMNS

## 7.1 INTRODUCTION

This chapter covers the design of *beam-columns*, members that carry both compression force and moment. Depending on the relative magnitude of the moment and the axial load, the behavior of a beam-column will vary from pure *beam* action at one extreme to pure *column* action at the other. The term *beam-column*, although more precise, is frequently shortened to *column* by practicing structural engineers, and the two terms will be used interchangeably.

If a column is slender, the determination of the maximum moment and the location at which it occurs becomes more complex than if a column is short and stocky. In a slender member, the deflections of the longitudinal axis by *primary moments* (member end moments or moments due to transverse loads) create *additional* or *secondary* moments equal to the product of the axial force $P$ and the deflection $\Delta$ of the centerline; this is sometimes referred to as the $P$-delta effect. When the secondary moments are large, a safe design requires that a cross section be sized for the sum of the primary and secondary moments.

Less stocky, more flexible members in which the secondary moments may be significant are called *long* or *slender* columns. Note that it is not the absolute length of a member that categorizes a column as long but the member's flexibility, a function of its length-to-thickness ratio, and the end restraint. If a member is very stocky (called a *short* column), its bending deformations will be very small because of its large bending stiffness; the resulting secondary moments, small in magnitude, can be neglected. It is important to emphasize that most reinforced concrete columns classified as long columns are still relatively stocky. Although these columns may develop significant secondary moments, they will normally fail by overstress, not by buckling.

Since failure of a column often produces extensive damage as the load, formerly supported by the column, falls toward the ground, smashing everything in its path, a high potential for loss of life is associated with failure; therefore, columns are designed with a higher factor of safety than beams. By designing beams to fail while columns remain intact, total collapse of a structure can be averted.

Examples of structures that contain beam-columns and free-body diagrams of individual beam-columns are shown in Fig. 7.1.

### Forces in Building Columns

Because of the monolithic nature of reinforced concrete construction, most concrete structures are highly indeterminate continuous frames. As a result of rigid joints, load applied to any member of a continuous frame will produce bending deformations and moments in all other members of the frame (see the dashed line in Fig. 7.1c.) These deformations and moments are largest in the loaded

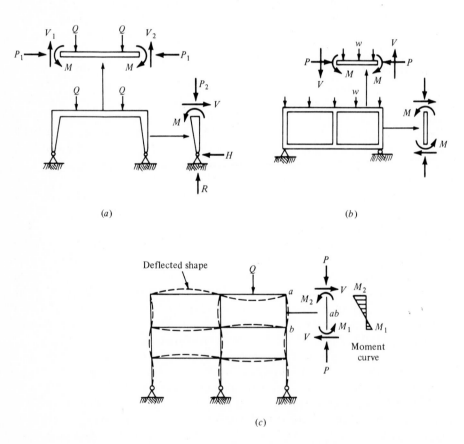

(a)

(b)

(c)

**Figure 7.1** Examples of structures containing beam-columns: (a) rigid frame, (b) Vierendeel truss, (c) building frame.

member and decrease rapidly with distance from the point of application of the load.

Since all members of a frame are bent either directly by applied loads or indirectly by joint rotations produced by the deflections of other members, three internal forces (shear, moment, and axial force) can develop in all members— columns as well as beams. No columns can carry only axial load and be completely free of moment. Moreover, moments develop in columns due to accidental eccentricities of the axial load with respect to the centroid of the column's cross section. Small variations in cross section produced by voids, honeycombing, misalignment of reinforcement, and crookedness of forms produce variations in the position of the centroid and contribute to the eccentricity of the axial force. Therefore, building columns are always designed for a minimum value of moment as well as for axial load even when loading conditions indicate that a column will be subject only to axial load.

Columns in multistory buildings are designed on the basis of their floor-to-floor dimensions, not their overall length from roof to basement. To size columns, the structure is analyzed as a unit. Then free bodies of the columns are cut out (see Fig. 7.1c) and the member sized at the section where the greatest value of moment occurs.

At any level, the total axial force in a column is approximately proportional to the number of floors that lie above the column; therefore columns in the upper stories, carrying the lightest axial loads, can be smaller in cross section or more lightly reinforced than those in the lower stories. Often the cross section of a column is held constant throughout the height of a building to simplify formwork, and the percentage of reinforcement and the strength of the concrete are increased in the lower floors.

Figure 7.2 shows a two-bay continuous frame whose girders all carry uniform load. In Fig. 7.2b free-body diagrams of the beams and columns framing into an interior and an exterior joint show that the axial force transmitted to a column at a floor is equal to the shears from the beams on either side of the joint.

If the beams framing into an interior column are similar in span and carry approximately the same load (the usual case), the end moments applied to each side of the joint are approximately equal and largely balance each other, resulting in a small net unbalanced moment applied to the columns. This moment is distributed to the columns above and below the joint in proportion to the bending stiffness of the columns.

At exterior joints, however, beams frame into one side of a column only; therefore, the total beam moment is applied to the columns framing into the joint. As a result, exterior columns (normally bent in double curvature) must be designed for higher moments and lower axial loads than interior columns.

Figure 7.2c shows that the axial force in a beam is a function of the difference in column shears. Since the column shears are moderate for gravity loads, their difference will be small and the axial force in beams will be negligible. As a result, floor beams are typically designed only for moment, and the small axial force is neglected.

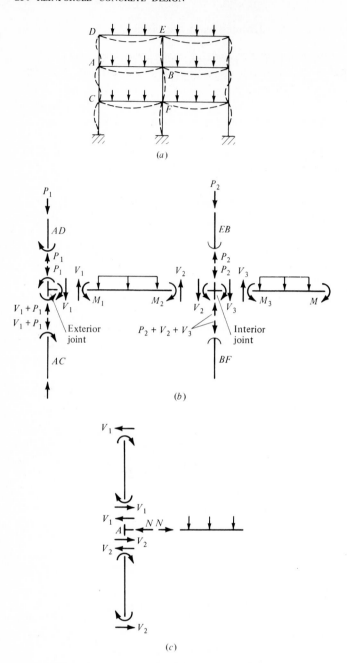

**Figure 7.2** Deformations and forces in multistory frames: (*a*) bending deformations of a two-bay frame, uniformly loaded girders; (*b*) free-body diagrams showing the transfer of forces from beams into columns through joints (column shears, axial forces in beams, and interior column moments omitted for clarity); (*c*) axial forces in beams produced by column shears (axial forces in columns and shears and moments in girders omitted for clarity); at joint $A$ $\Sigma F_x = 0$ and $N = V_2 - V_1$.

## Types of Columns

Most reinforced concrete columns have circular or rectangular cross sections and are reinforced with longitudinal bars. To ensure that the longitudinal reinforcement will be securely placed so that bars will not be pushed out of position when concrete is poured and compacted in column forms, the longitudinal steel is wired to lateral reinforcement to form a rigid cage. If individual hoops of steel, called *ties*, are used to position the longitudinal steel, the column is called a *tied column*. If the longitudinal steel is placed in a closely spaced continuous spiral and wired to it, the column is termed a *spiral column*. The area contained within the spiral is termed the *core area*. Both types are illustrated in Fig. 7.3.

Because thin columns are easier to conceal in walls and occupy less space, thereby allowing more rentable floor area, the designer is often under pressure to keep the cross section as small as possible. Although the ACI Code does not specify a minimum cross-sectional area, construction clearances require that the minimum width or diameter be not less than 8 to 10 in (203 to 254 mm).

To provide ductility, to reduce creep and shrinkage, and to ensure some bending strength, ACI Code §10.9.1 requires a minimum area of longitudinal reinforcing steel equal to 1 percent of the gross column area. An upper limit for the area of steel is set at 8 percent of the gross cross-sectional area. As a practical matter, it becomes difficult to fit more than 5 or 6 percent of steel reinforcement into a column form and still maintain sufficient space for concrete to flow between bars. As the amount of reinforcement is increased, less width is available to work concrete into the spaces between bars and the sides of the form, and the possibility of developing voids and honeycombing (porous concrete produced by a lack of mortar between aggregates next to the sides of the form) is increased. These imperfections can lead to a significant reduction in column strength.

If a high percentage of steel is required, reinforcing bars can be bundled and wired together to form a unit. ACI Code §7.6.6.1 specifies that up to four bars may

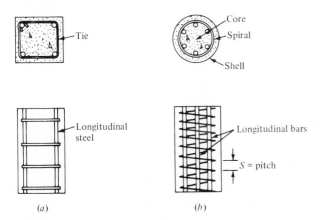

(a)　　　　(b)

**Figure 7.3** Types of columns: (*a*) tied column, minimum of four longitudinal bars required in a rectangular column; (*b*) spiral column, minimum of six longitudinal bars required.

**Figure 7.4** Column reinforced with bundled bars.

be placed in a given bundle. Figure 7.4 shows a column cross section reinforced with three-bar bundles in each corner.

To allow concrete to flow freely between the longitudinal steel in order to fill all portions of the form and reduce the likelihood of voids, ACI Code §7.6.3 requires that the clear distance between bars be not less than either 1.5 bar diameters or $1\frac{1}{2}$ in (38 mm).

If cast-in-place concrete is not exposed to weather or in contact with soil, a minimum of $1\frac{1}{2}$ in (38 mm) of concrete cover is required to protect the longitudinal steel, ties, and spirals from fire and corrosion. When concrete is exposed to earth or weather, the minimum cover must be increased to 2 in (51 mm) for bars greater than no. 6s.

### Long versus Short Columns: Slenderness Effects

The difference between the primary and the secondary moments in a beam-column is illustrated in Fig. 7.5, where a column is loaded by an axial force $P$ and two equal end moments that produce single-curvature bending. For this special case of equal end moments, no shears develop in the column. Cutting a section through the deflected member a distance $x$ from the top and summing moments about the centroid of the cross section to evaluate the internal moment $M_i$ gives

$$M_i = M + P\Delta$$

where $M$ represents the primary moment and $P\Delta$ equals the secondary moment.

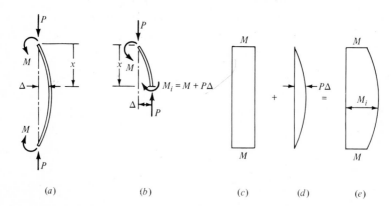

**Figure 7.5** Primary and secondary moments in a beam-column with equal end moments: (a) beam-column, (b) free-body diagram of column in deflected position, (c) primary moment, (d) secondary moment, (e) total moment.

For this case of equal end moments, the secondary moment attains its maximum value at midheight.

Columns in which the value of the secondary moment is less than 5 percent of the total moment are classified as *short* columns. If the secondary moment is greater than 5 percent of the primary moment, the column is classified as a *long* column.

## 7.2 FUNDAMENTALS OF COLUMN BEHAVIOR

Before discussing the ACI procedure for classifying columns, we review the fundamentals of column behavior to establish the parameters on which the ACI procedure is based. Most of this review is general and applies equally well to columns of other materials such as steel.

Since the student will have covered this material in a basic strength-of-materials course, derivations will be omitted and only major topics emphasized. For the student with a solid background on column behavior only a quick review of this section is required before starting Sec. 7.3.

### Buckling of Perfect Columns

Figure 7.6 shows a slender column supported at the bottom by a frictionless pin and restrained laterally at the top by a roller. Since the pins are frictionless, no moments can develop at either end to provide restraint against rotation. For simplicity, the column is assumed to be perfectly straight, free of residual stresses, and made of a perfect elastoplastic material. In the elastic region, the modulus of elasticity is constant and equal to $E$. Beyond the elastic region, the stress-strain curve is horizontal, and the modulus of elasticity $E_t$ drops to zero (see Fig. 7.6b).

If the column is slender, it will fail by buckling into the shape of a sine wave when the load reaches a particular value $P_c$, called the *Euler buckling* load or the *critical load*. Although the buckled column can support the critical load, it has no additional bending strength to support even the smallest additional increment of vertical load. The buckling load therefore represents the maximum theoretical load a slender column can support.

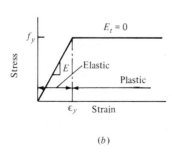

**Figure 7.6** Buckling of a pin-ended column: (*a*) Euler buckling load acts, (*b*) idealized stress-strain curve.

By considering the equilibrium of the pin-ended column in the bent configuration, the smallest load that produces buckling is given as

$$P_c = \frac{\pi^2 E I_{min}}{l^2}$$

(7.1)

where $E$ = modulus of elasticity
$I_{min}$ = minimum moment of inertia
$l$ = length of column between pin supports

If $P_c$ is divided by the cross-sectional area $A$ of the column, Eq. (7.1) can be converted into an equation that gives the average stress $f_c$ on the cross section of a buckled column

$$f_c = \frac{\pi^2 E I_{min}}{l^2 A}$$

(7.2)

Substituting $I_{min} = Ar^2$, where $r$ equals the minimum radius of gyration, gives the stress form of the Euler equation

$$f_c = \frac{\pi^2 E}{(l/r)^2} \qquad \text{but } f_c \le f_y$$

(7.3)

The nondimensional parameter $l/r$, called the *slenderness ratio* of the column, contains the geometric properties of the members that determine the buckling stress. Since $\pi^2$ is a constant and $E$ is a constant as long as $f_c$ does not exceed $f_y$, Eq. (7.3) plots as a hyperbola (see Fig. 7.7). Figure 7.7 shows that the average stress at which buckling occurs increases as the slenderness ratio decreases. The smallest slenderness ratio at which elastic buckling can occur can be established from Eq. (7.3) by setting $f_c$ equal to $f_y$ and solving for the corresponding slenderness ratio $(l/r)_{lim}$

$$\left(\frac{l}{r}\right)_{lim} = \sqrt{\frac{\pi^2 E}{f_y}}$$

(7.4)

As shown in Fig. 7.7, columns whose slenderness ratio is equal to or less than $(l/r)_{lim}$ fail by crushing at a stress of $f_y$.

Columns whose slenderness is greater than $(l/r)_{lim}$ fail by buckling.

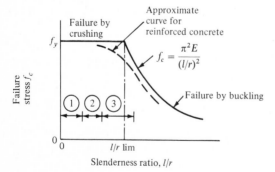

Figure 7.7 Influence of slenderness ratio on the magnitude of the failure stress of an axially loaded column; 1-compression blocks, 2-short columns, and 3-long columns.

A plot showing the variation of the failure stress with slenderness ratio for reinforced concrete columns would be similar in shape to the curve in Fig. 7.7, but calculations to determine points on the curve would be extremely complex because of the need to account for creep, the nonlinear stress-strain curve of concrete, imperfections, and the influence of steel. The approximate shape of the curve for an axially loaded reinforced concrete column is shown by the dashed line in Fig. 7.7.

Since most reinforced concrete columns are typically stocky, failure by buckling is unlikely. Most concrete columns, both long and short, have slenderness ratios between 0 and $(l/r)_{\text{lim}}$ (Fig. 7.7). In this region columns can be classified into one of three categories in order of increasing slenderness ratio:

*Short compression blocks or piers*: Members whose height does not exceed 2 or 3 times their smallest cross-sectional dimension.
The ACI Code does not define this category.
*Short column*: Stiff, short columns that are typical of most building columns. Secondary moments are small.
*Long columns*: Still relatively stocky, but bending deformations may create significant secondary moments for larger values of slenderness ratio.

## Influence of Boundary Conditions: Effective-Length Concept

Columns supported by frictionless pins and rollers do not exist in real structures. The ends of real columns are restrained against rotation by their supports, and end moments always develop; moreover, the ends of columns are sometimes free to displace laterally. Equations (7.1) and (7.3), derived for pin-ended columns, can be extended to cover columns with other boundary conditions by the substitution of an *effective length* of column in place of the actual length. The effective length is some proportion of the actual length.

The effective-length concept can be developed by considering the deformations of a fixed-end column whose top is supported by a roller (Fig. 7.8a). A free-body diagram of the column between the point of inflection (PI) and the top support is shown in Fig. 7.8c. This length of column behaves exactly like an axially loaded, pin-supported column of length $l_e$. The point of inflection is equivalent to a pin support for two reasons: (1) the moment is zero, and (2) like a frictionless pin, the buckled column below the point of inflection has no bending stiffness to resist rotation.

The effective length $l_e$ is typically expressed as the product of the actual length $l$ times a factor $k$, called the *effective length factor*. Substituting $kl$ for $l$ in Eqs. (7.1) and (7.3) gives

$$P_c = \frac{\pi^2 EI}{(kl)^2} \tag{7.5}$$

and

$$f_c = \frac{\pi^2 E}{(kl/r)^2} \tag{7.6}$$

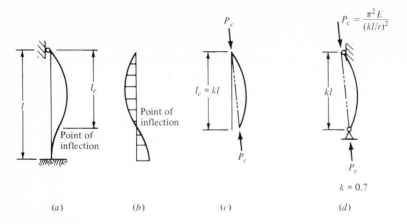

**Figure 7.8** Effective length of a column fixed at one end and pin-supported at the other end: (a) buckled shape, (b) moment curve, (c) free-body diagram of column above the point of inflection, (d) equivalent pin-ended column.

which predict the buckling load or the buckling stress for columns with any type of boundary conditions. These equations apply as long as behavior is elastic.

Values of $k$ are available in handbooks for a variety of supports, or they can be evaluated by using the design aids shown in Fig. 7.12. For the column in Fig. 7.8, an exact solution of the differential equation of the elastic curve gives a $k$ value of 0.7. Values of $k$ for other idealized support conditions are given in Fig. 7.9.

A comparison of the effective-length factors in Fig. 7.9 shows that columns whose supports prevent lateral displacement have $k$ values ranging from 0.5 to 1. The capacity of a given column is greatest when the supports provide both rotational and lateral restraint. If either end of a column is free to displace laterally, $k$ will be equal to or greater than 1.

### Braced and Unbraced Frames

Structural frames whose joints are restrained against lateral displacement by attachment to rigid elements or by bracing are called *braced frames*. Floors of buildings are usually braced by attachment to rigid elements such as structural walls (shear walls), elevator shafts, or reinforced masonry walls. The ACI commentary states that to be considered sufficiently stiff to brace a building frame a structural bracing element must have a bending stiffness at least 6 times larger than the sum of the stiffness of all the columns being braced. If a structural frame is not attached to an effective bracing element but depends on the bending stiffness of its columns and girders to provide lateral resistance, it is termed an *unbraced frame*. Examples of braced and unbraced frames are shown in Fig. 7.10.

Given two identical frames, one braced and the other unbraced, the effective length of the columns will always be greater in the unbraced frame than in the

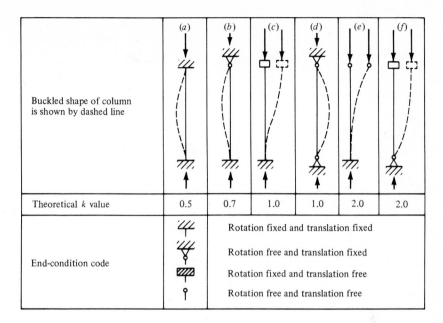

| | (a) | (b) | (c) | (d) | (e) | (f) |
|---|---|---|---|---|---|---|
| Buckled shape of column is shown by dashed line | | | | | | |
| Theoretical $k$ value | 0.5 | 0.7 | 1.0 | 1.0 | 2.0 | 2.0 |
| End-condition code | | Rotation fixed and translation fixed | | | | |
| | | Rotation free and translation fixed | | | | |
| | | Rotation fixed and translation free | | | | |
| | | Rotation free and translation free | | | | |

**Figure 7.9** Values of effective length factor for a number of idealized boundary conditions. (*Adapted from Ref. 11.*)

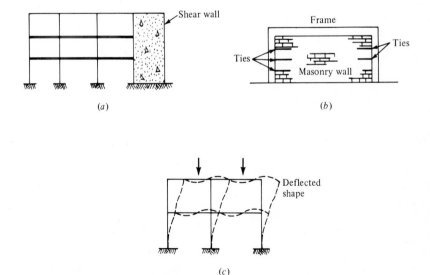

*(a)*

*(b)*

*(c)*

**Figure 7.10** Examples of braced and unbraced frames: (*a*) frame braced by a shear wall, (*b*) rigid frame braced by connection to masonry wall, (*c*) unbraced frame.

braced frame. Since the strength of a column, like the stiffness of a structure, decreases as the effective length increases, the designer should ensure that bracing elements are incorporated into a structure.

### Effective-Length Factors for Columns of Rigid Frames

In a reinforced concrete frame, columns are rigidly attached to girders and adjacent columns. The effective length of a particular column between stories will depend on how the frame is braced and on the bending stiffness of the girders. As a column bends in response to applied loads, the ends of the attached girders must rotate with the column because of the rigid joint. If the girders are stiff and do not bend significantly, they will provide full rotational restraint to the column, like a fixed support (Fig. 7.11a). If the girders are flexible and bend easily, as in Fig. 7.11b, they provide only a small degree of rotational restraint and the end conditions for the column approach those of a pin support that allows unrestrained rotation.

The Jackson and Moreland alignment charts[13] (Fig. 7.12) can be used to evaluate the influence of girder bending stiffness on the effective-length factor of a column that is part of a rigid frame. The charts are entered with values of $\psi$ for the joints at each end of a column. For a rigid frame whose members are prismatic, $\psi$, the ratio of the sum of the relative bending stiffnesses of the columns to that of the girders, is defined as

$$\psi = \frac{\Sigma(E_c I_c/L_c)}{\Sigma(E_g I_g/L_g)} \tag{7.7}$$

where  $I_c$ = moment of inertia of column
$I_g$ = moment of inertia of girder
$L_c$ = length of column, center to center of joints
$L_g$ = length of girder, center to center of joints
$E_g, E_c$ = modulus of elasticity of girders and columns respectively.

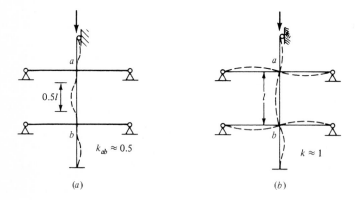

(a)                                             (b)

**Figure 7.11** Influence of girder stiffness on the effective length of a column in a braced frame: (a) rigid girders, (b) flexible girders.

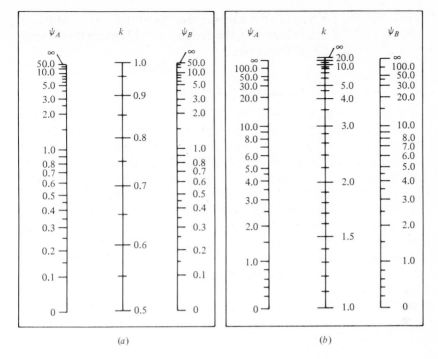

**Figure 7.12** Alignment charts for the effective-length factor $k$: ($a$) braced frames, ($b$) unbraced frames. $\psi$ = ratio of $\Sigma(EI/L_c)$ of compression members to $\Sigma(EI/L_g)$ of flexural members in a plane at one end of a compression member, $k$ = effective-length factor.[13]

The intersection of the straight line connecting the two $\psi$ values with the vertical line labeled $k$ gives the value of the column's effective-length factor. Since the strength of a column is influenced by the presence or absence of lateral support, charts are given for both *braced* and *unbraced* frames.

The value of the $k$ factor is based on the assumption that all columns in a braced frame buckle simultaneously and that the girders bend into single curvature with equal but opposite rotations at each end (Fig. 7.13$a$).

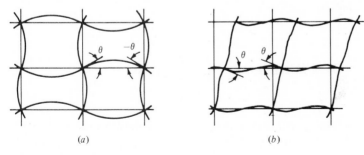

**Figure 7.13** Frame geometry on which $k$ factors in Fig. 7.12 are based: ($a$) bending deformations of a braced frame, ($b$) bending deformations of an unbraced frame.

For the *unbraced* frame, the k factors are based on the assumption that the girders and the buckled columns are bent into double curvature (Fig. 7.13b). If the boundary conditions for the girder are not consistent with the above assumptions, their stiffness may be modified to account for the influence of the boundary restraint. For an *unbraced* frame, the girder stiffness should be multiplied by 0.5 if the far end of a girder connects to a hinge and by 0.67 if the far end of the girder is fixed. For a *braced* frame, the girder stiffness should be multiplied by 1.5 if the far end of the girder connects to a hinge and by a factor of 2 if the far end of the girder is connected to a fixed support.

ACI Code §10.11.2.2 requires that the influence of both cracking and the area of the reinforcement on the stiffness of the beams and the columns be considered in the computation of k for an *unbraced* frame. This specification can be satisfied by basing the computation of girder moment of inertia on the properties of the cracked transformed section. For columns, the numerator of ACI equation 10.9 given as Eq. (7.8) is recommended for the moment of inertia:

$$I_c = \frac{I_{gr}}{5} + nI_{se} \tag{7.8}$$

$I_{gr}$ = moment of inertia of column based on gross area
$n$ = modular ratio $E_s/E_c$
$I_{se}$ = moment of inertia of steel about centroidal axis of cross section

As an alternate procedure if $kl_u/r$ does not exceed 60, the ACI Commentary indicates that reasonable results can also be secured by using half the girder's moment of inertia based on the gross area and basing the column's moment of inertia on the gross column area. This procedure accounts for the fact that the cracking is less severe in columns because compression in the columns closes bending cracks.

Since the value of k for braced frames varies from 0.5 to 1, ACI Code §10.11.2.1 permits the effective-length factor k to be taken conservatively as 1 in order to speed design.

The k factor for columns in unbraced frames will always be greater than 1; the ACI Commentary indicates the designer should not expect a value of k less than 1.2 for unbraced columns.

## 7.3 THE ACI PROCEDURE FOR CLASSIFYING BEAM-COLUMNS

The effective-slenderness ratio $kl_u/r$, a measure of a member's overall flexural stiffness, can be used to divide beam-columns into two categories: *short* members, in which the secondary moments are negligible, and *long* or *slender* columns, in which secondary moments are significant and must be considered. Studies[1] have shown that a column can be classified as short when its effective-slenderness ratio satisfies the following criteria

$$\frac{kl_u}{r} \leq \begin{cases} 22 & \text{unbraced members†} & (7.9) \\ 34 - 12\dfrac{M_1}{M_2} & \text{braced members} & (7.10) \end{cases}$$

where $k$ = effective-length factor (see Fig. 7.12)

$l_u$ = unsupported length of member; defined in ACI §10.11.1.1 as clear distance between floor slabs, beams, or other members capable of providing lateral support

$r$ = radius of gyration of cross section of column associated with axis about which bending is occurring

$M_1$ = value of the smaller end moment on the column calculated from a conventional elastic analysis, e.g., moment distribution; $M_1$ is positive for single curvature and negative for double curvature

$M_2$ = value of larger design end moment on a compression member, always positive

For rectangular sections the ACI Code specifies that $r$ can be taken as 0.3 times the depth of the section in the direction of bending. For a circular section $r$ equals 0.25 times the diameter. For other shapes, computation of $r$ is based on the gross section (see ACI Code §10.11.3).

## Special End Conditions

If a column is supported by a frictionless hinge, no rotational restraint is exerted on the end of the column. Therefore a hinge can be treated like a joint into which girders of zero stiffness frame. For this condition $\psi = \infty$ (infinity).

A fixed-end support may be considered equivalent to a joint into which a girder of infinite stiffness frames. By being too stiff to bend, such a girder would restrain the joint against rotation. For this condition, $\psi$ is taken equal to zero.

**Example 7.1:** Evaluate the effective-length factor for the cantilever column in Fig. 7.14a.

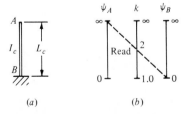

(a)  (b)

**Figure 7.14** (a) Fix-ended column, (b) evaluation of $k$ by Fig. 7.12b.

---

† A study of over 20,000 typical building columns found that 90 percent of the columns in braced frames and over 40 percent of the columns in unbraced frames were short. Since most structural frames are braced, secondary moments can be neglected in the design of most columns.

SOLUTION At top use $I_g = 0$; at bottom use $I_g = \infty$

$$\psi = \begin{cases} \dfrac{\Sigma(I_c/L_c)}{\Sigma(I_g/L_g) = 0} = \infty & \text{top} \\[4mm] \dfrac{\Sigma(I_c/L_c)}{\Sigma(I_g/L_g) = \infty} = 0 & \text{bottom} \end{cases}$$

Use Fig. 7.12b and read $k = 2$ (Fig. 7.14b).

**Example 7.2:** Evaluate the effective-length factor for the column in Fig. 7.15.

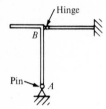

Figure 7.15

SOLUTION At both top and bottom rotational restraint is zero; use $I_g = 0$

$$\psi = \begin{cases} \dfrac{\Sigma(I_c/L_c)}{\Sigma(I_g/L_g) = 0} = \infty & \text{top} \\[4mm] \dfrac{\Sigma(I_c/L_c)}{\Sigma(I_g/L_g) = 0} = \infty & \text{bottom} \end{cases}$$

Enter Fig. 7.12a and read $k = 1$.

**Example 7.3:** The frame in Fig. 7.16 is composed of members with rectangular cross sections. All members are constructed of the same strength concrete ($E$ same for both beams and columns). Considering bending in the plane of the frame only, classify column $bc$ as long or short if the frame is (a) braced and (b) unbraced. All girders 12 × 24 in, $I_g = 13,824$ in⁴, $I_{bc} = 2,744$ in⁴, $I_{ab} = 4,096$ in⁴.

SOLUTION (a) For the column to be short we must have using Eq. (7.10)

$$\frac{kl_u}{r} \leq 34 - 12\frac{M_1}{M_2} = 34 - \frac{12(-20)}{30} = 42$$

where $M_1$ is negative because of double-curvature bending. Since the frame is braced against sidesway, $k$ may be taken as 1 to compute the slenderness ratio.

$$l_u = (12.5 - 2)(12) = 126 \text{ in} \qquad r = 0.3h = 0.3(14) = 4.2 \text{ in}$$

$$\frac{kl_u}{r} = \frac{1(126)}{4.2} = 30 < 42$$

Therefore, the column is short.

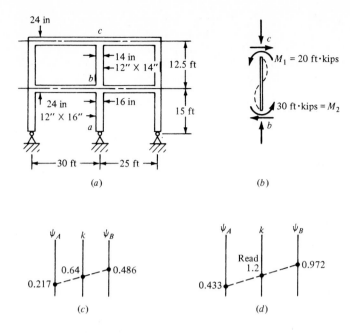

**Figure 7.16** (a) Building frame, (b) free-body diagram of column bc, (c) evaluation of k by Fig. 7.12a, (d) evaluation of k by Fig. 7.12b.

Although a smaller value of $k$ is not required to demonstrate that the column is short, the influence of girder stiffness on $k$ will be illustrated by the use of Fig. 7.12

$$\text{Top:} \quad \psi_A = \frac{\Sigma(I_c/L_c)}{\Sigma(I_g/L_g)} = \frac{2744/12.5}{(13,824/30) + 13,824/25} = 0.217$$

$$\text{Bottom:} \quad \psi_B = \frac{\Sigma(I_c/L_c)}{\Sigma(I_g/L_g)} = \frac{(2744/12.5) + 4096/15}{(13,824/30) + 13,824/25} = 0.486$$

Using Fig. 7.12a, read $k = 0.64$ (see Fig. 7.16c); then $kl_u/r = 19.2$.

(b) For the column to be short we must have $kl_u/r \le 22$. To compute the effective-length factor $k$, use Fig. 7.12b. If $kl_u/r \le 60$, the effect of cracking on the moment of inertia of the girder can be taken into account when the $\psi$ factors are computed by using $I_g = \frac{1}{2}I_{gr}$

$$\text{Top:} \quad \psi_A = \frac{\Sigma(I_c/L_c)}{\Sigma(I_g/L_g)} = \frac{2744/12.5}{[\frac{1}{2}(13,824)/30] + \frac{1}{2}(13,824)/25} = 0.433$$

$$\text{Bottom:} \quad \psi_B = \frac{\Sigma(I_c/L_c)}{\Sigma(I_c/L_c)} = \frac{(2744/12.5) + 4096/15}{[\frac{1}{2}(13,824)/30] + \frac{1}{2}(13,824)/25} = 0.972$$

Enter Fig. 7.12b with $\psi_A$ and $\psi_B$ and read $k = 1.2$ (see Fig. 7.16d).

$$\frac{kl_u}{r} = \frac{1.2(126)}{4.2} = 36 > 22$$

Therefore the column is long.

**Example 7.4:** In Fig. 7.17 consider the behavior of the frame in the plane of the paper. Compute $kl/r$ of columns $ab$ and $bc$. All members are 18 by 12 in. Frame unbraced $E$ is constant for beams and columns. Since the nomographs in Fig. 7.12 are based on the assumption that the ends of columns are rigidly joined to girders (see Fig. 7.13), we shall assume that an imaginary girder with zero flexural stiffness ($I_g = 0$) runs between joints $c$ and $g$ (Fig. 7.17b).

$$I_c = I_g = \frac{bh^3}{12} = \frac{12(18)^3}{12} = 5832 \text{ in}^4$$

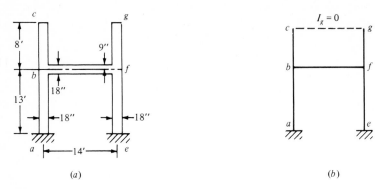

(a)                                                                 (b)

**Figure 7.17** (a) Rigid frame, (b) an imaginary girder with $I_g = 0$ added from points $c$ to $g$.

SOLUTION  Column $ab$: compute $k$. Since joint $a$ is fixed use $I_g = \infty$

$$\psi_a = \frac{\Sigma(I_c/L_c)}{\Sigma(I_g/L_g) = \infty} = 0 \qquad \psi_b = \frac{\Sigma(I_c/L_c)}{\Sigma(I_g/L_g)} = \frac{(5832/8) + 5832/13}{(5832(\frac{1}{2})/14)} = 5.65$$

where the $\frac{1}{2}$ factor reduces the girder stiffness to account for cracking. Using Fig. 7.12b, unbraced chart, read $k = 1.5$

$$\frac{kl_u}{r} = \frac{1.5[(13 \text{ ft} \times 12) - 9 \text{ in}]}{0.3(18 \text{ in})} = 41$$

Column $bc$: compute $k$. For member $cg$, $I_g = 0$

$$\psi_b = 5.65 \text{ (see above)} \qquad \psi_c = \frac{\Sigma(I_c/L_c)}{\Sigma(I_g/L_g) = 0} = \infty$$

Using Fig. 7.12b, unbraced chart, read $k = 3.5$

$$\frac{kl_u}{r} = \frac{3.5[(8 \text{ ft} \times 12) - 9 \text{ in}]}{0.3(18 \text{ in})} = 56.4$$

## 7.4 STRENGTH OF SHORT AXIALLY LOADED COLUMNS

To begin the study of column design, we examine the special case of a short uniformly compressed column loaded to failure by an axial load. This case establishes both the strength of a reinforced column and the influence of ties and spirals on the failure mode.

The strength of a short axially loaded tied or spiral column can be established by considering a free-body diagram of the upper section of the short column shown in Fig. 7.18$a$. The failure load $P_0$ is applied through a rigid block at the centroid of the section to produce uniform deformation $\Delta$ of the member and equal values of strain in the steel and the concrete. Failure occurs when both the steel and concrete have been strained to their maximum strength in the horizontal regions of the stress-strain curves. This strain will be about 0.002 (Fig. 7.18$b$). At this level of strain the stress in the steel will be equal to $f_y$ and the stress in the concrete equal to $f_c'$. From vertical equilibrium of the free-body diagram of the column in Fig. 7.18$c$, the axial force $P_0$ can be expressed in terms of the internal forces in the steel and in the concrete. Summing forces in the vertical direction gives

$$P_0 = f_y A_{st} + f_c'(A_g - A_{st}) \qquad (7.11)$$

where $A_g$ is the gross area of the cross section and $A_{st}$ is the total area of steel.

Experimental studies show that the strength predicted by Eq. (7.11) is too high. If the compressive strength of the concrete $f_c'$ is reduced by a factor of 0.85, the

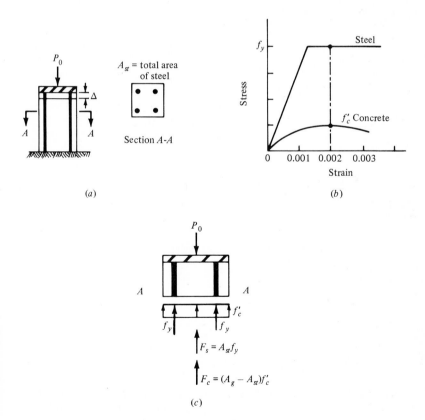

(a)

(b)

(c)

**Figure 7.18** Strength of a short axially loaded column: ($a$) axially loaded short column, ($b$) uniaxial stress-strain curves showing magnitude of stress at failure, ($c$) forces in steel and concrete at failure.

column's predicted strength will compare closely with test results. Making this adjustment in Eq. (7.11) gives

$$P_0 = A_{st} f_y + (A_g - A_{st})0.85 f_c' \qquad (7.12)$$

The 0.85 factor in Eq. (7.12) accounts for the fact that concrete poured into forms is not as strong as concrete poured into test cylinders used to establish the stress-strain curve of concrete. Better compaction and curing of concrete in test cylinders than in forms probably accounts for the difference in strength. There is also a tendency for the excess water in the concrete to bleed, i.e., rise to the upper portion of the form. This water increases the water-cement ratio of the concrete in the top of the form and reduces its compressive strength.

## Behavior of Columns Loaded to Failure by Axial Load

Tests verify that short tied and spiral columns with identical areas of steel and concrete have the same capacity for axial load; however, once the ultimate load has been applied, the failure mode for each varies (Fig. 7.19). As the ultimate load is reached in the tied column, the highly compressed concrete becomes heavily cracked and falls out, leaving the longitudinal reinforcement unsupported between ties. Lacking support, the slender, highly stressed bars buckle, and the column fails rapidly (Fig. 7.20a).

When a spiral column is stressed to its full capacity, total collapse of the member does not take place. Although the outer shell of concrete covering the spiral cracks up and falls off as the ultimate load is initially reached, the reduced cross section of the column consisting of concrete and longitudinal reinforcement within the spiral is able to develop the additional strength needed to support the load that produces failure. Total collapse of the spiral column does not occur because the closely spaced turns of the continuous spiral and the longitudinal reinforcement form a cage that confines the concrete core and holds it together (Fig. 7.20b).

The magnitude of the additional load that can be supported by the column is a function of the strength of the spiral. The spiral eventually fails when the outwardly expanding core stretches the spiral steel to its yield point. The dashed lines in Fig. 7.19 compare the strength of short axially loaded columns reinforced with a

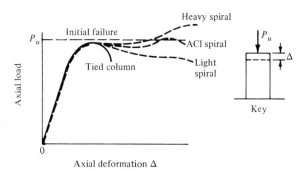

**Figure 7.19** Load-deflection curves for tied and spiral columns.

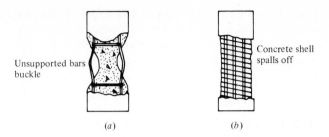

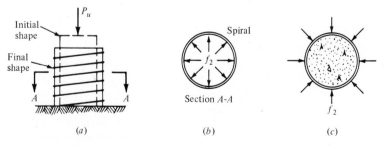

**Figure 7.20** Appearance of columns at failure: (*a*) tied column, (*b*) spiral column.

light, a heavy, and a standard ACI spiral. Notice that the ACI spiral is sized so the value of the final failure load (after large vertical deflections have taken place) is slightly greater than the initial failure load. The ability to undergo large deformations without failure gives spiral columns a degree of toughness and ductility the standard tied column lacks.

Recognizing the difference in failure modes between tied and spiral columns, the ACI Code specifies a reduction factor of 0.7 for tied columns and for the more ductile spiral column a reduction factor of 0.75. For both columns the relatively large reduction factors reflect the difficulty of producing dense, well-compacted concrete when the plastic concrete is poured into long narrow column forms and worked into the spaces between the column reinforcement.

## Design of Spirals

The ability of the concrete confined within the spiral to carry additional load is due to the lateral pressure exerted on the concrete core by the heavy coils of the spiral (Fig. 7.21). Experimental studies have shown that the compressive strength of concrete increases when the test cylinders are loaded by lateral pressure. In terms of the concrete's 28-day compressive strength $f'_c$ and the lateral pressure $f_2$ the compressive strength of concrete can be expressed by

$$f_f = f'_c + 4.1f_2 \tag{7.13}$$

where $f_f$ is the ultimate compressive strength of laterally loaded concrete.

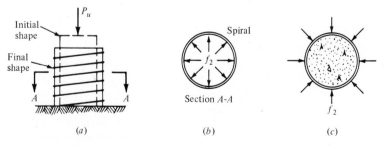

**Figure 7.21** Stressing of spiral: (*a*) lateral expansion of compressed core stressing spiral, (*b*) lateral pressure on spiral applied by expanding core, (*c*) lateral pressure on core exerted by spiral.

The spiral is proportioned so that the compressive strength lost by spalling of the concrete shell covering the spiral is equal to the additional compressive strength that can be developed by the core due to the lateral pressure exerted by the spiral when its coils are stressed to the yield point. Expressing the above criteria in terms of material properties gives

$$(A_g - A_{core})(0.85f'_c) = A_{core}(4.1f_2) \tag{7.14}$$

where $A_{core}$, the area of core, is based on the outside diameter of the spiral, and $f_2$ is given by Eq. (7.13).

To express $f_2$ in terms of the area and yield point of the spiral steel, consider a cylinder of concrete equal in depth to the pitch of the spiral (Fig. 7.22a). The slight slope of the spiral is neglected. Due to symmetry the lateral pressure is equal in all directions. Cutting a vertical section through the cylinder along a diameter gives the free-body diagram shown in Fig. 7.22b. Summing forces perpendicular to the vertical plane through the diameter gives

$$h_{core} S f_2 = 2A_b f_y$$

$$f_2 = \frac{2A_b f_y}{h_{core} S} \tag{7.15}$$

where  $S$ = spiral pitch
$A_b$ = area of spiral bar

Substituting Eq. (7.15) into (7.14) and dividing both sides by $A_{core}$ gives

$$\left(\frac{A_g}{A_{core}} - 1\right)(0.85f'_c) = \frac{4.1(2A_b f_y)}{h_{core} S} \tag{7.16}$$

After defining

$$\rho_s = \frac{\text{volume of spiral steel in one turn}}{\text{volume of core}}$$

we have

$$\rho_s = \frac{A_b \pi h_{core}}{(\pi h_{core}^2/4)S} = \frac{4A_b}{h_{core} S} \tag{7.17}$$

Core

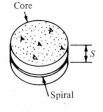

Spiral

(a)

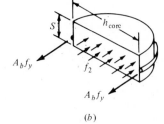

(b)

**Figure 7.22** Stresses in spiral induced by compression in core: (a) free-body diagram of the cylindrical core restrained by one turn of the spiral, (b) free-body diagram of core and spiral cut along a diameter.

Solving Eq. (7.17) for $A_b$ in terms of $\rho_s$, substituting into Eq. (7.16), and simplifying gives

$$\rho_s = \frac{0.42f'_c}{f_y}\left(\frac{A_g}{A_{core}} - 1\right)$$

Rounding the coefficient 0.42 to 0.45 gives

$$\rho_s = \frac{0.45f'_c}{f_y}\left(\frac{A_g}{A_{core}} - 1\right) \tag{7.18}$$

which is ACI equation 10.5. $f_y$ is not to exceed 60 kips/in$^2$ (ACI Code §10.9.3).

ACI Code §7.10.4 contains provisions for detailing spirals that produce columns with a high degree of toughness. For cast-in-place concrete, spirals must be fabricated from bars that are at least $\frac{3}{8}$ in (9.5 mm) in diameter. To ensure that sufficient clearance will be available for concrete to flow into the spaces between turns of the spiral, the minimum clear spacing between turns must be at least 1 in (25.4 mm). By specifying that the clear spacing between turns of the spiral not exceed 3 in (76 mm) proper confinement of the core within the spiral is assured.

Considerable information on placing reinforcement is contained in Ref. 12, which contains many details that can be incorporated into structural drawings and specifications.

**Example 7.5:** Establish the size and required pitch of a spiral for the 16 in-diameter column in Fig. 7.23. $f_y = 60$ kips/in$^2$ (413.7 MPa), and $f'_c = 4$ kips/in$^2$ (27.58 MPa).

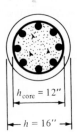

$h_{core} = 12''$

$\leftarrow h = 16'' \rightarrow$    **Figure 7.23**

SOLUTION Using Eq. (7.18), we have

$$\rho_s = 0.45\frac{4000}{60,000}\left[\frac{\pi(8^2)}{\pi(6^2)} - 1\right] = 0.0233$$

Try a $\frac{3}{8}$ in-diameter spiral, $A_b = 0.11$ in$^2$. Using Eq. (7.17), we have

$$\rho_s = \frac{4A_b}{h_{core}S}$$

$$0.0233 = \frac{4(0.11)}{12S}$$

and

$$S = 1.574 \text{ in} \qquad \text{use } S = 1.5 \text{ in (38 mm)}$$

$$1 \text{ in} < 1.5 \text{ in} < 3 \text{ in} \qquad \text{OK}$$

## Design of Ties

Tests on full-sized columns indicate the spacing between ties has no significant influence on the ultimate strength of columns. However, once a column begins to fail due to overload, closely spaced ties reduce the tendency of the longitudinal steel to buckle and help hold the column together.[10]

As noted in ACI Code §7.10.5, all longitudinal bars in tied columns must be enclosed by lateral ties. For longitudinal bars no. 10 or smaller, ties must be at least no. 3s. For larger bars (nos. 11, 14, 18, and bundled bars) ties must be at least no. 4s.

ACI Code §10.7.5.2 specifies that the spacing between ties must not exceed 16 longitudinal bar diameters, 48 tie diameters, or the least dimension of the column. Also, ties must be so arranged that every corner and alternate longitudinal bar has lateral support provided by the corner of a tie having an included angle of not more than 135°, and no bar shall be farther than 6 in (152 mm) clear on either side from such a laterally supported bar (Fig. 7.24).

## Plastic Centroid

The plastic centroid of a cross section is the point through which the resultant axial force on a cross section must pass to produce uniform strain at failure. If the applied load does not pass through the plastic centroid, it will create bending as well as compression. The moment created will be equal to the product of the axial force times the distance between the force and the plastic centroid.

If the cross section is symmetrical about both its principal axes, the plastic centroid coincides with the intersection of the two axes of symmetry. For unsymmetrical cross sections, the plastic centroid will coincide with the position of the resultant internal force associated with a uniform strain distribution on the cross section.

**Example 7.6:** Locate the position of the plastic centroid of the section in Fig. 7.25; $f_y = 60 \text{ kips/in}^2$, $f'_c = 3 \text{ kips/in}^2$, and the cross section is reinforced with four no. 10 bars.

SOLUTION The plastic centroid will lie on the $x$ axis because of symmetry. Apply the axial load $P_n$ at the plastic centroid, creating uniform strain. Using statics, equate the internal and external forces.

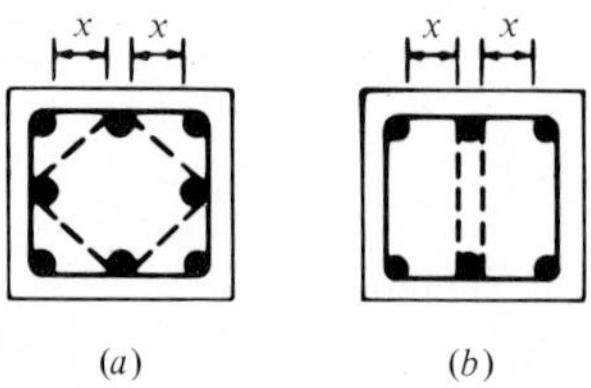

(a)

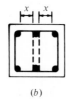

(b)

**Figure 7.24** Tie details: (a) eight-bar pattern, (b) six-bar pattern (the middle bars must be tied by the dashed stirrups if $x$ is greater than 6 in).

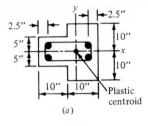

(a)

(b)

(c)

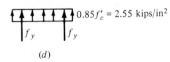

(d)

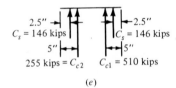

(e)

**Figure 7.25** Unsymmetrical column: (a) cross section, (b) elevation, (c) strain distribution, (d) stresses, (e) internal forces.

Break the concrete areas into two rectangular portions 10 by 10 and 10 by 20 in.

$$C_{c_1} = (200 \text{ in}^2)(2.55 \text{ kips/in}^2) = 510 \text{ kips}$$

$$C_{c_2} = (100 \text{ in}^2)(2.55 \text{ kips/in}^2) = 255 \text{ kips}$$

The force in the steel adjusted for concrete stress is

$$C_s = A_s(f_y - 0.85f'_c) = 2.54(57.45) = 146 \text{ kips}$$

The resultant vertical force is evaluated by $\Sigma F_y$

$$P_n = 255 \text{ kips} + 510 \text{ kips} + (146 \text{ kips})(2) = 1057 \text{ kips}$$

Locate $P_n$ by summing moments about the left edge of the column

$$P_n \bar{x} = \Sigma F_n x_n$$

$$(1057 \text{ kips})(\bar{x}) = (146 \text{ kips})(2.5 \text{ in}) + (255 \text{ kips})(5 \text{ in}) + (510 \text{ kips})(15 \text{ in}) + (146 \text{ kips})(17.5 \text{ in})$$

$$\bar{x} = 11.2 \text{ in}$$

## 7.5 STRENGTH OF SHORT COLUMNS FOR AXIAL LOAD AND MOMENT

The basic problem in column design is to establish the proportions of a reinforced concrete cross section whose theoretical strength, multiplied by a reduction factor, is just adequate to support the axial load and maximum moment in the column produced by factored design loads. This criterion can be summarized as

$$P_u \le \phi P_n \quad \text{and} \quad M_u \le \phi M_n$$

where $P_u$ and $M_u$ are the axial load and moment produced by factored service loads and $P_n$ and $M_n$ are the theoretical axial and bending strengths, also referred to as the nominal strengths.

Since the stress distribution produced by axial load and moment depends on the cross section's proportions, which are not initially known, column design cannot be carried out directly. Instead, the proportions of a cross section must be estimated and then investigated to determine whether its capacity is adequate for the design loads. Although various combinations of steel and concrete areas are possible to support a particular set of loads, Code restrictions on the percentage of steel and architectural limitations on dimensions usually result in similar cross-sectional dimensions regardless of who designs the column.

### Representation of Axial Load and Moment

To enable the designer to proportion a reinforced concrete column, the response of a cross section to various combinations of axial load and moment will be established. Axial load and moment can be represented in two ways. Figure 7.26a shows a cross section acted upon by an axial load $P_u$ and a moment $M_u$. $P_u$, acting through the centroid of the section, represents the load producing uniform compression of the member. $M_u$ represents the pure moment tending to cause rotation of the cross section about the centroidal axis. As an alternate statically equivalent representation (see any statics text) the load and moment can be replaced by a single load $P_u$ shifted a distance $e$ to the right of the centroidal axis so that $P_u e = M_u$ (Fig. 7.26b). In this second representation the relative size of the axial force and the moment is clearer. If $M_u$ is small relative to $P_u$, $e$ will be small. When $e$ is small, most of the cross section will be in compression and column behavior will dominate. On the other hand, if $e$ is large and $P_u$ lies well outside the cross section, compression is small and beam action dominates.

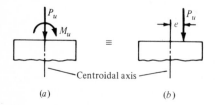

**Figure 7.26** Alternate representations of axial load and moment: (a) as a load and a moment, (b) as an eccentric load where $e = M_u/P_u$.

## Deformations Produced by Axial Load and Moment

As shown in Fig. 7.27a, a cross section compressed by an axial force acting through its centroid undergoes a uniform shortening. Under the action of a pure moment a cross section undergoes a linear rotation about its neutral axis; plane sections are assumed to remain plane. Therefore, when a cross section is loaded by both axial load and moment, superposition of the two linear deformations produces a resultant linear displacement and the variation of strain is correspondingly linear (Fig. 7.27c).

The strain distribution in part 1 of Fig. 7.27c represents the case in which the bending deformations on the tension side exceed the axial deformations. As a result, tension strains develop on part of the cross section. Wherever tensile strains develop, the concrete is assumed to be cracked.

If the axial load is large and the moment small, the compressive strains due to axial load will exceed the tensile strains produced by moment and compressive strains will exist over the entire section (see part 2 of Fig. 7.27c).

## Interaction Curves

As a first step in the development of a design procedure for columns, the simpler problem of establishing the axial load and moment associated with any strain distribution producing *failure* of a given cross section will be outlined. This procedure will permit the construction of an *interaction curve* that shows the variation of axial capacity with the magnitude of the load's eccentricity.

To illustrate the relationship between the strain distribution at failure and the applied loads, we shall assume in Example 7.7 that the axial force $P_n$ and the moment $M_n$ represent one of the many possible combinations of forces that produce failure. Failure of the cross section is assumed to occur when the compressive strain on any part of the cross section reaches 0.003. In this example, we arbitrarily consider the case where a tensile strain of 0.001 has developed on the left edge of the cross section when the compressive strain on the right edge reaches 0.003.

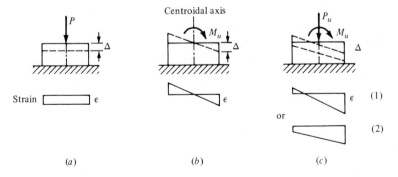

**Figure 7.27** Strains produced by axial load and moment: (*a*) axial load, (*b*) pure moment, (*c*) axial load and moment, (1) moment large relative to axial load; (2) axial load large relative to moment.

The stresses in the steel and concrete on a horizontal section are established from the stress-strain curves of the materials. Multiplying these stresses by the areas on which they act gives the internal forces. As was done for beams, computations are simplified by assuming that the actual nonlinear variation of compressive stress at failure can be represented by an equivalent rectangular stress block in which the concrete is uniformly stressed to $0.85f_c'$. Once the internal forces in the steel and in the concrete have been established, the two equations of equilibrium, $\Sigma F_y = 0$ and $\Sigma M = 0$, can be used to compute the applied force $P_n$ and moment $M_n$. The eccentricity of the load can then be computed from $e = M_n/P_n$.

**Example 7.7:** If the strain distribution at failure is specified for the column in Fig. 7.28, compute $P_n$ and $M_n$, the theoretical values of axial load and moment that produced the specified strain distribution. The column is reinforced with four no. 9 bars; $f_c' = 3$ kips/in², and $f_y = 60$ kips/in².

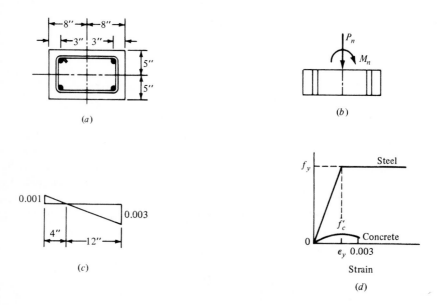

**Figure 7.28** Computation of internal forces when strain distribution at failure is specified: (a) cross section, (b) elevation, (c) specified strain, (d) stress-strain curves.

SOLUTION To establish the stresses in the steel on each side of the cross section of Fig. 7.28a we first compute the strains in the reinforcement from the geometry of the strain distribution (see Fig. 7.29a) and then determine the stresses by using the stress-strain curve of Fig. 7.28d.

$$\frac{\epsilon_{s_1}}{9 \text{ in}} = \frac{0.003}{12 \text{ in}}$$

$$\epsilon_{s_1} = 0.00225$$

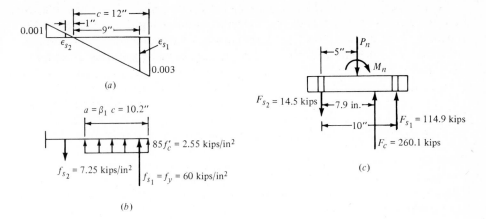

$$\frac{\epsilon_{s_2}}{1 \text{ in}} = \frac{0.001}{4 \text{ in}}$$

**Figure 7.29** (*a*) Strain, (*b*) stresses, (*c*) internal forces.

Since $\epsilon_{s_1} > \epsilon_y = 0.00207$, $f_{s_1} = f_y = 60$ kips/in$^2$. Also

$$\frac{\epsilon_{s_2}}{1 \text{ in}} = \frac{0.001}{4 \text{ in}}$$

$$\epsilon_{s_2} = 0.00025$$

Since $\epsilon_{s_2} < \epsilon_y$,

$$f_{s_2} = \epsilon_{s_2} E_s = 0.00025(29{,}000 \text{ kips/in}^2) = 7.25 \text{ kips/in}^2$$

Now we compute the force in the concrete $F_c$. The stress distribution in the concrete can be represented by a uniform stress of $0.85f'_c = 2.55$ kips/in$^2$ acting over a depth of stress block $a = \beta_1 c = 0.85(12) = 10.2$ in

$$F_c = ab(0.85f'_c) = 10.2(10)(2.55 \text{ kips/in}^2) = 260.1 \text{ kips}$$

The force in the tension steel $F_{s_2}$ is

$$F_{s_2} = A_s f_s = (2 \text{ in}^2)(7.25 \text{ kips/in}^2) = 14.5 \text{ kips}$$

To compensate for the additional force created by the inclusion of the area occupied by compression steel in the computation of $F_c$, the stress in the compression steel is reduced by $0.85f'_c$. This adjustment eliminates the need for deducting the area occupied by the steel when the centroid of the concrete compression zone is located to establish the position of $F_c$

$$F_{s_1} = A_s(f_{s_1} - 0.85f'_c) = 2(60 - 2.55) = 114.9 \text{ kips}$$

From the equilibrium of the free body of the column segment in Fig. 7.29*c*, compute $P_n$ and $M_n$

$$\overset{+}{\uparrow}\Sigma F = 0$$

$$0 = -P_n - 14.5 + 260.1 + 114.9$$

$$P_n = 360.5 \text{ kips}$$

Sum moments about any point; say the tension steel

$$\Sigma M = 0$$

$$0 = 360.5(5 \text{ in}) + M_n - 260.1(7.9 \text{ in}) - (114.9 \text{ kips})(10 \text{ in})$$

$$M_n = 1401.29 \text{ in} \cdot \text{kips} = 116.77 \text{ ft} \cdot \text{kips}$$

By holding the strain constant at 0.003 on one edge of the cross section and allowing the strain on the opposite edge to vary, other combinations of axial load and moment that produce failure of the cross section in Example 7.7 can be determined. When these pairs of axial load and moment producing failure are plotted, a curve showing the variation of axial capacity with moment results. This curve, termed an *interaction curve*, represents all possible combinations of axial load and moment that produce failure of the given cross section. Figure 7.30 shows the interaction curve for the cross section of Example 7.7. The values of $P_n$ and $M_n$ computed in Example 7.7 are represented by point $D$.

Points $A$, $B$, and $C$, at the extremities of the curve, closely define the shape of the interaction curve. Point $A$ represents a column carrying axial load only, $M_n = 0$. The magnitude of the axial force $P_n$ at point $A$ can be computed as follows:

$$P_n = P_0 = A_{st} f_y + (A_g - A_{st})(0.85 f_c') \tag{7.12}$$

$$= 4(60 \text{ kips/in}^2) + (160 - 4)(0.85)(3 \text{ kips/in}^2) = 637.8 \text{ kips}$$

Point $B$, the most distant point to the right, represents the combination of axial load and moment that produces a balanced failure; the concrete crushes at a strain of 0.003 just as the steel on the tension side reaches its yield point. A balanced failure is not a function of the area of reinforcing steel, as in a beam, but depends on the eccentricity of the axial load. Any combination of axial load and moment on the interaction curve between points $A$ and $B$, where the eccentricity of the load is less

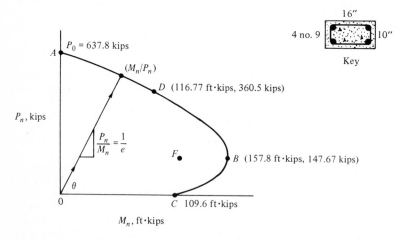

**Figure 7.30** Interaction curve for cross section of Example 7.7.

than the eccentricity for a balanced failure $e_b$, produces a failure that initiates by crushing the concrete on the compression side of the member. Between points $B$ and $C$, where moment is high and the eccentricity of the axial load is greater than the balanced eccentricity, beam behavior dominates and failure starts by yielding of the tension steel, followed by a shift of the neutral axis and a secondary compression failure in the concrete. Values of $P_n$ and $M_n$ producing a balanced failure are termed $P_b$ and $M_b$ and computed in Example 7.8.

Point $C$, located on the $M_n$ axis, represents the capacity of the cross section for moment only; the axial force equals zero. The value of moment represented by point $C$ is computed in Example 7.9.

In Fig. 7.30 a point such as $F$, located within the interaction curve and the reference axes, represents a combination of axial load and moment that can be supported by a column without producing failure.

If the load on a cross section of a short column increases from zero to its ultimate value while the eccentricity $e$ remains constant, the plot showing the variation of load and moment can be represented on the $P_n M_n$ coordinate plane by a straight line that slopes upward from the origin. Where the line intersects the interaction curve, the coordinates of the point of intersection $P_n$ and $M_n$ give the combination of force and moment producing failure. The slope of the line $\theta$ equals $P_n/M_n$ or $1/e$.

**Example 7.8:** (a) Compute the values of axial load and moment that produce a balanced failure of the cross section in Fig. 7.31a. (b) Determine the eccentricity of the load at failure. $f'_c = 3$ kips/in$^2$, and $f_y = 60$ kips/in$^2$. Reinforcement consists of four no. 9 bars; $A_s = 2$ in$^2$, and $A'_s = 2$ in$^2$.

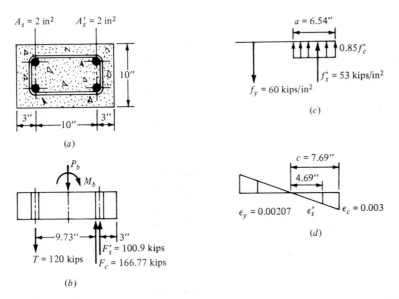

**Figure 7.31** Balanced failure: (a) cross section, (b) elevation with forces, (c) stresses, (d) strain at failure.

# 278 REINFORCED CONCRETE DESIGN

SOLUTION (a) From the strain distribution compute $c$ using similar triangles (Fig. 7.31d)

$$\frac{0.00207}{13-c} = \frac{0.003}{c} \quad \text{and} \quad c = 7.69 \text{ in}$$

$$a = \beta_1 c = 0.85(7.69 \text{ in}) = 6.54 \text{ in}$$

Compute strain in compression steel

$$\frac{0.003}{7.69 \text{ in}} = \frac{\epsilon'_s}{4.69 \text{ in}} \quad \text{and} \quad \epsilon'_s = 0.00183$$

$$f'_s = E_s \epsilon'_s = (29{,}000)(0.00183)$$

$$= 53 \text{ kips/in}^2$$

$$F_c = ab(0.85f'_c)$$

$$= 6.54(10)(0.85)(3) = 166.77 \text{ kips}$$

$$F'_s = A'_s(f'_s - 0.85f'_c)$$

$$= 2(53 - 2.55) = 100.9 \text{ kips}$$

$$T = f_y A_s = 60(2) = 120 \text{ kips}$$

Compute $P_b$; $\Sigma F_y = 0$

$$0 = -P_b - 120 \text{ kips} + 100.9 \text{ kips} + 166.77 \text{ kips}$$

$$P_b = 147.67 \text{ kips}$$

Compute $M_b$; $\Sigma M = 0$ about $T$ (Fig. 7.31b)

$$0 = M_b + (147.67 \text{ kips})(5 \text{ in}) - (100.9 \text{ kips})(10 \text{ in}) - (166.77 \text{ kips})(9.73 \text{ in})$$

$$M_b = 1893 \text{ in} \cdot \text{kips} = 157.8 \text{ ft} \cdot \text{kips}$$

(b) Compute eccentricity of load

$$e_b = \frac{M_b}{P_b} = 12.82 \text{ in}$$

**Example 7.9:** Compute the theoretical value of moment $M_n$ that will cause the 10 by 16-in cross section of Example 7.7 to fail. The section is reinforced with four no. 9 bars, $f'_c = 3$ kips/in$^2$, and $f_y = 60$ kips/in$^2$ (see Fig. 7.32).

SOLUTION Analyze by the trial method (see Sec. 3.11). Guess $c = 4$ in; then

$$a = \beta_1 c = 0.85(4) = 3.4 \text{ in}$$

Using the strain diagram, compute $\epsilon'_s$ and then $f'_s$ (Fig. 7.32d)

$$\epsilon'_s = \tfrac{1}{4}(0.003) = 0.00075 \text{ in/in}$$

$$f'_s = \epsilon'_s E_s = 0.00075(29{,}000) = 21.8 \text{ kips/in}^2$$

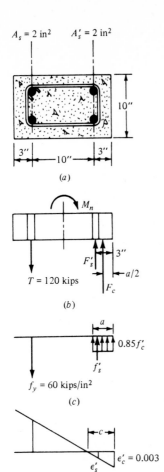

$A_s = 2$ in$^2$     $A'_s = 2$ in$^2$

10"

3"    3"

10"

(a)

$M_n$

$F'_s$    3"

$T = 120$ kips

$a/2$

$F_c$

(b)

a

$0.85 f'_c$

$f'_s$

$f_y = 60$ kips/in$^2$

(c)

c

$\epsilon'_c = 0.003$

$\epsilon'_s$

(d)

**Figure 7.32** Moment capacity: (a) cross section, (b) elevation, (c) stress, (d) strain at failure.

Compute $F'_s$ and $F_c$; compare with $T$

$$F'_s = (21.8 - 2.55)(2) = 38.5 \text{ kips}$$

$$F_c = 10(3.4)(2.55) = 86.7 \text{ kips}$$

$$F'_s + F_c = 125.2 \text{ kips} > T = 120 \text{ kips}$$

Therefore reduce $c$ slightly and repeat. Try $c = 3.8$ in.
      The second trial gives

$$F'_s + F_c = 119 \text{ kips} \approx 120 \text{ kips}$$

Using $F'_s = 37$ kips and $F_c = 83$ kips, compute the internal moment by summing forces about $T$

$$M_n = F'_s 10 + F_c\left(13 - \frac{a}{2}\right)$$

$$= 37(10 \text{ in}) + (83 \text{ kips})\left(13 - \frac{3.23}{2}\right) = 1315 \text{ in} \cdot \text{kips} = 109.6 \text{ ft} \cdot \text{kips}$$

**The shape of the interaction curve**    As previously noted, the interaction curve has two distinct regions (Fig. 7.30). In the region between points $A$ and $B$, failure initiates by crushing of the concrete on the compression side of the member. In this region the capacity of the member to carry axial load decreases almost linearly with increase in moment or equivalently with increase in eccentricity of the applied load. Such a response seems logical. If a portion of the member's internal strength is used to carry moment, less is available to support axial load.

Between points $B$ and $C$ bending dominates, and a member with symmetrical reinforcement behaves like an underreinforced beam. Since failure initiates by yielding of the tension steel while the strain in the concrete is below its ultimate value, the application of a small compression force to the cross section as the steel is about to yield in tension prevents yielding by reducing the tensile strain in the steel. Now additional moment can be added until the steel is strained again to its yield point; therefore in this region of the interaction curve, the capacity of the cross section to carry moment increases moderately with an increase in the axial load. This aspect of behavior is reflected in the shape of the interaction curve, which now slopes upward and to the right from point $C$ to the balanced failure point at $B$ (Fig. 7.30).

The interaction curve in Fig. 7.30 represents the strength of one particular reinforced cross section for all possible combinations of axial load and moment that produce failure. If the cross section had been reinforced with a larger area of steel, say four no. 11 bars, the general shape of the interaction curve would be the same but the strength of the cross section would increase. If plotted to the same scale as the cross section reinforced with the four no. 9 bars, its interaction curve would lie above the curve representing the strength of the section reinforced with four no. 9 bars (Fig. 7.33).

ACI Code §9.3.2 specifies that a reduction factor of $\phi = 0.7$ for tied and $\phi = 0.75$ for spiral columns be applied to the coordinates of the theoretical interaction curve to produce a design interaction curve. For example, in Fig. 7.34 a typical point $D'$ on the design curve is established by multiplying both coordinates of point $D$ on the theoretical curve by the appropriate reduction factor. Recognizing that a beam-column behaves like a beam when the axial load is small, the ACI Code specifies that the reduction factor $\phi$ may be increased linearly to 0.9 as $\phi P_n$ decreases from $0.1 f'_c A_g$ to zero. This increase in $\phi$ applies to members in which $f_y$ does not exceed

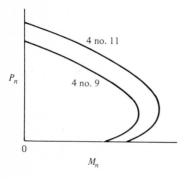

**Figure 7.33** Influence of the reinforcement ratio on the proportions of the interaction curve.

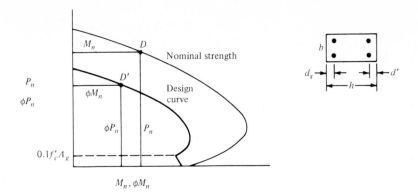

**Figure 7.34** The design interaction curve.

60 kips/in², the reinforcement pattern is symmetrical, and $h - d' - d_s > 0.7$, ($d_s$ defined in Fig. 7.34). For other members which do not satisfy the above requirements, $\phi$ may be increased linearly to 0.9 as $\phi P_n$ decreases from $0.1f'_c A_g$ or $\phi P_b$, whichever is smaller, to zero. The effect of this modification is to produce a break in the slope of the design interaction curve near the bottom of the curve.

A final modification is made to the design interaction curve by cutting off the upper part of it with a horizontal line (line $ab$ in Fig. 7.35). The position of the line $\phi P_{n, \max}$ is given by

$$\text{Tied columns:} \quad \phi P_{n, \max} = 0.8\phi[0.85f'_c(A_g - A_{st}) + f_y A_{st}]$$

$$\phi = 0.7$$

$$\text{Spiral columns:} \quad \phi P_{n, \max} = 0.85\phi[0.85f'_c(A_g - A_{st}) + f_y A_{st}]$$

$$\phi = 0.75$$

The purpose of this cutoff is to reduce the axial load permitted on columns that carry pure compression or compression and a small moment. This reduction is made for two reasons: (1) All columns are subjected to moment produced by accidental eccentricities not considered in their analysis. By limiting the axial load permitted on the cross section, a portion of the column's strength is reserved for

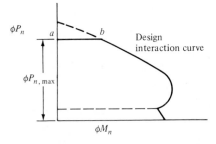

**Figure 7.35** Reduction in axial capacity of columns supporting loads with small eccentricities: Tied column, $\phi P_{n, \max} = .8\phi[0.85f'_c(A_g - A_{st}) + f_y A_{st}]$ and $\phi = .7$; spiral column, $\phi P_{n, \max} = .85\phi[0.85f'_c(A_g - A_{st}) + f_y A_{st}]$ and $\phi = .75$.

these moments. (2) The strength of columns that carry pure compression for long periods of time is lower than that of columns loaded for short durations. The cutoff line reduces the pure axial load permitted on spiral and on tied columns by an additional 15 and 20 percent respectively.

Column design is normally carried out with interaction curves. Curves for a large variety of rectangular and circular cross sections are published in Ref. 7. Examples of them are shown in Figs. 7.36 to 7.43.

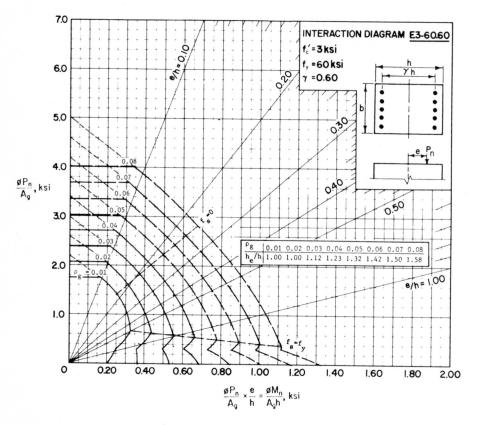

**Figure 7.36** Interaction diagram.[7]

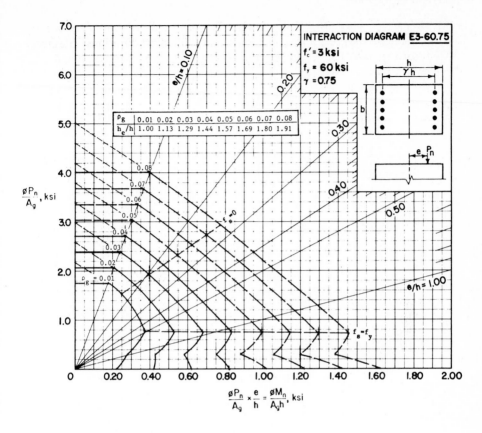

**Figure 7.37** Interaction diagram.[7]

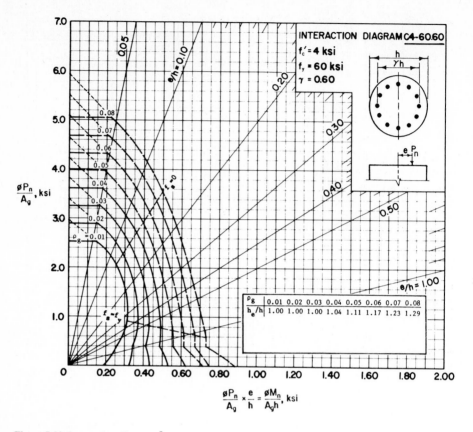

**Figure 7.38** Interaction diagram.[7]

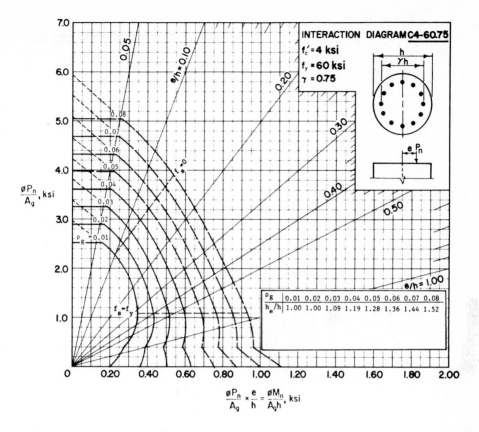

**Figure 7.39** Interaction diagram.[7]

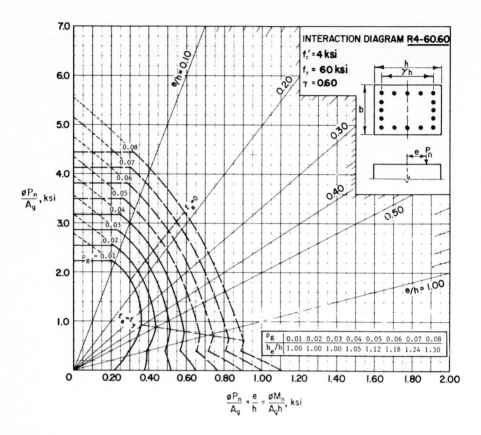

**Figure 7.40** Interaction diagram.[7]

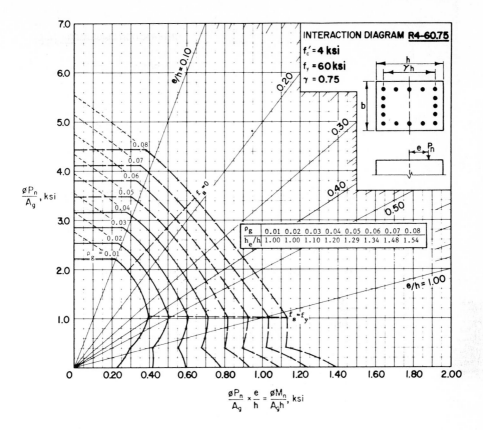

**Figure 7.41** Interaction diagram.[7]

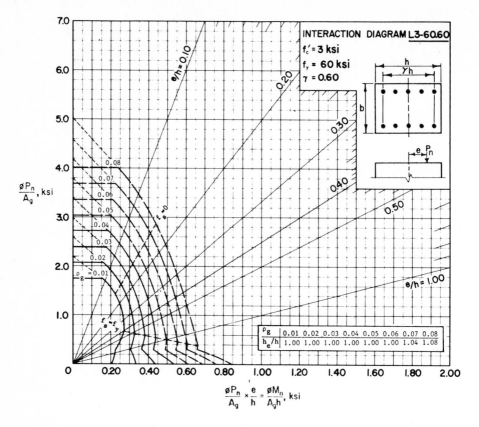

**Figure 7.42** Interaction diagram.[7]

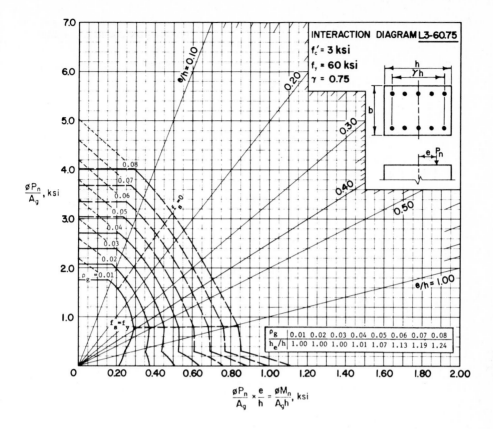

**Figure 7.43** Interaction diagram.[7]

## 7.6 DESIGN PROCEDURE FOR A SHORT COLUMN

For columns that are components of rigid-jointed structures the analysis and design are interrelated. Since the distribution of forces in an indeterminate structure depends on the relative stiffness of the members, the dimensions of members must be estimated before the structure can be analyzed. After both the axial load and the maximum moment in the column due to factored loads are determined, the reinforcement is selected. If the area of steel falls within the 1 to 8 percent limits established by the ACI Code, the design is complete, but if the initial dimensions prove to be either unacceptably small or large, the dimensions must be readjusted and the analysis and design repeated. For most building frames, small changes in column dimensions do not produce significant changes in the column forces; therefore, an acceptable cross section can usually be established in one or two trials. To complete the design, the ties enclosing the steel are sized and the shear strength of the cross section is checked. Although shear forces are typically small in interior columns, they may be significant in exterior columns bent into double curvature.

## Use of ACI Interaction Curves to Proportion Columns

The dimensions of a column cross section and the required area of reinforcement required to support a specific combination of axial load and moment due to factored loads can be quickly established by using the interaction curves contained in Ref. 7, which contains interaction curves for tied and spiral columns with common patterns of reinforcement. To permit these interaction curves to be used for cross sections of any dimension, the ordinates are plotted in terms of $\phi P_n/A_g$ and the abscissas in term of $\phi P_n e/A_g h$ instead of $\phi P_n$ and $\phi M_n$.

The properties of the materials $f'_c$ and $f_y$ and the parameter $\gamma$, a function of the distance between rows of reinforcement, determine the appropriate design aid. To compute the coordinates required to enter the design aid, $\phi P_n$, the axial strength of the cross section, is set equal to $P_u$, $e$ is computed by $M_u/P_u$, and $A_g$ (the gross area) is estimated. The designer then reads the required value of the reinforcement ratio $\rho_g$ from the curve that lies at the intersect of the coordinates. After $\rho_g$ has been established, $A_{st}$, the total area of steel, is computed from $A_{st} = \rho_g A_g$. The design is then completed by selecting bars, checking clearances, and sizing the ties or the spiral. The use of design aids for column design is illustrated in Examples 7.10 and 7.11.

**Example 7.10: Design of a rectangular column** Rectangular tied columns support the building frame shown in Fig. 7.44. Design the column between the second and third floors. The clear distance $l_u$ between floors equals 10 ft. The effective-length factor $k$ may be taken as 0.8 for bending in the plane of the frame and 0.75 for bending out of plane; $f'_c = 3$ kips/in² and $f_y = 60$ kips/in². The in-plane moments are shown in Fig. 7.44. Moments are zero perpendicular to the plane of the frame. Design the column for $P_u = 320$ kips, $M_u = 100$ ft · kips.

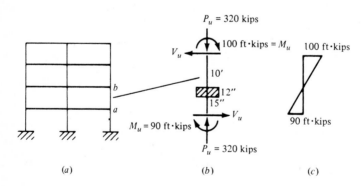

**Figure 7.44** (a) Building frame, (b) column ab, (c) moment curve.

SOLUTION Select a trial area. Estimate average stress at 2 kips/in²

$$\text{Trial } A_g = \frac{P_u}{2 \text{ kips/in}^2} = \frac{320}{2} = 160 \text{ in}^2$$

Try a 12- by 15-in cross section, giving $A_g = 180$ in². See Fig. 7.44b for orientation.

Check the slenderness ratios about both axes to verify that the column is short. For bending in the plane of the frame, $h = 15$ in, $M_1 = 90$ ft·kips, and $M_2 = 100$ ft·kips

$$\frac{kl_u}{r} = \frac{0.8(10)(12)}{0.3(15)} = 21.33$$

$$34 - 12\frac{M_1}{M_2} = 34 - 12\frac{-90}{100} = 44.8$$

$$21.33 < 44.8 \qquad \text{column short}$$

For bending perpendicular to the frame, $h = 12$ in, $M_1 = 0$, and $M_2 = 0$

$$\frac{kl_u}{r} = \frac{0.75(10)(12)}{0.3(12)} = 25$$

$$34 - 12\frac{M_1}{M_2} = 34 - \frac{0}{0} = 34$$

$$25 < 34 \qquad \text{column short}$$

Compute coordinates for the ACI interaction curves. Set $\phi P_n = P_u$

$$\frac{\phi P_n}{A_g} = \frac{P_u}{A_g} = \frac{320}{180} = 1.78 \text{ kips/in}^2$$

$$e = \frac{M_u}{P_u} = \frac{(100 \text{ ft·kips})(12)}{320 \text{ kips}} = 3.75 \text{ in}$$

$$\frac{P_n}{A_g}\frac{e}{h} = 1.78\frac{3.75 \text{ in}}{15 \text{ in}} = 0.445$$

Estimate $\gamma h = 9$ in, then $\gamma = 0.6$ (see Fig. 7.45). From Fig. 7.36 read $\rho_g = 0.034$

$$A_{st} = \rho_g A_g = 0.034(180) = 6.12 \text{ in}^2$$

Use four no. 11 bars

$$A_{st,\,\text{sup}} = 6.25 \text{ in}^2$$

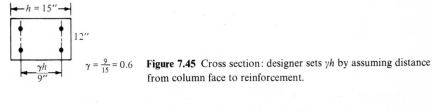

$\gamma = \frac{9}{15} = 0.6$   **Figure 7.45** Cross section: designer sets $\gamma h$ by assuming distance from column face to reinforcement.

Maximum spacing of ties. Use no. 4 ties because of no. 11 bars. Use smallest value of

$$16 \times \text{longitudinal } d_b = 16(\tfrac{11}{8}) = 22 \text{ in}$$

$$48 \times \text{tie diameters} = 48(\tfrac{1}{2}) = 24 \text{ in}$$

$$\text{Least dimension} = 12 \text{ in controls}$$

Check shear to determine if it controls tie spacing. Shear will normally be small in interior columns but may be significant in exterior columns. Compute the column shear $V_u$ by summing moments about the bottom of the column

$$\Sigma M = 0$$

$$V_u(10 \text{ ft}) = 100 \text{ ft} \cdot \text{kips} + 90 \text{ ft} \cdot \text{kips}$$

$$V_u = 19 \text{ kips}$$

Compute the shear capacity of concrete $V_c$ with Eq. (4.11)

$$V_c = 2\left(1 + \frac{N_u}{2000A_g}\right)\sqrt{f'_c}b_w d$$

Allowing 3 in for cover, $d = h - 3 = 12$ in

$$= 2\left(1 + \frac{320{,}000}{2000(180)}\right)\frac{\sqrt{3000}}{1000}\,12(12) = 29.8 \text{ kips}$$

$$V_u > \frac{\phi V_c}{2} = \frac{0.85(29.8)}{2} = 12.7 \text{ kips}$$

Use no. 4 at $d/2 = 6$ in.

**Example 7-11. Design of a spiral column.** Design a short circular column to carry $P_u$ = 424 kips (1.886 kN) and $M_u$ = 127.2 ft · kips (172.5 kN · M); $f'_c$ = 4 kips/in² (27.58 MPa), and $f_y$ = 60 kips/in² (413.7 MPa). The steel is to be contained within a $\frac{3}{8}$-in (9.5-mm) spiral (see Fig. 7.46).

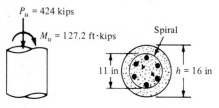

Cross section          **Figure 7.46**

SOLUTION Compute a trial area. Estimate the average concrete stress at 2 kips/in² (13.79 MPa)

$$A_g = \frac{P_u}{f_c} = \frac{424}{2 \text{ kips/in}^2} = 212 \text{ in}^2 \ (136.782 \text{ mm}^2)$$

Use a 16-in-diameter section; $A_g$ = 201 in²; allowing 2.5 in from face of column to steel, $\gamma = \frac{11}{16} = 0.69$.

Compute variables for ACI interaction curves

$$e = \frac{M_u}{P_u} = \frac{(127.2)12}{424} = 3.6 \text{ in} \ (91.44 \text{ mm})$$

$$\frac{e}{h} = \frac{3.6}{16} = 0.225$$

Set $\phi P_n = P_u$; then

$$\frac{\phi P_n}{A_g} = \frac{424}{201} = 2.11 \text{ kips/in}^2 \qquad \frac{\phi P_n}{A_g}\frac{e}{h} = 0.475 \text{ kips/in}^2$$

Interpolating between Figs. 7.38 and 7.39, read $\rho_g = 0.041$

$$A_{st} = \rho_g A_g = 0.041(201) = 8.24 \text{ in}^2$$

Use seven no. 10 = 8.86 in$^2$ (5716.5 mm$^2$). A $\frac{3}{8}$-in-diameter spiral with a pitch of 1.5 in is required (see Example 7.5).

## Trial Method

When an interaction curve is not available because a cross section has an unusual shape or a nonstandard pattern of reinforcement, cross sections can be designed by a method of trials, as described below.

**Step 1** Select a trial section and analyze the structure to establish the design forces $P_u$ and $M_u$ produced by factored service loads. An approximate trial area can be established by dividing $P_u$ by $0.5f'_c$ when the eccentricity does not exceed half the depth of the section. For larger eccentricities use 0.3 to $0.4f'_c$.

**Step 2** Establish a reinforcement pattern. The area of steel $A_{st}$ might be taken initially as 3 or 4 percent of the gross area of the cross section.

**Step 3** Compute the eccentricity of the axial force $e = M_u/P_u$.

**Step 4** Guess a strain distribution associated with failure, compute the internal forces, and solve for $P_n$ and $M_n$ using the two equations of statics

$$\Sigma F_y = 0 \qquad \Sigma M = 0$$

(see Example 7.7).

**Step 5** Compute $e = M_n/P_n$ and $\phi P_n$. If $P_u < \phi P_n$ and the eccentricity of $P_n$ is equal to or slightly greater than the eccentricity of $P_u$, the cross section is adequate. If a large difference in eccentricities exists between $P_u$ and $P_n$, modify the strain distribution and repeat the analysis. If the axial strength of the section $\phi P_n$ is too large or too small, modify the cross section by changing the area of steel or the dimensions and repeat the analysis and design. The method is illustrated in Example 7.12.

**Example 7.12: Design of a column by the trial method** Select a square cross section for a short tied column that is to support design loads of $P_u = 372$ kips and $M_u = 124$ ft · kips; $f'_c = 5$ kips/in$^2$, and $f_y = 60$ kips/in$^2$.

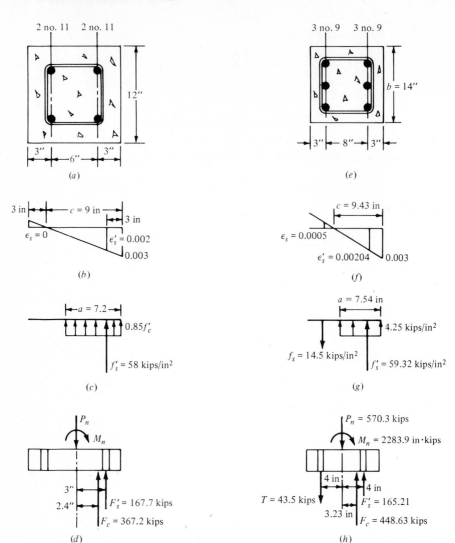

**Figure 7.47** Design by the trial method: (a) initial trial section (designer estimates dimensions and area of reinforcement); (b) strain distribution assumed by designer; (c) stresses based on assumed strain distribution; (d) forces in equilibrium, $P_n$ and $M_n$ computed from $\Sigma F_y = 0$ and $\Sigma M = 0$; (e) revised section, dimensions increased to raise capacity; (f) new strain distribution assumed by designer; (g) stresses; (h) forces.

SOLUTION

1. Assuming that $P_u$ produces an average compressive stress on the cross section of $0.5f'_c$, select a trial area

$$A_g = \frac{P_u}{0.5f'_c} = \frac{372 \text{ kips}}{2.5 \text{ kips/in}^2} = 148.8 \text{ in}^2$$

Try a 12- by 12-in section, giving $A_g = 144 \text{ in}^2$.

2. Compute the eccentricity of $P_u$

$$e = \frac{M_u}{P_u} = \frac{(124 \text{ ft} \cdot \text{kips})(12)}{372 \text{ kips}} = 4 \text{ in}$$

3. Guess the area of reinforcement. Try 4 percent steel

$$A_{st} = 0.04(144 \text{ in}^2) = 5.76 \text{ in}^2$$

Use four no. 11; then $A_{st} = 6.25 \text{ in}^2$. For trial section see Fig. 7.47a.
4. Assume a strain distribution producing failure (Fig. 7.47b).

$$\epsilon_s' = \tfrac{6}{9}(0.003) = 0.002$$

Establish the internal stresses from the strain distribution. Compute the internal forces in the steel and concrete

$$a = \beta_1 c \quad \text{and} \quad \beta_1 = 0.85 - 0.05(f_c' - 4) = 0.8$$

$$a = 0.8(9 \text{ in}) = 7.2 \text{ in}$$

$$f_s' = \epsilon_s' E_s = 0.002(29,000) = 58 \text{ kips/in}^2$$

$$F_c = 0.85 f_c' ab = 0.85(5)(7.2)(12) = 367.2 \text{ kips}$$

$$F_s' = (f_s' - 0.85 f_c')\dagger A_s' = (58 - 4.25)(3.12) = 167.7 \text{ kips}$$

Compute $P_n$ from $\Sigma F_y = 0$ (see Fig. 7.47d)

$$P_n = F_c + F_s' = 367.2 \text{ kips} + 167.7 \text{ kips} = 534.9 \text{ kips}$$

Compute $M_n$ from $\Sigma M = 0$ about centerline

$$0 = M_n - F_c(2.4 \text{ in}) - F_s'(3 \text{ in})$$

$$M_n = 1384.4 \text{ in} \cdot \text{kips}$$

5. Compute eccentricity of $P_n$

$$e = \frac{M_n}{P_n} = \frac{1384.4 \text{ in} \cdot \text{kips}}{534.9 \text{ kips}} = 2.59 \text{ in}$$

The capacity of the cross section for axial load is

$$\phi P_n = 0.7(534.9 \text{ kips}) = 374.4 \text{ kips}$$

The solution in step 5 indicates that the assumptions in step 4 do not produce a satisfactory design. Although the capacity of the cross section, 374.4 kips, exceeds the design load $P_u = 372$ kips, the eccentricity of $\phi P_n$ is less than the 4-in eccentricity of $P_u$; therefore a new strain distribution must be assumed that implies a larger moment. If a larger portion of the capacity must be used to carry moment, less is available to support axial load. For a second trial the capacity of the cross section will be raised by increasing the dimensions to 14 by 14 in and the area of the steel to three no. 9 bars on each side. For the next trial, the section and adjusted strain distribution are shown in Fig. 7.47e and f.

† Reduce stress in the steel to account for the inclusion of $0.85 f_c' A_s'$ that was included in $F_c$.

6. Repeating the analysis gives (see Fig. 7.47g and h for stresses and forces)

$$P_n = 570.3 \text{ kips} \qquad M_n = 2283.9 \text{ in} \cdot \text{kips}$$

$$\phi P_n = 0.7(570.3 \text{ kips}) = 399.2 \text{ kips} > P_u = 372 \text{ kips}$$

$$e = \frac{M_n}{P_n} = \frac{2283.9 \text{ in} \cdot \text{kips}}{570.3 \text{ kips}} = 4.0 \text{ in} \qquad \text{OK}$$

Since the axial strength $\phi P_n$ exceeds the factored load $P_u$, the cross section is adequate for the design loads. The comparison of axial strength requires, of course, that $P_n$ act at an eccentricity equal to or greater than $P_u$.

7. Verify that

$$P_u \leq 0.8\phi[A_{st} f_y + 0.85f'_c(A_g - A_{st})] \qquad \text{ACI Code §10.3.5.1}$$

$$\leq 0.8(0.7)[6(60) + 0.85(5)(196 - 6)]$$

$$372 \text{ kips} < 653.8 \text{ kips} \qquad \text{OK}$$

## 7.7 DESIGN OF LONG COLUMNS

### Introduction

As shown in Fig. 7.5, the internal moment in a long column is the sum of the secondary moment (the $P\Delta$ component) and the primary moment produced by transverse loads or end moments. To evaluate the secondary moment the deflection of the column's longitudinal axis must be evaluated. Once the deflected shape is known, the $P\Delta$ moment can be evaluated at any section.

One method for determining the deflected shape of the column's axis is to integrate the basic differential equation of the elastic curve given by

$$\frac{d^2 y}{dx^2} = \frac{M(x)}{EI} \tag{7.19}$$

This *double-integration method* requires that the internal moment $M(x)$ be expressed in terms of the applied loads. Deflections can also be evaluated by applying a variety of numerical procedures to solve Eq. (7.19).

Although various procedures to analyze elastic beam-columns for a variety of loads and boundary conditions exist, only an approximate analysis of a reinforced concrete beam-column is possible because properties of the cross section (such as $E$ and $I$) cannot be established with precision. Flexural cracking produces variations in effective cross section; creep due to the sustained portion of the applied loads causes deflections and curvature to increase with time; and the modulus of elasticity $E_c$ varies with the level of the compressive stress because the stress-strain curve for concrete is nonlinear (Fig. 7.48).

While ACI Code §10.10.1 suggests that the moments in slender columns be determined by a second-order analysis that accounts for the influence of axial load, creep, and variations in cross section on the member's bending stiffness,

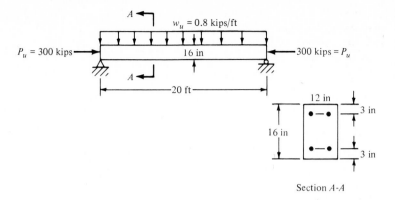

Section A-A

**Figure 7.57**

The column is long; magnify moments. The maximum moment equals the midspan moment multiplied by the magnification factor

$$M_c = \frac{w_u l^2}{8}\,\delta = \frac{0.8(20^2)}{8}(2.46) = 98.4 \text{ ft}\cdot\text{kips}$$

where

$$\delta = \frac{C_m}{1 - P_u/\phi P_c} = \frac{1}{1 - 300/[0.7(721.6)]} = 2.46$$

and

$$P_c = \frac{\pi^2 EI}{l_u^2} = 721.6 \text{ kips}$$

where

$$EI = \frac{EI_g/2.5}{1 + \beta} = \frac{57,000\sqrt{4000}(\frac{12}{12})(16^3)/2.5}{1 + 0.4} = 4.21 \times 10^6 \text{ kips}\cdot\text{in}^2$$

$$l_u = 20 \times 12 \text{ in} = 240 \text{ in}$$

**Example 7.18: Design of a slender column** The tied column in Fig. 7.58 is a member of a braced frame. The factored moments shown produce bending about the y-y axis.

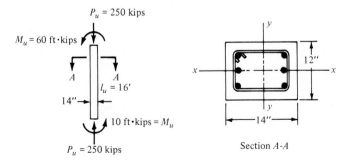

Section A-A

**Figure 7.58**

Moments about the $x$-$x$ axis are zero. The effective length factor equals 0.9 with regard to bending about the $y$-$y$ axis and 0.85 with respect to bending about the $x$-$x$ axis. Determine the longitudinal reinforcement required to support the factored loads. $f'_c = 3$ kips/in$^2$, $f_y = 60$ kips/in$^2$, and $\beta_d = 0.6$. Assume that a preliminary analysis has indicated a 12- by 14-in trial cross section.

SOLUTION The strength of the column will be checked independently for bending about both principal axes. Classify the column as long or short.

Bending about the $y$-$y$ axis

$$\frac{kl_u}{r} = \frac{0.9(16 \text{ ft})(12)}{0.3(14 \text{ in})} = 41.14$$

$$34 - 12\frac{M_1}{M_2} = 34 - 12\frac{-10}{60} = 36 \qquad 41.14 > 36$$

The column is long. Consider secondary moment.

Bending about the $x$-$x$ axis

$$\frac{kl_u}{r} = \frac{0.85(16 \text{ ft})(12)}{0.3(12 \text{ in})} = 45.3$$

$$34 - 12\frac{M_1}{M_2} = 34 - 12\frac{0}{M_2} = 34 \qquad 45.3 > 34$$

The column is long. Consider secondary moment.

Analysis for bending about the $y$-$y$ axis. Design for

$$M_1 = -10 \text{ ft} \cdot \text{kips} \qquad M_2 = 60 \text{ ft} \cdot \text{kips}$$

Verify that $M_2$ exceeds the minimum moment specified in ACI Code §10.11.5.4. This moment equals

$$P_u(0.6 + 0.03h) \text{ in} = 250 \text{ kips} \frac{[0.6 + (0.03 \times 14 \text{ in})]}{12} = 21.25 \text{ ft} \cdot \text{kips}$$

$$60 \text{ ft} \cdot \text{kips} > 21.25 \text{ ft} \cdot \text{kips} \qquad \text{use } 60 \text{ ft} \cdot \text{kips}$$

Compute $M_c = \delta M_2$

$$\delta = \frac{C_m}{1 - P_u/\phi P_c} \geq 1$$

$$C_m = 0.6 + 0.4\frac{M_1}{M_2} = 0.6 + 0.4\frac{-10}{60} = 0.533 > 0.4 \qquad \text{OK}$$

$$P_c = \frac{\pi^2 EI}{(kl_u)^2}$$

$$E_c = \frac{57,000\sqrt{3000}}{1000} = 3.12 \times 10^3 \text{ kips/in}^2$$

$$EI = \frac{E_c I_g/2.5}{1 + \beta_d} \qquad \text{Eq. (7.27)}$$

$$P_c = \frac{[\pi^2(3.12 \times 10^3)(2744 \text{ in}^4)/2.5]/(1 + 0.6)}{[0.9(16)(12)]^2} = 707.4 \text{ kips}$$

$$\delta = \frac{C_m}{1 - P_u/\phi P_c} = \frac{0.533}{1 - (250 \text{ kips})/(0.7)(707.4)} = 1.08 > 1 \qquad \text{OK}$$

$$M_c = \delta M_2 = 1.08(60 \text{ ft} \cdot \text{kips}) = 64.8 \text{ ft} \cdot \text{kips}$$

Using ACI interaction curves, determine $\rho_g$

$$e = \frac{M_c}{P_u} = \frac{64.8(12)}{250 \text{ kips}} = 3.1 \text{ in}$$

Set $\phi P_n = P_u$

$$\frac{\phi P_n}{A_g} = \frac{250 \text{ kips}}{12(14)} = 1.49 \text{ kips/in}^2 \qquad \frac{\phi P_n}{A_g} \frac{e}{h} = 1.49 \frac{3.1}{14} = 0.33$$

Estimate $\gamma h = 9$ in. $\gamma = \frac{9}{14} = 0.64$. Interpolate between Figs. 7.36 and 7.37. Read $\rho_g = 0.02$, $A_{st} = \rho_g A_g = 0.02(12)(14) = 3.36$ in$^2$. Use six no. 7. $A_{st} = 3.61$ in$^2$.

Analyze for bending about the x-x axis. $M_u = 0$; use

$$M_u = P_u(0.6 + 0.03h) = (250 \text{ kips})[0.6 + 0.03(12)](\tfrac{1}{12}) = 20 \text{ ft} \cdot \text{kips}$$

Compute $M_c = \delta M_2$

$$\delta = \frac{C_m}{1 - P_u/\phi P_c} \qquad C_m = 0.6 + 0.4 \frac{M_1}{M_2}$$

Since $M_1 = M_2 = 0$, use $M_1/M_2 = 1$ (ACI Code §10.11.5.4); then

$$C_m = 0.6 + 0.4 = 1 \qquad P_c = \frac{\pi^2 EI}{(kl)^2}$$

$$I = \frac{bh^3}{12} = \frac{14(12^3)}{12} = 2016 \text{ in}^4$$

$$E_c = 57{,}000\sqrt{3000} = 3.12 \times 10^6 \text{ lb/in}^2$$

$$EI = \frac{E_c I_g/2.5}{1 + \beta_d} = 1572.5 \times 10^3 \text{ kips} \cdot \text{in}^2$$

$$P_c = \frac{\pi^2(1572.5 \times 10^3)}{[0.85(16)(12)]^2} = 582.71 \text{ kips}$$

$$\delta = \frac{1}{1 - 250/[0.7(582.71)]} = 2.58$$

$$M_c = 2.58(20) = 51.6 \text{ ft} \cdot \text{kips}$$

Compute eccentricity of load

$$e = \frac{M_u}{P_u} = \frac{51.6(12)}{250} = 2.48 \text{ in}$$

Design aids (see Figs. 7.42 and 7.43) indicate that six no. 8 bars are required.

**Example 7.19** The plan of a multistory building is shown in Fig. 7.59. In Table 7.1 the column loads and the buckling loads for each of the three sizes of columns used are given. Determine the moment for which column no. 1 must be designed if the frame is unbraced.

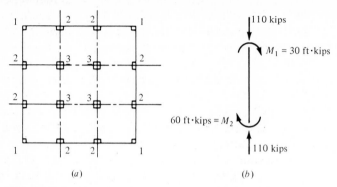

**Figure 7.59** (a) Floor plan, (b) free-body diagram of column 1, showing applied loads.

SOLUTION    Compute the magnification factor for sidesway of all columns. Because of symmetry, stiffness the same in both principal directions

$$\delta = \frac{C_m}{1 - \dfrac{\Sigma P_u}{\phi \Sigma P_c}} = \frac{1}{1 - \dfrac{4(110) + 8(200) + 4(400)}{0.7[4(420) + 8(715) + 4(1055)]}} = 1.81$$

Compute the magnification factor for column no. 1 if no sidesway occurs because other columns provide lateral restraint

$$\delta = \frac{C_m}{1 - P_u/\phi P_c}$$

$$C_m = 0.6 + \frac{0.4 M_1}{M_2} = 0.6 + 0.4 \frac{-30}{60} = 0.4$$

$$\delta = \frac{0.4}{1 - 110/[0.7(1170)]} = 0.46 \text{ use } 1$$

Sidesway governs; use $\delta = 1.81$.

$$M_c = \delta M_2 = 1.81(60 \text{ ft} \cdot \text{kips}) = 108.6 \text{ ft} \cdot \text{kips}$$

**Table 7.1    Properties of columns**

| No. | $P_u$ kips | $I$ by Eq. (7.27) $\beta_d = 0$ | $P_c$, kips Sidesway | No sidesway |
|---|---|---|---|---|
| 1 | 110 | 1000 | 420† | 1170† |
| 2 | 200 | 1920 | 715 | |
| 3 | 400 | 3200 | 1055 | |

† The effective-length factor $k$ in the sidesway case is greater than 1. For the braced case $k$ is less than 1.

## 7.8 BIAXIAL BENDING

### Introduction

As the position of live load on a floor varies, building columns may be subject to loading patterns that produce biaxial bending, i.e., bending about both principal axes of the cross section. Nevertheless, to simplify design, code provisions specify loading patterns that produce uniaxial bending in most building columns. Corner columns, routinely designed for biaxial bending, are the exception.

 Columns with a circular cross section that are subject to biaxial bending are excluded from this discussion because the applied moments can be combined vectorially to produce a single resultant moment and the cross section designed as if bending existed in one direction only. This procedure is possible because the bending strength of a circular cross section is the same in all directions thanks to its symmetry with regard to any centroidal axis (Fig. 7.60).

### Design of Rectangular Sections for Biaxial Bending and Axial Load

Designing a rectangular column cross section for biaxial bending and axial load is complicated because the position and direction of the neutral axis are difficult to establish (Fig. 7.61). Therefore the trial method used to design columns for uniaxial bending, as in Example 7.12, would not lead to an expedient solution. Furthermore, since the strain on the cross section varies linearly in both directions, considerable computation time is required to establish the relationship between stress and strain at all points on the cross section.

 The design procedure presented in this section is based on expressing the required uniaxial bending strength about each principal axis in terms of the design moments about both principal axes. The relationship can be developed by using the three-dimensional failure surface generated by interaction curves for various ratios of biaxial eccentricity. In Fig. 7.62 $ECG$, the line of intersection of the failure surface with the $\phi P_n - \phi M_{nx}$ plane represents the interaction curve for uniaxial bending about the x-x axis. Similarly, curve $EAF$ represents the interaction curve for uniaxial bending about the y-y axis. The variation of biaxial bending strength with axial load for a cross section loaded by a constant ratio of $M_{ux}/M_{uy}$ is given by the interaction curve $EDH$.

 In Fig. 7.62 the distance $OF$ is equal to the distance $OG$ only when the cross section is square and is reinforced with the same number of bars on each of its four sides.

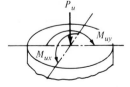

**Figure 7.60** Resultant moment for a circular cross section loaded by moments from two perpendicular directions given as

$$M_u = \sqrt{M_{ux}^2 + M_{uy}^2}.$$

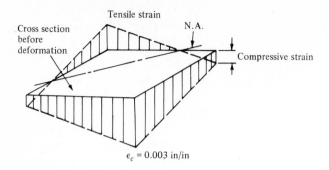

Figure 7.61 Variation of strains on the cross section of a column carrying axial load and moments acting about both principal axes.

A design procedure[6] using the failure surface can be developed by considering the variation of the biaxial moment capacity at a particular value of axial load. For example, curve $ADC$ in Fig. 7.62, cut by a horizontal plane that lies a vertical distance $P'_u$ above the plane of the horizontal reference axes, represents combinations of moment $M_{ux}$ and $M_{uy}$ that produce failure when the cross section is loaded by the axial force $P'_u$. A plan view of curve $ADC$, referred to as a *load contour*, is shown in Fig. 7.63a, where $M_{ux}$ and $M_{uy}$ represent factored design moments and $\phi M''_{nx}$ and $\phi M''_{ny}$ represent the uniaxial bending strengths about the $x$ and $y$ axes, respectively, of the cross section with an axial compression force $P'_u$ acting. If the cross section is to supply the required strength to support $P'_u$, $M_{ux}$, and $M_{uy}$, it must have uniaxial bending strengths of $\phi M''_{nx}$ and $\phi M''_{ny}$. (Note that many cross sections can be designed to support a given set of biaxial moments and axial load.) To simplify the design procedure, curve $ADC$ in Fig. 7.63a can be approximated by a straight line that makes an angle $\theta$ with the $\phi M_{nx}$ axis (see Fig. 7.63b). As can be seen, the straight-line approximation increases the required uniaxial bending

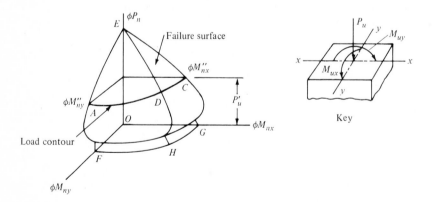

Figure 7.62 Interaction failure surface for a rectangular column subject to axial load and biaxial bending.

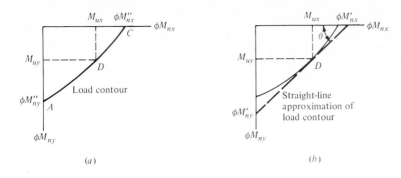

**Figure 7.63** Combinations of $M_{ux}$ and $M_{uy}$ that produce failure of an axially loaded cross section: (a) load contour formed by the intersection of a horizontal plane and an interaction failure surface, (b) straight-line approximation of a curved load contour.

strengths to $\phi M'_{nx}$ and $\phi M'_{ny}$ from $\phi M''_{nx}$ and $\phi M''_{ny}$ and produces a conservative design. By dividing all distances on the $\phi M_{nx}$ axis by $\phi M'_{nx}$ and all distances on the $\phi M_{ny}$ axis by $\phi M'_{ny}$ the equation of the straight line can be nondimensionalized. With this transformation, the equation of the line in Fig. 7.63b can be represented by

$$\frac{M_{ux}}{\phi M'_{nx}} + \frac{M_{uy}}{\phi M'_{ny}} = 1 \qquad (7.28)$$

(See Fig. 7.64.) Multiplying each term of Eq. (7.28) by $\phi M'_{nx}$ gives

$$M_{ux} + M_{uy}\frac{\phi M'_{nx}}{\phi M'_{ny}} = \phi M'_{nx} \qquad (7.29)$$

If the restriction is imposed that the rectangular cross section be reinforced with the same area of steel along each of the four sides, the moment capacities about each principal axis will be approximately proportional to the dimensions of the sides

$$\frac{\phi M'_{nx}}{\phi M'_{ny}} = \frac{b}{h} \qquad (7.30)$$

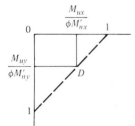

**Figure 7.64** Nondimensional plot of linearized load contour.

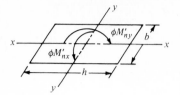

**Figure 7.65** Relationship between moments and dimensions for Eq. (7.30); $b$ = short side, $h$ = long side, $\phi M'_{nx}$ = bending capacity, $x - x$ axis, and $\phi M'_{ny}$ = bending capacity, $y - y$ axis.

Figure 7.65 shows the relationship between sides and moments in Eq. (7.30). Substituting Eq. (7.30) into Eq. (7.29) gives

$$\phi M'_{nx} = M_{ux} + M_{uy}\frac{b}{h} \tag{7.31}$$

Because of the restriction of equal areas of steel on all sides, the cross section is completely designed when analyzed for bending about the minor principal axis. A design aid such as Fig. 7.40 or 7.41 must be used to select the area of steel. If a square column is designed, $b = h$ and Eq. (7.31) reduces to

$$\phi M'_{nx} = M_{ux} + M_{uy} \tag{7.32}$$

A more accurate estimate of $\phi M'_{nx}$ is possible if the assumption is made that the load contour is a curve, the result is given by

$$\phi M'_{nx} = M_{ux} + M_{uy}\frac{b}{h}\frac{1-\beta}{\beta} \tag{7.33}$$

where $\beta$ is related to the curvature of the nondimensionalized load contour (see Fig. 7.66). Charts[7] for evaluating $\beta$ give values between 0.55 and 0.65 for most columns. For a value of $\beta = 0.5$, Eq. (7.33) reduces to Eq. (7.31).

## The Bresler Equation

Since the design of columns by the procedure discussed in Sec. 7.8 is approximate, the capacity of the column should be verified to determine if the steel area is

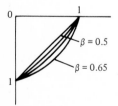

$\beta = 0.5$

$\beta = 0.65$

**Figure 7.66** Variation of load-contour curvature with the parameter $\beta$; curves associated with larger values of $\beta$ not shown.

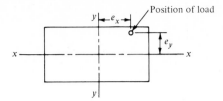

Figure 7.67

adequate. The axial capacity of columns can be checked by the use of the Bresler equation[9].

$$\frac{1}{P_n} = \frac{1}{P_{nx}} + \frac{1}{P_{ny}} + \frac{1}{P_0} \tag{7.34}$$

where $P_n$ = theoretical axial capacity of column under biaxial bending
$P_{nx}$ = theoretical axial capacity if bending occurs about $x$-$x$ axis only; $e_x = 0$ (see Fig. 7.67)
$P_{ny}$ = theoretical axial capacity if bending occurs about the $y$-$y$ axis only; $e_y = 0$ (see Fig. 7.67)
$P_0$ = theoretical axial capacity of column carrying pure axial load only; use Eq. (7.12)

Tests have shown that Eq. (7.34) predicts the strength of columns under biaxial bending with excellent accuracy. The equation is limited to cases where $P_n$ is equal to or greater than $0.1P_0$

**Example 7.20:** (*a*) Using the straight-line approximation of the load contour given by Eq. (7.32), design a short square column to support design loads of $P_u = 432$ kips, $M_{ux} = 108$ ft · kips, and $M_{uy} = 144$ ft · kips. (*b*) Check results with Eq. (7.34); $f'_c = 4$ kips/in², and $f_y = 60$ kips/in². Use equal areas of steel on all four sides (see Fig. 7.68).

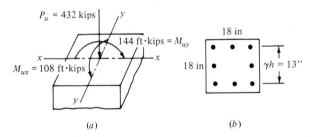

(*a*)                    (*b*)

**Figure 7.68** (*a*) Design loads, (*b*) arrangement of reinforcement ($\gamma h$ is set by the designer to provide required cover for reinforcement); $\gamma = 13/18 = 0.722$.

SOLUTION (*a*) Estimate a trial area. Assume the average compressive stress to be $0.3 \, f'_c$

$$A_g = \frac{P_u}{0.3f'_c} = \frac{432 \text{ kips}}{1.2} = 360 \text{ in}^2$$

Try 18 by 18 in, giving $A_g = 324$ in$^2$. Then assuming $\gamma h = 13$ in

$$\gamma = \frac{13}{18} = 0.722$$

Using Eq. (7.32), select the steel based on the required bending strength about the y-y axis

$$\phi M'_{nx} = M_{ux} + M_{uy} = 108 + 144 = 252 \text{ ft} \cdot \text{kips}$$

Using Fig. 7.41, select $\rho_g$; set $\phi P_n = P_u$

$$\frac{\phi P_n}{A_g} = \frac{432}{324} = 1.333$$

$$e = \frac{\phi M'_{nx}}{P_u} = \frac{252(12)}{432} = 7 \text{ in} \qquad \frac{e}{h} = \frac{7}{18} = 0.389$$

Read $\rho_g = 0.025$; then

$$A_{st} = \rho_g A_g = 0.025(324) = 8.1 \text{ in}^2$$

Use eight no. 9; $A_{st,\,\text{sup}} = 8$ in$^2$

(b) Use Eq. (7.34) to check the capacity of the cross section

$$\frac{1}{P_n} = \frac{1}{P_{nx}} + \frac{1}{P_{ny}} - \frac{1}{P_0}$$

For $P_{ny}$ consider bending about y-y axis, $M_{uy} = 144$ ft $\cdot$ kips

$$e = \frac{144(12)}{432} = 4 \text{ in} \qquad \frac{e}{h} = \frac{4}{18} = 0.222$$

for $\rho_g = 0.025$ and $e/h = 0.222$, read from Fig. 7.41

$$\frac{\phi P_{nx}}{A_g} = 2 \qquad \text{then} \qquad P_{nx} = \frac{2(324)}{0.7} = 925.7 \text{ kips}$$

For $P_{nx}$, consider bending about the x-x axis, $M_{ux} = 108$ ft $\cdot$ kips

$$e = \frac{108(12)}{432} = 3 \text{ in} \qquad \frac{e}{h} = \frac{3}{18} = 0.167$$

For $\rho_g = 0.025$ and $e/h = 0.167$, read from Fig. 7.41

$$\frac{\phi P_{ny}}{A_g} = 2.22 \qquad \text{then} \qquad P_{ny} = \frac{2.22(324)}{0.7} = 1027.5 \text{ kips}$$

$$P_0 = A_{st} f_y + (A_g - A_{st})(0.85 f'_c) = (8 \text{ in}^2)(60 \text{ kips/in}^2) + (324 - 8)(3.4 \text{ kips/in}^2)$$
$$= 1554.4 \text{ kips}$$

Substituting into the Bresler equation gives

$$\frac{1}{P_n} = \frac{1}{925.7} + \frac{1}{1027.5} - \frac{1}{1554.4}$$

$$P_n = 709 \text{ kips} \qquad P_n > 0.1 P_0$$

$$\phi P_n = 0.7(709 \text{ kips}) = 496.3 \text{ kips} > 432 \text{ kips}$$

The design is OK, but $A_{st}$ can be reduced.

**Example 7.21:** (a) Repeat the design of the column in Example 7.20 using Eq. (7.33) and a value of $\beta = 0.65$ (the value recommended by Ref. 7). (b) Check results with Eq. (7.34).

SOLUTION (a) Compute the required moment capacity $\phi M'_{nx}$

$$\phi M'_{nx} = M_{ux} + \frac{b}{h} \frac{1 - \beta}{\beta} M_{uy}$$

For a square column $h = b$ and

$$\phi M'_{nx} = 108 + \frac{1 - 0.65}{0.65} 144 = 185.5 \text{ ft} \cdot \text{kips}$$

Using Fig. 7.41 select $\rho_g$

setting $\phi P_n = P_u$ $\qquad \dfrac{\phi P_n}{A_g} = \dfrac{432}{324} = 1.333 \text{ kips/in}^2 \qquad \gamma = \dfrac{13}{18} = 0.722$

$$e = \frac{\phi M'_{nx}}{P_u} = \frac{185.5(12)}{432} = 5.15 \text{ in} \qquad \frac{e}{h} = \frac{5.15 \text{ in}}{18 \text{ in}} = 0.29$$

Read $\rho_g = 0.018$; then

$$A_{st} = \rho_g A_g = 5.83 \text{ in}^2$$

Use eight no. 8 bars

$$A_{st, \, \sup} = 6.28 \text{ in}^2 \qquad \rho_g = \frac{A_{st}}{A_g} = 0.0193$$

(b) Check using the Bresler equation. Compute $P_{nx}$

$$e = \frac{108(12)}{432} = 3 \text{ in} \qquad \text{and} \qquad \frac{e}{h} = 0.167$$

with $\rho_g = 0.0193$ and $e/h = 0.167$, enter Fig. 7.41. Read $\phi P_{nx}/A_g = 2.1$; then $P_{nx} = 972$ kips. Similarly $P_{ny} = 833.1$ kips.

$$P_0 = A_{st} f_y + (A_g - A_{st})(0.85 f'_c) = 1457 \text{ kips}$$

$$\frac{1}{P_n} = \frac{1}{972} + \frac{1}{833.1} - \frac{1}{1457}$$

$$P_n = 648 \text{ kips}$$

$$\phi P_n = 453.7 \text{ kips} > 432 \text{ kips} \qquad \text{OK}$$

**Example 7.22:** (a) Using Eq. (7.33) with a trial value of $\beta = 0.65$, redesign the column of Example 7.20 with a 16- by 20-in cross section. (b) Verify capacity with Eq. (7.34).

SOLUTION (a) Compute the bending capacity about the minor principal axis (Fig. 7.69).

$$\phi M'_{nx} = M_{ux} + \frac{b}{h} \frac{1 - \beta}{\beta} M_{uy} = 108 + \frac{16}{20} \frac{1 - 0.65}{0.65} 144 = 170 \text{ ft} \cdot \text{kips}$$

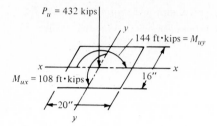

**Figure 7.69**

Using ACI design aids, compute $\rho_g = 0.016$. Then $A_{st} = 5.12$ in$^2$. Use four no. 8 and four no. 7; $A_{st,\,sup} = 5.55$ in$^2$.

(b) Equation (7.34) gives $P_n = 634$ kips

$$\phi P_n = 444 \text{ kips} > 432 \text{ kips} \qquad \text{OK}$$

## QUESTIONS

**7.1** What is the difference between a long and a short column?

**7.2** What is the $P\Delta$ effect?

**7.3** Why is the moment in most interior columns small?

**7.4** Why is the axial force in beams that are components of multistory rigid frames neglected?

**7.5** Give three reasons why the ACI Code requires a minimum area of steel in all columns.

**7.6** Why must the amount of steel in a column not exceed 8 percent?

**7.7** What is the primary advantage of a spiral column over a tied column?

**7.8** What is the primary function of ties?

**7.9** Do ties increase the axial capacity of a tied column?

**7.10** What is the effective length of a column? How is it used?

**7.11** What is the difference between a braced and an unbraced frame?

**7.12** What building elements are normally used to produce a braced frame?

**7.13** What do points on an interaction curve represent?

## PROBLEMS

**7.1** A perfect pin-ended column with a 2- by 4-in cross section has the stress-strain curve shown in Fig. P7.1. Determine both the mode of failure and the theoretical failure load if the length of the column is (a) 30 in and (b) 60 in.

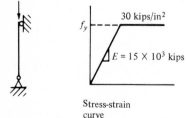

Stress-strain curve                     **Figure P7.1**

**7.2** Considering behavior in the plane of the frame (Fig. P7.2), determine the slenderness ratio of $AB$ and classify it as long or short. The frame is braced perpendicular to its own plane but unbraced in its own plane. All members are 10 in thick normal to the plane of the frame.

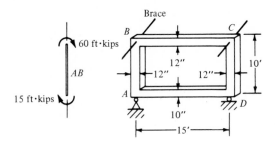

**Figure P7.2**

**7.3** Considering the behavior of the frame in its own plane only, classify columns $AB$ and $CD$ (Fig. P7.3) as long or short if all members are 12 in deep normal to the plane of the frame, $f'_c = 3$ kips/in² for beams, $f'_c = 5$ kips/in² for columns, and the frame is (a) braced and (b) unbraced.

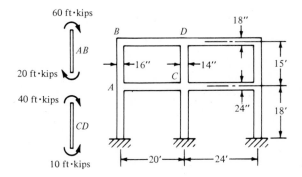

**Figure P7.3**

**7.4** Consider the behavior only in the plane of the frame (Fig. P7.4); $f'_c = 3$ kips/in², and $f_y = 60$ kips/in².
   (a) Determine the required steel in each row of the cross section.
   (b) If $AB$ is a short cantilever column, what is the maximum value $L_c$ can have?

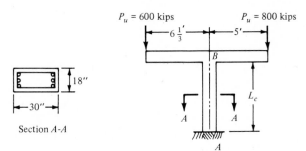

**Figure P7.4**

**7.5** Determine the theoretical capacity of the cross section in Fig. P7.5 for pure axial load. Eight no. 10 bars are used as reinforcement; $f'_c = 5$ kips/in², and $f_y = 60$ kips/in².

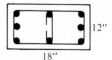

18"

**Figure P7.5**

**7.6** Locate the plastic centroid of the cross section in Fig. P7.6 if $f'_c = 3$ kips/in² and $f_y = 50$ kips/in².

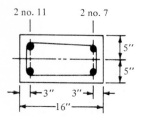

2 no. 11    2 no. 7

**Figure P7.6**

**7.7** Repeat Prob. 7.6 for $f'_c = 5$ kips/in² and $f_y = 50$ kips/in².

**7.8** Draw the interaction curve for the cross section in Fig. P7.8 by computing the coordinates associated with axial load, balanced failure, pure bending, and a triangular distribution of strain. Draw both the theoretical and design curves. $f'_c = 4$ kips/in², $f_y = 60$ kips/in², and there are four no. 8 bars.

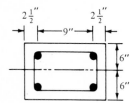

**Figure P7.8**

**7.9** Draw the interaction curve for the cross section in Fig. P7.9 for the points noted in Prob. 7.8 if load acts to the right of the y-y axis. $f'_c = 5$ kips/in², and $f_y = 60$ kips/in².

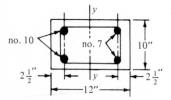

no. 10    no. 7

**Figure P7.9**

**7.10** Repeat Prob. 7.9 if the load acts to the left of the *y-y* axis.

**7.11** (*a*) Design a short square tied column with not more than 4 percent steel to carry an axial load, $P_u = 400$ kips. Note that for pure compression $P_u$ is not to exceed $0.8\phi(0.85f'_c A_c + f_y A_{st})$. (*b*) Design the ties. $f'_c = 4$ kips/in², and $f_y = 60$ kips/in².

**7.12** Repeat Prob. 7.11 if $P_u = 900$ kips.

**7.13** Design (*a*) a short circular column (Fig. 7.23) with spiral reinforcement to carry an axial load of 600 kips and (*b*) the spiral. $f'_c = 3.5$ kips/in², and $f_y = 60$ kips/in². Do not use more than 5 percent steel.

**7.14** Repeat Prob. 7.13 if $P_u = 1200$ kips.

**7.15** Using the ACI interaction curves, design a short tied column to carry loads of $P_u = 400$ kips and $M_u = 80$ ft · kips if $f'_c = 3$ kips/in² and $f_y = 60$ kips/in². Bending about *y-y* axis (see Fig. P7.15).

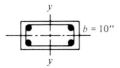

*y*

$b = 10''$

*y*

**Figure P7.15**

**7.16** Using the ACI interaction curves, design a short circular spiral column to carry an axial force $P_u = 500$ kips and a moment of 96 ft · kips. $f'_c = 4$ kips/in², and $f_y = 60$ kips/in².

**7.17** Repeat Prob. 7.16 for $P_u = 950$ kips and $M_u = 45$ ft · kips.

**7.18** Using the trial method, determine the capacity of the short column with the cross section in Fig. P7.18 for an axial load with an eccentricity of 5 in; $f'_c = 5$ kips/in², and $f_y = 60$ kips/in². Three no. 8 bars are on each side.

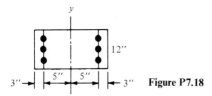

*y*

12''

3'' — 5'' — 5'' — 3''    **Figure P7.18**

**7.19** Repeat Prob. 7.18 if the load acts at an eccentricity of 10 in from the *y* axis.

**7.20** Using the trial method, determine whether the cross section in Fig. P7.20 can safely carry an axial load of $P_u = 300$ kips and $M_u = 180$ ft · kips if $f'_c = 4$ kips/in² and $f_y = 60$ kips/in².

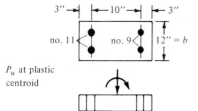

3'' — 10'' — 3''

no. 11    no. 9    12'' = b

$P_u$ at plastic centroid

**Figure P7.20**

**7.21** Repeat Prob. 7.20 for $P_u = 250$ kips and $M_u = 140$ ft · kips.

**7.22** Determine whether the 12- by 14-in tied column (Fig. P7.22) can be reinforced to support the indicated loads. Analysis shows bending about the y-y axis only, and 50 percent of the factored moment is due to dead load. Use $k = 0.85$, $f'_c = 3$ kips/in$^2$, and $f_y = 60$ kips/in$^2$.

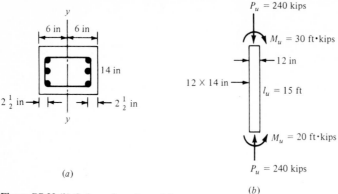

(a)

**Figure P7.22** (b) Column from braced frame.

**7.23** Consider bending in the plane of the paper only; a beam-column (Fig. P7.23) with a 12- by 16-in cross section carries an axial load of 240 kips and a uniform load of 0.8 kip/ft. Is the reinforcement of four no. 10 bars adequate? $f'_c = 3$ kips/in$^2$, and $f_y = 60$ kips/in$^2$.

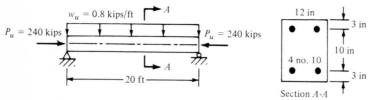

**Figure P7.23**

**7.24** Using the approximate design procedure of Sec. 7.8, design a 16-in square cross section with equal reinforcement on all four sides to carry an axial force $P_u = 315$ kips and moments $M_{ux} = 90$ ft · kips and $M_{uy} = 120$ ft · kips. Check the capacity of the cross section with the Bresler equation and modify the reinforcement until the axial capacity is within 5 percent of the design load. $f'_c = 4$ kips/in$^2$, and $f_y = 60$ kips/in$^2$.

**7.25** Repeat the design in Prob. 7.24 using a 14- by 18-in cross section to carry the specified forces. The 18-in dimension is to be oriented in the direction of the larger moment; $f'_c = 4$ kips/in$^2$, and $f_y = 60$ kips/in$^2$.

**7.26** Use Eq. (7.34) to determine the maximum load the cross section in Fig. P7.26 can support if the design load is applied with the eccentricities shown. $f'_c = 3$ kips/in$^2$, and $f_y = 60$ kips/in$^2$.

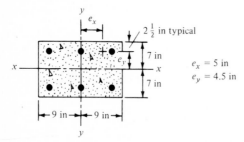

**Figure P7.26**

# REFERENCES

1. J. G. MacGregor, J. Breen, and E. O. Pfrang: Design of Slender Concrete Columns, *ACI, Proc.*, vol. 67, pp. 6–28, January 1970.
2. J. Breen, J. G. MacGregor, and E. O. Pfrang: Determination of Effective Length Factors for Slender Columns, *ACI Proc.*, vol. 69, 669–672, 1972.
3. B. Bresler: "Reinforced Concrete Engineering," vol. I, Wiley, New York, 1974.
4. S. P. Timoshenko and J. M. Gere: "Theory of Elastic Stability," 2d ed., McGraw-Hill, New York, 1961.
5. R. Furlong: Ultimate Strength of Square Columns under Biaxial Eccentric Loads, *ACI Proc.*, vol. 57, p. 1129, March 1961.
6. M. Fintel, (ed.): "Handbook of Concrete Design," Van Nostrand, New York, 1974.
7. Design Handbook, vol. 2, Columns, *ACI Publ.* SP-17a(78), Detroit, 1978.
8. A. L. Parme, J. M. Nieves, and A. Gouwens: Capacity of Rectangular Columns Subject to Biaxial Bending, *ACI.*, vol. 63, p. 911, September 1966.
9. B. Bresler: Design Criteria for Reinforced Columns under Axial Load and Biaxial Bending, *ACI Proc.*, vol. 57, pp. 481–490, November 1960.
10. R. Park and T. Paulay: "Reinforced Concrete Structures," p. 28, Wiley-Interscience, New York, 1975.
11. "Manual of Steel Construction," 8th ed., American Institute of Steel Construction, 1980.
12. "Manual of Standard Practice for Detailing Reinforced Concrete Structures," 6th ed., *ACI* 315–74, Detroit, 1974.
13. "Commentary on Building Code Requirements for Reinforced Concrete (ACI 318–77)," *ACI*, Detroit, 1977.

Palazetto dello Sport, Rome, a reinforced-concrete domed roof supported on Y-shaped columns. Designed and constructed by the master builder Pier Luigi Nervi, an engineer known for the inventiveness and logic of his structural forms. (*Photograph by the Italian Government Travel Offices.*)

# EIGHT

## FOOTING DESIGN

## 8.1 INTRODUCTION

Loads from buildings, bridges, and other structures must be transmitted into the ground through *foundations*; these structural elements support the main load-bearing members, i.e., columns and walls, of the structure. Since the bearing pressures that soils can sustain are much smaller than the compressive stresses in columns or walls, foundations must be used to reduce the pressures applied directly to the soil by spreading the supported loads over an area large enough to prevent rupture or excessive deformation of the soil.

Foundation design involves both a *soil study*, to establish the most suitable type of foundation, and a *structural design*, to determine the proportions of the foundation elements. When foundations are supported on soil whose properties are well known and no unusual design conditions are present, the structural engineer may feel qualified to carry out both the soils study and the structural design. Conversely, if soil conditions are complex, if installation of the foundations may be difficult, or if settlements must be closely controlled, a soils specialist (geotechnical engineer) and a structural engineer may work as a team to design the foundations. The geotechnical engineer will carry out the site investigation, establish the physical properties of the soil, and study the feasibility of various types of foundations. The structural designer is responsible for sizing and detailing the components of each foundation, a step required before the cost of a particular design can be accurately estimated.

## 8.2 TYPES OF FOUNDATIONS

The type of foundation selected for a particular structure is influenced by the following factors:

1. The strength and compressibility of the various soil strata at the site
2. The magnitude of the column loads
3. The position of the water table
4. The need for a basement
5. The depth of foundations of adjacent buildings

When soil of adequate bearing capacity and stiffness lies near the ground surface, columns and walls are often supported on pads of concrete, termed *footings* (Fig. 8.1). Walls are typically supported on long narrow footings that bend into single

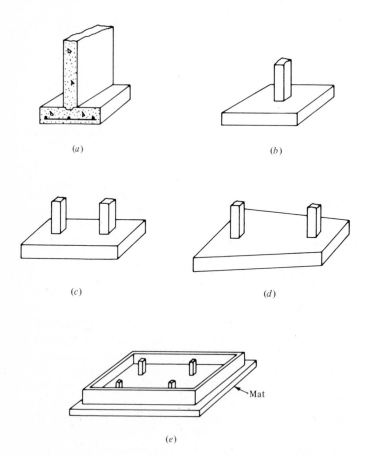

*(a)*  *(b)*

*(c)*  *(d)*

*(e)*

**Figure 8.1** Types of footings: (*a*) wall footing, (*b*) two-way column footing, (*c*) rectangular combined footing, (*d*) trapezoidal footing, (*e*) mat.

curvature perpendicular to the direction of the wall (Fig. 8.1*a*). The primary reinforcement is placed in the bottom of the footing in the short direction, i.e., transverse to the direction of the wall. Temperature and shrinkage steel, running in the direction of the wall, is often added to position the transverse reinforcement and to provide some bending strength so that the footing can bridge over local areas of weakness in the soil. As shown in Fig. 8.1*b*, individual columns are placed on rectangular pads of concrete that bend in two directions. These *two-way footings* or *spread footings* must be reinforced in both directions of bending with bars of steel placed in the bottom of the footing parallel to its sides. If clearances permit, square two-way footings are used wherever possible to reduce the bending moments.

If footings overlap because columns are closely spaced, or if the extension of a footing in a particular direction is limited by property lines, two or more columns may be placed on one common footing, called a *combined footing* (Fig. 8.1*c*). Although combined footings are usually rectangular, a trapezoid may produce a more economical design when large differences exist in the magnitudes of the column loads or clearances do not permit a rectangular footing (Fig. 8.1*d*).

When the combined area of individual footings exceeds one-half of the area contained within the perimeter of the plan of a building, a condition associated with a weak soil, columns and walls may be more effectively supported on one large footing, called a *mat* (Fig. 8.1*e*). Although many mats are designed as slabs of constant depth, economy often results when mats are designed as a system of beams and slabs.

In addition to saving the cost of forming a large number of individual footings, the continuity of the mat also reduces differential settlements between columns by a factor of approximately 100 percent over that of columns placed on individual footings. If the water table lies well above the base of the foundation and a basement is required, a reinforced concrete mat combined with the perimeter walls can be formed into a watertight box to produce a dry basement. To prevent flooding of the basement by groundwater, a heavily reinforced mat of well-compacted concrete is essential to resist cracking by the high water pressures on the base of the mat.

Although locating foundations as close as possible to the ground surface reduces excavation costs, the designer must take care to place the base of foundations below the zone of soil that is subject to freezing. If the soil below a foundation freezes, the foundation is subject to *heaving*, i.e., vertical movement due to the increase in volume that occurs as the moisture in the soil turns to ice. In northern regions of the United States a depth of 4 to 6 ft (1.22 to 1.83 m) of soil cover is usually adequate to prevent freezing of the soil during the coldest winters. In warm regions, the minimum depth of cover is controlled by the depth at which good bearing material is located.

If the upper soil strata lack the strength to carry shallow foundations, building loads must be transmitted deeper into the ground until a stratum of adequate bearing material, e.g., dense sand or rock, is encountered. When a suitable bearing stratum is located far below the surface, foundations are often supported on piles; these are columns of steel, concrete, or wood that have been driven into the bearing

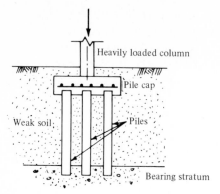

Figure 8.2 Pile foundation.

stratum. When a group of piles is required to support a large column load, a block of reinforced concrete, called a *pile cap*, is used to distribute the column load to the supporting piles (see Fig. 8.2).

## 8.3 SOIL PRESSURES

### Bearing Pressures under Axially Loaded Footings

Bearing pressures under the base of a footing are determined by treating the footing as a rigid element and the soil directly under the footing as a homogeneous elastic material that is isolated from the surrounding soil. Since soil pressures are assumed to be directly proportional to the deformation of the soil, uniform pressure is assumed to develop under the base of an axially loaded footing because the soil is uniformly compressed (see Fig. 8.3*a*).

Actually, the pressure distribution on the base of a footing is not uniform but varies. The variation is a function of (1) the flexibility of the footing, (2) the depth of the footing below the ground surface, and (3) the type of soil, e.g., clay or sand. For example, the downward displacement of a loaded footing into a *cohesionless soil*, a sand or a gravel, produces a lateral movement of the soil from under the edge of the footing. For a footing located at or near the ground surface, the small depth of soil covering the base of the footing offers little resistance to the escape of soil from under the base. The loss of support to the outer edges of the footing which accompanies the movement of the soil particles reduces the bearing pressures at the perimeter of the footing and produces the parabolic pressure distribution shown in Fig. 8.3*b*. If a footing on cohesionless soil lies well below the surface, however, the soil pressure on the base becomes more nearly uniform. Now the weight of the large depth of soil over the base confines the soil under the footing and restricts the lateral movement of the soil from under the edges of the footing. As a result of the confinement, the soil is able to provide vertical support to the outer edges of the footing.

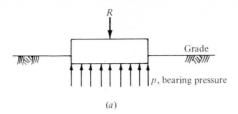

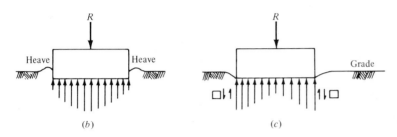

**Figure 8.3** Bearing pressures under axially loaded footings: (*a*) assumed pressure distribution, (*b*) deformations and approximate pressure under a rigid surface footing on a cohesionless soil, (*c*) deformations and approximate pressure distribution under a rigid footing on cohesive soil.

If an axially loaded footing is founded on a clay, which is a cohesive soil, a uniform settlement of a rigid footing creates the approximate stress distribution shown in Fig. 8.3*c*. In a cohesive soil, the shear stresses induced in the soil surrounding the base of the footing create additional vertical support at the edge of the footing and produce higher stress at the edge than at the center of the footing.

Although the magnitude of the soil pressures under the base of axially loaded footings is not uniform, footing design, regardless of the soil type, is usually based on the assumption of a uniform pressure distribution under the base to simplify the analysis. Experimental studies and the satisfactory performance of large numbers of building foundations indicate that this assumption produces conservatively designed foundations.

## Eccentrically Loaded Footings

Although it is always desirable to load footings axially to ensure uniform settlements and to minimize soil pressures, footings must frequently be designed for both axial load and moment. Moment may be caused by lateral forces due to wind or to earthquake, by lateral soil pressures, or by thrust at the base of a column (see Fig. 8.4).

If the eccentricity of the vertical load is small, compression stresses develop over the entire base of the foundation. For this case the soil pressure can be determined by superimposing the direct stress $P/A$ due to the axial load and the

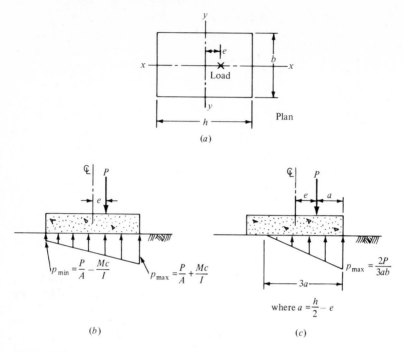

**Figure 8.4** Soil pressure under an eccentrically loaded rectangular footing: (a) plan; (b) small eccentricity of load, $e < h/6$; (c) large eccentricity of load, $e > h/6$.

bending stress $Mc/I$ created by the moment. Since tensile stresses cannot be transmitted between soil and concrete, superposition of stresses is valid only when the tensile bending stresses do not exceed the direct compression stresses. For a rectangular footing, the maximum eccentricity for which superposition holds can be established from the limiting case of a triangular stress distribution on the base of a foundation. Although compression stresses develop over the entire base for this case, the stress is zero at the edge, where the tensile bending and direct compression stresses are equal. Expressing the moment as $Pe$ and setting $p_{min}$ in Fig. 8.4b equal to zero gives

$$\frac{P}{A} = \frac{Mc}{I} = \frac{Pec}{I}$$

Solving for the eccentricity $e$ gives

$$e = \frac{I}{Ac} \tag{8.1}$$

For a rectangular footing of length $h$ and width $b$, Eq. (8.1) becomes

$$e = \frac{bh^3/12}{bh(h/2)} = \frac{h}{6} \tag{8.2}$$

If the eccentricity of the vertical load is large and the tensile bending stresses exceed the direct stress, a triangular stress distribution will develop over a portion of the base. The maximum pressure associated with this distribution can be established by recognizing that the centroid of the soil pressure is located directly under the vertical component of the applied load. With the dimensions of the footing established and with the eccentricity of the vertical load known, the distance between the resultant of the applied load $P$ and the outside edge (denoted by $a$ in Fig. 8.4c) can be established. The length of base on which the triangular distribution of soil pressure acts is then $3a$. Equating the resultant of the soil pressure to the applied force gives

$$\frac{p_{max}}{2} \, 3ab = P$$

solving for $p_{max}$, we get

$$p_{max} = \frac{2P}{3ab} \qquad \text{where } a = \frac{h}{2} - e \qquad (8.3)$$

Example 8.1 shows the computations required to find the soil pressure under the base of eccentrically loaded footings and illustrates the importance of keeping the eccentricity as small as possible to prevent overstressing the soil.

**Example 8.1:** In Fig. 8.5a a 6- by 4-ft (1.83- by 1.22-m) footing is loaded by a resultant force $P = 200$ kips (889.6 kN) applied on the x-axis 0.5 ft (0.15 m) to the right of the centroid of the footing. Determine the soil pressure on the base; repeat the computation if the eccentricity is 1.25 ft (0.38 m).

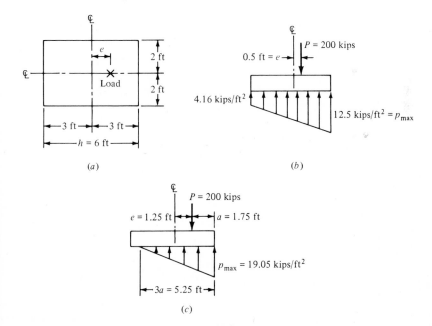

Figure 8.5 (a) Plan, (b) $e = 0.5$ ft, (c) $e = 1.25$ ft.

SOLUTION   Since $e = 0.5$ ft $< h/6 = 1$ ft, direct and bending stresses can be superimposed (see Fig. 8.5b)

$$p_{max} = \frac{P}{A} + \frac{Pec}{I} = \frac{200}{24} + \frac{200(0.5)(3)}{\frac{4}{12}(6^3)}$$

$$= 8.33 + 4.17 = 12.5 \text{ kips/ft}^2 \text{ (598.5 kPa)}$$

$$p_{min} = 8.33 - 4.17 = 4.16 \text{ kips/ft}^2 \text{ (199.2 kPa)}$$

Since $e = 1.25$ ft $> h/6 = 1$ ft, stress acts over a length $3a$. Use Eq. (8.3) (see Fig. 8.5c)

$$p_{max} = \frac{2P}{3ab} = \frac{2(200)}{3(1.75)4} = 19.05 \text{ kips/ft}^2 \text{ (912.1 kPa)}$$

## Allowable Soil Pressures

To be properly designed, a foundation must (1) have an adequate factor of safety against punching into the soil and (2) must not settle excessively. Large differential settlements between adjacent footings produce uneven floor surfaces and may crack masonry walls or overstress the structural frame. Unless a structure contains equipment that must be aligned with great precision, a maximum settlement of 1 in can be tolerated without producing damage to either the structure or the attached or supported nonstructural elements.

Values of allowable soil pressure for various types of soils are usually specified in building codes. These allowable pressures are based both on theoretical studies of soil strength and on field measurements of actual building settlements. Typically, these pressures provide a factor of safety of 3 against the foundation punching into the ground but do not necessarily ensure that settlements will be limited to 1 in (25.4 mm). To ensure that settlements are within the design requirements, charts are available in various reference books that relate settlement to soil strength as measured by the standard penetration test.[1]

In building codes, the allowable soil pressure for various classes of soil may be stated as either the gross or net pressure permitted on the soil directly under the base of the footing. The gross pressure represents the total stress in the soil created by all loads above the base of the footing. These loads include (1) service loads, (2) the weight of the footing, and (3) the weight of the soil between the top of the footing and the ground surface. The net pressure is that pressure over and above the pressure created by the weight of the soil. It is computed by subtracting from the gross pressure the weight of a 1-ft-square column of soil whose height extends from the base of the footing to the ground surface.

**Example 8.2:** Compute the gross and the net soil pressure at the base of the 5-ft (1.52-m) square footing in Fig. 8.6. The unit weight of soil is $\gamma_s = 130$ lb/ft$^3$ (2082 kg/m$^3$), and $\gamma_c = 150$ lb/ft$^3$ (2403 kg/m$^3$).

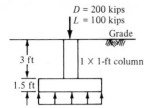

**Figure 8.6**

SOLUTION  Total load above the base of the footing

$$\text{Weight of footing: } 5(5)(1.5)(0.15) = \quad 5.63 \text{ kips}$$

$$\text{Weight of column: } 1(1)(3)(0.15) = \quad 0.45$$

$$\text{Weight of soil: } 3(25 - 1)(0.13) = \quad 9.36$$

$$\underline{\text{Service loads: } 200D + 100L = 300.00}$$

$$315.44 \text{ kips } (1403.08 \text{ kN})$$

The gross soil pressure is

$$q_{gr} = \frac{P}{A} = \frac{315.44 \text{ kips}}{25 \text{ ft}^2} = 12.62 \text{ kips/ft}^2 \text{ (604.25 kPa)}$$

and the net soil pressure is

$$q_n = q_{gr} - \text{weight of 4.5 ft of soil} = 12.62 - 4.5(0.13) = 12.04 \text{ kips/ft}^2 \text{ (576.48 kPa)}$$

## 8.4 DESIGN OF FOOTINGS FOR VERTICAL LOAD

Footings, which are designed as shallow beams, must be proportioned for the internal forces created by the pressures acting on their top and bottom surfaces as well as their dead weight. When a footing covered by a constant depth of soil supports a load that passes through its centroid, the weights of both the soil and the footing may be omitted from the analysis and the footing proportioned only for the pressures created by the factored column or wall loads. This simplification is possible because the portion of the pressure induced on the base of a footing by the weight of the footing and the soil has the same magnitude and distribution as the forces applied by the two materials. In other words, neither shear nor moment is induced by the dead weight of the footing or by a constant depth of soil (Fig. 8.7).

With few exceptions, the proportions of beams and slabs in buildings, which carry loads of several *hundred* pounds per square foot, are determined by the bending strength required to carry moment or by the minimum depth required to limit deflections. Since large soil pressures—4000 to 10,000 lb per square foot or more are common—under footings create high shear forces even when the footings are small, and since shear reinforcement is not normally used, shear rather than moment typically determines the minimum required depth of footings: the depth

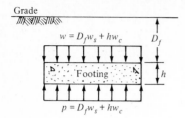

Figure 8.7 Uniform pressure created on the base of a footing by the weight of concrete and soil; $w_c$ and $w_s$ are the unit weights of concrete and soil.

of the footing must be set so that the shear capacity of the concrete $\phi V_c$ on all potential failure planes equals or exceeds the shear produced by factored loads.

Shear reinforcement is not generally used in shallow footings because (1) establishment of an effective pattern of shear reinforcement is difficult when footings bend in two directions and (2) the depth of the compression zone may not be adequate in a shallow footing to anchor shear reinforcement that is designed to yield at failure.

To prevent crushing of the concrete along the plane of contact between the column and the top surface of the footing, the ACI Code limits the bearing stresses on the base of the column and the surface of the footing. Although the bearing pressure on the base of the column is not to exceed $\phi(0.85f_c')$, the bearing pressure on the footing surface under a column may go as high as

$$\phi(0.85f_c')\sqrt{A_2/A_1} \leq \phi(0.85f_c')2 \tag{8.4}$$

where $\phi = 0.7$, $A_1$ is the loaded area (the column cross section), and $A_2$ is the maximum area of that portion of the supporting surface that is geometrically similar to and concentric with area $A_1$. The factor $\sqrt{A_2/A_1}$, which is greater than 1, accounts for the increase in bearing capacity produced by the lateral confinement of the concrete under the loaded area by the surrounding concrete (see Fig. 8.15).

In the next two sections the design procedure for a wall footing carrying a uniformly distributed load and an axially loaded rectangular column footing will be discussed. The same principles that apply to the design of wall and single column footings can be readily extended and applied to the design of more complicated foundations such as the combined footings and the mats shown in Fig. 8.1.

## 8.5 DESIGN OF A WALL FOOTING WITH A UNIFORM LOAD

Under the action of a uniform soil pressure, the segment of a wall footing that extends beyond the face of a wall acts like a cantilever beam and bends upward under the forces created by the unbalanced pressures on the base of the footing (Fig. 8.8b). When a wall carries a uniformly distributed line load, all sections along the length of the footing behave identically; therefore, the design of the footing can be based on the analysis of a typical 1-ft slice cut by transverse planes normal to the longitudinal axis of the wall (Fig. 8.8a). The steps required to design a wall footing are summarized below.

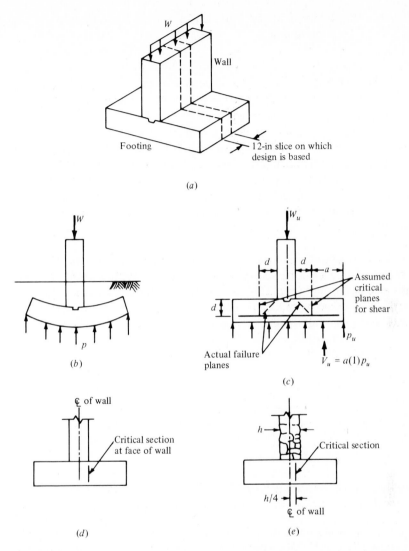

**Figure 8.8** Details of wall footing design: (*a*) a uniformly loaded wall; (*b*) bending deformations; (*c*) critical planes for beam shear; (*d*) critical section for moment, concrete wall; (*e*) critical section for moment, masonry wall.

**Step 1**  Establish the required width of the footing by dividing the total *service* load by the allowable soil pressure. The width of the footing is typically rounded off to the nearest even inch.

**Step 2**  Estimate the depth of the footing. ACI Code §15.7 requires that the depth of the footing above the reinforcement be at least 6 in (152 mm) for footings on soil and at least 12 in (305 mm) for footings supported on piles.

**Step 3** Increase the service loads by load factors and compute a fictitious soil pressure for which the footing is sized.

**Step 4** Verify that the shear capacity of the concrete is adequate to prevent a beam-shear failure. The critical section for shear is assumed to be located out from the face of the wall a distance $d$, the effective depth of the footing (see Fig. 8.8c); i.e., on the critical section $V_u \leq \phi 2\sqrt{f'_c}\,b_w d$, where $\phi = 0.85$, $b_w = 12$ in, $d$ is the effective depth of the footing, and $V_u$ equals the shear produced by the factored soil pressure $p_u$ that acts on the portion of the footing base located between the critical section for shear and the outside edge. In the center region of the footing, between the critical sections, load is transmitted into the soil directly (see the dashed lines in Fig. 8.8c). If a significant difference exists between the required and actual shear capacity, the dimensions must be modified and the analysis repeated until a safe design is achieved.

**Step 5** Compute the reinforcement required for moment. If a footing carries a concrete wall, ACI Code §15.4.2 specifies that the critical section for moment be taken at the face of the wall (Fig. 8.8d). If a footing supports a masonry wall, the critical section is assumed to be located out from the center of the wall a distance equal to one-fourth the wall thickness (Fig. 8.8e). Since the relatively large depth required to carry shear reduces the area of flexural steel, minimum steel should always be checked;

i.e.,

$$A_s \geq \begin{cases} \dfrac{200b_w d}{f_y} & \text{USCU} \\[2ex] \dfrac{1.4b_w d}{f_y} & \text{SI} \end{cases}$$

**Step 6** Verify that the anchorage length between the critical section for moment and the end of the bar equals or exceeds the required development length.

**Example 8.3: Design of a wall footing** Design the reinforced concrete footing in Fig. 8.9 to support a concrete wall that carries a dead load $D$ of 10 kips/ft (145.9 kN/m) and a

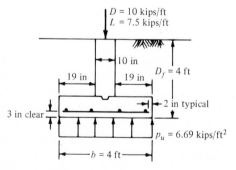

Figure 8.9

live load of 7.5 kips/ft (109.5 kN/m). The maximum pressure on the soil under the foundation is not to exceed 5 kips/ft² (239.4 kPa); $f'_c = 3$ kips/in² (20.68 MPa), $f_y = 60$ kips/in² (413.7 MPa), and the unit weight of the soil is $\gamma_s = 130$ lb/ft³ (2082 kg/m³).

SOLUTION Base the analysis on a 1-ft-thick slice of wall and foundation. Using service loads, determine the width of footing. Establish the net bearing capacity of the soil by deducting the weight of the footing and the soil above the base of the foundation from the maximum allowable soil pressure of 5 kips/ft². Assume footing 12 in thick. Neglect the small difference between the weight of the wall and the soil. Computing the weight of material above the footing base gives

$$q_n = 5.0 - [0.15 + 3(0.13)] = 4.46 \text{ kips/ft}^2$$

The width of footing is

$$b = \frac{D + L}{q_n} = \frac{10 + 7.5}{4.46} = 3.92 \text{ ft} \qquad \text{use 4 ft (1.22 m)}$$

Pressure produced by factored loads

$$p_u = \frac{1.4(10) + 7.5(1.7)}{4} = \frac{26.75 \text{ kips}}{4} = 6.69 \text{ kips/ft}^2 \text{ (320.3 kPa)}$$

Check shear at $d$ in out from the face of the wall (see Fig. 8.10a). Allowing 3 in (76 mm) of clear cover for reinforcement in bottom of footing plus 0.5 in (13 mm) for half a bar

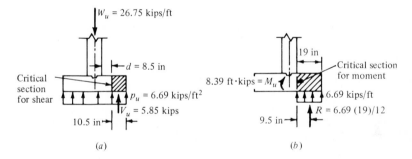

(a)    (b)

**Figure 8.10** (a) Design shear, (b) design moment.

diameter gives $d = 8.5$ in (215.9 mm). Shear on a vertical section $d$ in out from the face of the wall is

$$V_u = \frac{10.5}{12} (1)(6.69 \text{ kips/ft}^2) = 5.85 \text{ kips (26 kN)}$$

Capacity of the concrete

$$\phi V_c = 0.85(2)\sqrt{f'_c}b_w d = 0.85(2)\frac{\sqrt{3000}}{1000}(12)(8.5)$$

$$= 9.5 \text{ kips (42.3 kN)}$$

Since the shear strength $\phi V_c$ is considerably greater than $V_u$, the depth of the footing can be reduced. Try $h = 10$ in and $d = 6.5$ in

$$V_u = \frac{12.5}{12}(1)(6.69 \text{ kips/ft}^2) = 6.97 \text{ kips (31 kN)}$$

$$\phi V_c = 0.85(2)\frac{\sqrt{3000}}{1000}(12)(6.5) = 7.26 \text{ kips (32.29 kN)} \qquad \text{shear OK}$$

Find the area of reinforcement required for moment. For the critical section at the face of the wall (see Fig. 8.10$b$),

$$M_u = 6.69 \frac{19}{12}\frac{9.5}{12} = 8.39 \text{ ft} \cdot \text{kips}$$

$A_s$ may be established from Table 3.5 or by the trial method. Using the trial method gives

$$M_u = \phi T\left(d - \frac{a}{2}\right) \qquad \text{guess } a = 1 \text{ in}$$

$$8.39(12) = 0.9 A_s(60)(6.5 - 0.5)$$

$$A_s = 0.31 \text{ in}^2 \qquad \text{controls}$$

Use no. 5 bars at 12 in on center. If $a$ is checked by equating $T = C$, we find $a = 0.6$ in; therefore the initial assumption is OK. Compute $A_{s,\min}$ using requirements for beams

$$A_{s,\min} = \frac{200}{f_y}b_w d = \frac{200}{60,000}(12)(6.5) = 0.26 \text{ in}^2$$

Check development length. Since spacing exceeds 6 in, the basic $l_d$ can be reduced by 0.8

$$l_d = \frac{0.04 A_b f_y}{\sqrt{f'_c}} = \frac{0.04(0.31)(60,000)(0.8)}{\sqrt{3000}} = 10.87 \text{ in}$$

$$= 0.0004 d_b f_y = 0.0004(\tfrac{5}{8})(60,000)(0.8) = 12.0 \text{ in}$$

but not less than 12 in (controls). If reinforcement extends to within 2 in of side of the footing, the length available to anchor is 17 in

$$17 \text{ in} > 12 \text{ in} \qquad \text{anchorage OK}$$

## 8.6 DESIGN OF AN AXIALLY LOADED TWO-WAY FOOTING

Axially loaded rectangular footings that support a single column must be designed for punching shear (see Sec. 4.6), beam shear, and bending. When a footing bends in double curvature, beam shear and bending about both principal axes must be considered unless the footing is square. If it is, its analysis can be simplified by assuming its strength to be identical in both principal directions even though the effective depth is slightly different in each direction because a two-layer grid of steel is required. The procedure for designing an axially loaded rectangular footing supporting a single column is detailed below.

## Design Procedure

**Step 1**    Establish the required area of the footing by dividing the total service load by the allowable soil pressure. Then set the dimensions. If possible, use a square footing to achieve the greatest economy.

**Step 2**    Select a trial footing depth. As previously noted for wall footings, the ACI Code requires that the depth of the footing above the reinforcement be at least 6 in (152 mm) for footings on soil and at least 12 in (305 mm) for footings on piles; 3 in (76.2 mm) of clear cover over the reinforcement is required if concrete is cast directly against soil. A practical minimum depth is 10 in (254 mm).

**Step 3**    Increase the service loads by the appropriate load factors and compute the soil pressure $p_u$ by dividing the factored load by the area of the footing.

**Step 4**    Verify that the punching shear capacity is adequate. The perimeter of the assumed failure surface for punching shear (Fig. 8.11a) is located out from the face of the column a distance $d/2$ and has the same shape as that of the column. For a

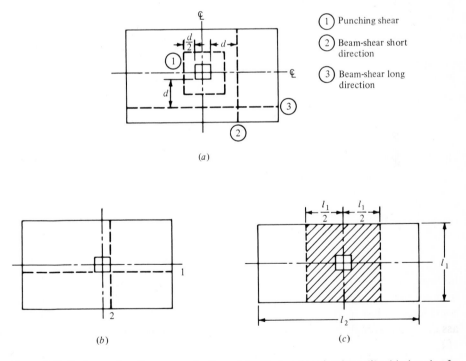

(a)

① Punching shear

② Beam-shear short direction

③ Beam-shear long direction

(b)    (c)

**Figure 8.11**  Design sections for two-way footings: (a) critical sections for shear, (b) critical section for moment, (c) location of band of concentrated reinforcement in the short direction.

satisfactory design the shear $V_u$ transmitted across the failure surface must not exceed

$$\phi V_c = 0.85\left(2 + \frac{4}{\beta_c}\right)\sqrt{f'_c}b_0 d \le 0.85(4)\sqrt{f'_c}b_0 d \qquad (4.24)$$

where $\beta_c$ = ratio of long to short sides of column
$b_0$ = perimeter of failure surface
$d$ = effective depth of footing

Since there are two layers of steel, an average value of $d$ may be used. If necessary, modify the punching-shear capacity by changing the depth.

**Step 5** Check beam-shear capacity in each direction on planes located a distance $d$ from the face of the support. Increase depth if additional shear strength is required. See sections 2 and 3 in Fig. 8.11$a$.

**Step 6** Compute the area of flexural steel required in each direction with respect to a critical section that crosses the footing at the face of the column. In the long direction, the reinforcement is distributed uniformly across the width of the footing. In the short direction a portion of the reinforcement is concentrated in a band centered about the column centerline. The width of the band is equal to that of the short dimension of the footing, and the ratio of the reinforcement in the band to the total reinforcement is equal to $2/(\beta + 1)$, where $\beta$ equals the ratio of the long to short sides of the footing. The reinforcement that is not placed in the band is spaced uniformly in the regions on either side of the center band. Of course, in each principal direction the area of steel must not be less than $200b_w d/f_y$. If the footing is square, the steel is not concentrated in bands but spaced uniformly throughout the width in both directions (see Fig. 8.11$c$).

**Step 7** Check anchorage to verify that the bond is adequate.

Examples 8.4 and 8.5 illustrate the design of a square and a rectangular footing for identical column loads. A comparison of the final designs shows that the square footing requires less material and is therefore more economical.

**Example 8.4: Design of a two-way square footing** Design the square two-way footing in Fig. 8.12 to support a 16-in (406-mm) square column located at the center of the footing.

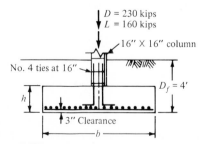

Figure 8.12

The column carries service loads of 230 kips (1023 kN) dead load and 160 kips (711.7 kN) live load. The allowable soil pressure on the base of the footing is 5.5 kips/ft² (273 kPa). The column is reinforced with six no. 10 bars, $f'_c = 3$ kips/in² (20.68 MPa) and $f_y = 50$ kips/in² (344.7 MPa). The weight of the soil is $\gamma_s = 130$ lb/ft³.

SOLUTION

1. Establish the dimensions of the base using the total service load of 390 kips. Assume a footing 2 ft (0.61 m) deep. The net allowable soil pressure is

$$q_n = 5.5 - [2(0.15) + 2(0.13)] = 4.94 \text{ kips/ft}^2$$

The required area is

$$\frac{P}{q_n} = \frac{390 \text{ kips}}{4.94} = 78.95 \text{ ft}^2$$

Make it 9 by 9 ft (2.74 by 2.74 m). Area supplied = 81 ft.²

2. Design pressure on base of footing due to factored column loads is

$$p_u = \frac{(230 \text{ kips})(1.4) + (160 \text{ kips})(1.7)}{81 \text{ ft}^2} = \frac{594 \text{ kips}}{81} = 7.33 \text{ kips/ft}^2$$

Since two rows of reinforcement, one on top of the other, will be used in the bottom of the footing, an average value of $d$, measured from the top surface of the footing to the middle of the two rows, will be used. Assuming 1-in-diameter bars, we get

$$d = 24 \text{ in} - (3 \text{ in cover} + 1 \text{ in bar diameter}) = 20 \text{ in (508 mm)}$$

3. Check punching shear on the critical plane $d/2$ in out from the face of the column (Fig. 8.13a). Shear $V_u$ on the critical plane is

$$V_u = 594 \text{ kips} - 3(3)(7.33 \text{ kips/ft}^2) = 528 \text{ kips (2349 kN)}$$

$$\phi V_c = \phi 4\sqrt{f'_c} b_0 d = 0.85(4)\frac{\sqrt{3000}}{1000}(4)(36)(20) = 536.33 \text{ kips (2386 kN)}$$

$$\phi V_c > V_u \qquad \text{OK}$$

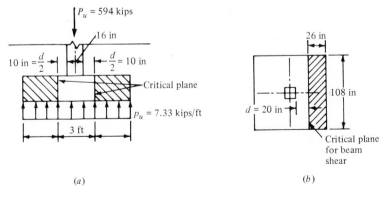

(a)                                                                 (b)

**Figure 8.13** (a) Failure plane for punching shear, (b) critical section for beam shear.

4. Check beam shear on the plane located $d$ in out from the face of the column, (see Fig. 8.13$b$)

$$V_u = \tfrac{26}{12}(9 \text{ ft})(7.33 \text{ kips/ft}^2) = 142.94 \text{ kips } (635.8 \text{ kN})$$

$$\phi V_c = \phi 2\sqrt{f'_c}b_w d = 0.85(2)\sqrt{3000}(108)(\tfrac{20}{1000}) = 201.1 \text{ kips } (894.5 \text{ kN})$$

$$\phi V_c > V_u \qquad \text{OK}$$

5. Compute the area of flexural steel in each direction. The critical section extends across the width of the footing at the face of the column (Fig. 8.14)

$$M_u = \tfrac{46}{12}(9 \text{ ft})(7.33 \text{ kips/ft}^2)(\tfrac{23}{12}) = 484.7 \text{ ft} \cdot \text{kips } (657.3 \text{ kN} \cdot \text{m})$$

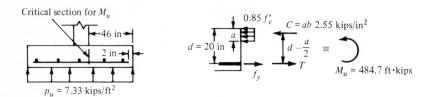

**Figure 8.14**

Use trial method to establish $A_s$

$$M_u = \phi T\left(d - \frac{a}{2}\right) \qquad \text{guess } a = 1.5 \text{ in}$$

$484.7(12) = 0.9T(20 - 0.75)$, and $T = 335.7$ kips. Then

$$A_s = \frac{T}{f_y} = 6.71 \text{ in}^2 \ (4329.3 \text{ mm}^2)$$

Check $\dot{a}$. Equate $T = C$

$$335.7 \text{ kips} = a(108)(2.55 \text{ kips/in}^2) \qquad \text{and} \qquad a = 1.22 \text{ in}$$

Therefore the initial estimate is OK. The minimum $A_s$ is

$$\frac{200}{f_y}b_w d = \frac{200}{50,000}(108)(20) = 8.64 \text{ in}^2 \ (5574.5 \text{ mm}^2) \qquad \text{controls}$$

Use 15 no. 7 bars; $A_{s,\,\text{sup}} = 9.02 \text{ in}^2 \ (5819.7 \text{ mm}^2)$.
6. Check anchorage. $l_d = 0.04 A_b f_y/\sqrt{f'_c}$ but not less than $0.0004 d_b f_y$, or 12 in

$$l_d = \frac{0.04(0.6)(50,000)}{\sqrt{3000}} = 21.9 \text{ in } (556.3 \text{ mm}) \qquad \text{controls}$$

$$= 0.0004(0.875)(50,000) = 17.5 \text{ in}$$

Adjusted $l_d$

$$l_d = (21.9 \text{ in})0.8(s > 6 \text{ in}) = 17.5 \text{ in } (445 \text{ mm})$$

If 2 in of cover is allowed, 44 in is available for anchorage from the face of the column to the edge of the footing; anchorage OK.

combined footing is treated like a beam in the longitudinal direction for flexure, the shear provisions that apply for beams should also be used; i.e., the maximum value of factored shear $V_u$ on the critical section, a distance $d$ from the face of the column, should not exceed $\phi V_c/2$ unless shear reinforcement is provided. Other designers, following the provisions of ACI Code §11.5.5.1, allow a factored shear of $\phi V_c$ on the unreinforced web. If the footing is wide and shallow, the latter assumption appears reasonable, particularly since neither footings in contact with soil nor buried footings are subject to large temperature and shrinkage strains that might produce cracking and lead to a reduction in shear capacity. If the footing is narrow and deep, using the beam provisions for shear seems more appropriate. Because the failure of a footing could lead to serious damage of the superstructure, footing design should be very conservative; therefore, the shear provisions for beams will be used in this text.

**Example 8.6: Design of a combined footing**   Design a rectangular combined footing to support the two columns shown in Fig. 8.20. Property lines require that the footing not extend beyond the face of column $A$ more than 15 in (381 mm). The allowable soil pressure at the base of the foundation must not exceed 4 kips/ft² (191.5 kPa); $f'_c = 3$ kips/in² (20.68 MPa), $f_y = 60$ kips/in² (413.7 MPa), the unit weight of soil is $\gamma_s = 130$ lb/ft³ (2082 kg/m³), and $\gamma_c = 150$ lb/ft³ (2403 kg/m³).

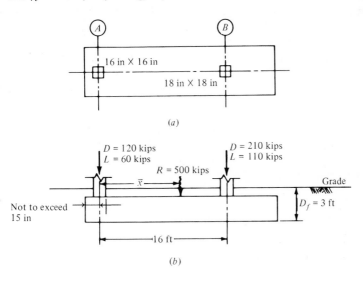

(a)

(b)

**Figure 8.20**

SOLUTION

   1. Locate the position of the resultant of the column loads by summing moments about the centerline of column $A$. The resultant $R$ of the service loads, is 180 kips + 320 kips = 500 kips (see Fig. 8.20b)

$$500\bar{x} = (320 \text{ kips})(16)$$

$$\bar{x} = 10.2 \text{ ft (3.1 m)}$$

2. *Length of footing* To produce a uniform distribution of soil pressure on the base of the footing, the centroid of the footing will be positioned under the resultant of the service loads carried by the columns. The maximum distance the footing can be extended to the left of the resultant is controlled by the limitation that the footing not extend more than 15 in beyond the outside face of column $A$; therefore, the maximum distance between the center and the left end of the footing (half the footing length) equals

$$10.2 \text{ ft} + \tfrac{8}{12} \text{ (half the column thickness)} + \tfrac{15}{12} = 12.2 \text{ ft} \quad \text{use 12 ft}$$

The distance the footing extends beyond the face of column $A$ is

$$12 - (10.2 + \tfrac{8}{12}) = 1.13 \text{ ft} < 15 \text{ in}$$

3. *Width of footing* Assume footing 2 ft deep; determine the net soil pressure $q_n$

$$q_n = 4 - [2(0.15) + 0.13] = 3.57 \text{ kips/ft}^2$$

The required area of footing (based on service loads) is

$$\frac{R}{q_n} = \frac{500 \text{ kips}}{3.57 \text{ kips/ft}^2} = 140.06 \text{ ft}^2$$

$$\text{Width of footing} = \frac{\text{area}}{\text{length}} = \frac{140.05 \text{ ft}^2}{24 \text{ ft}} = 5.84 \text{ ft} \quad \text{use 6 ft}$$

4. *Shear and moment curves for footing in the longitudinal direction* Treat the footing as a beam loaded by factored service loads. When the ratio of live to dead load in each column is different, the position of the resultant of the factored loads will deviate from the position of the resultant of the service loads. If the resultant of the factored loads is eccentric with respect to the centroid of the footing, which is located under the resultant of the service loads, the bending stresses created by the eccentricity and the direct pressure of the factored loads will combine to produce a trapezoidal variation of soil pressure (Fig. 8.4b). To retain a uniform distribution of soil pressure under factored loads (and thereby simplify the design computations), all loads can be multiplied by the same average-load factor, which is based on the ratio of the total factored load to that of the total service load. Since the service loads rather than the factored loads establish the distribution of soil pressure, the use of a uniform pressure seems logical.

Alternatively, as a slightly more conservative procedure, the average load factor can be based on the ratio of the factored load to the service load for that column in which the ratio of live to dead load is maximum. Following the first procedure, the design loads are computed as

$$\text{Average load factor} = \frac{D(1.4) + L(1.7)}{D + L}$$

$$= \frac{(120 + 210)(1.4) + (60 + 110)(1.7)}{500}$$

$$= 1.502 \quad \text{use 1.51}$$

The factored loads are

Column $A$: $\qquad P_u = 1.51(120 + 60) = 271.8 \text{ kips}$

Column $B$: $\qquad P_u = 1.51(210 + 110) = 483.2$

$$\text{Total} = 755.0 \text{ kips}$$

The factored soil pressure is

$$p_u = \frac{P_u}{A} = \frac{755 \text{ kips}}{(24)6 \text{ ft}^2} = 5.24 \text{ kips/ft}^2$$

The factored load per foot in the longitudinal direction is

$$w_u = \frac{755 \text{ kips}}{24} = 31.46 \text{ kips/ft}$$

While the column loads in Fig. 8.21 are assumed to act uniformly over the width of the column in the longitudinal direction (a standard design assumption), calculations can be simplified slightly without much change in results by treating the column loads as concentrated forces applied at the centerline of the column.

5. *Required depth of footing* Once the shear and moment curves are drawn for the longitudinal direction, the depth of the footing can be established based on the maximum values of shear and moment (see Fig. 8.21).

In the longitudinal direction the depth required for maximum moment $M_u = 709.38$ ft · kips may be established with Eq. (3.46).

$$M_u = \phi \rho f_y b d^2 \left( 1 - \frac{\rho f_y}{1.7 f'_c} \right) \tag{3.46}$$

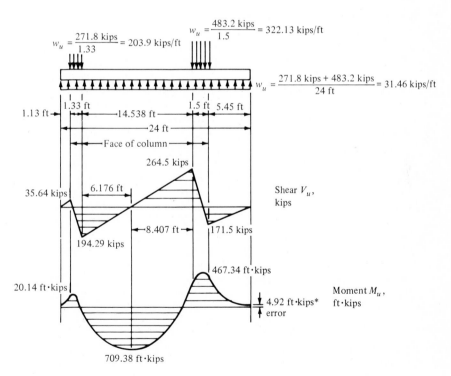

*Since numbers are carried to two significant figures in calculations, a small roundoff error results.

**Figure 8.21** Shear and moment curves in the longitudinal directions.

where $\rho \leq \frac{3}{4}\rho_b = 0.016$. For $\rho = 0.014$, $b = 72$ in, and the specified values of $f'_c$ and $f_y$, Eq. (3.46) indicates a required $d = 14$ in.

*Depth required for punching shear* Several trials indicate that a depth $h = 21$ in ($d = 17$ in) is required to provide adequate strength for punching shear. The failure surfaces are shown in Fig. 8.22.

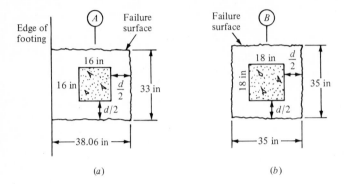

**Figure 8.22** Failure planes for punching shear: (a) column A, (b) column B.

At column $A$ the computation for the last trial follows.

$$V_u = 271.8 \text{ kips} - \left[\frac{38.06(33)}{144}(5.24 \text{ kips/ft}^2)\right] = 226.1 \text{ kips}$$

Punching shear capacity at column $A$ on the $U$-shaped failure surface is

$$\phi V_c = \phi(4)\sqrt{f'_c}b_0 d = 0.85(4)\frac{\sqrt{3000}}{1000}[38.06(2) + 33](17) = 345.5 \text{ kips}$$

At column $B$ check punching shear.

$$V_u = 483.2 \text{ kips} - \tfrac{35}{12}(\tfrac{35}{12})(5.24 \text{ kips/ft}^2) = 438.6 \text{ kips}$$

$$\phi V_c = 0.85(4)\frac{\sqrt{3000}}{1000}(35)(4)(17) = 443.22 \text{ kips}$$

For the depth required for beam shear if no stirrups are used, check shear at a distance $d$ to the left of column $B$. Assume footing 22 in deep ($d = 18$ in). The depth required for punching shear has been rounded off to the nearest even larger inch.

$$V_u = 264.5 \text{ kips} - \tfrac{18}{12}(31.46 \text{ kips/ft}) = 217.31 \text{ kips}$$

$$\phi V_c = 0.85(2)\frac{\sqrt{3000}}{1000}(72)(18) = 120.67 \text{ kips}$$

Since $V_u > \phi V_c$, shear capacity is not adequate; increase depth. Additional trials indicate that a 32-in-deep footing ($d = 28$ in) is required. As an alternate design, select $h = 22$ in ($d = 18$ in), and design stirrups to supply the required shear capacity.

6. *Design of stirrups*

$$V_u = \phi(V_c + V_s)$$

where $\phi V_c = 120.67$ kips, $V_u = 217.31$, and $\phi = 0.85$

$$217.31 \text{ kips} = 120.67 \text{ kips} + \phi V_s$$

$$V_s = 113.69 \text{ kips}$$

Since $V_s < 4\sqrt{f_c'}b_w d$, the maximum spacing is $s = d/2 = 9$ in. Compute $A_v$ with $s = 9$ in

$$A_v = \frac{V_s s}{f_y d} = \frac{113.69(9)}{60(18)} = 0.95 \text{ in}^2$$

$$A_{v,\min} = \frac{50b_w s}{f_y} = \frac{50(72)(9)}{60,000} = 0.54 \text{ in}^2$$

Use no. 5 stirrups with four legs (see Fig. 4.4*f*) at 9 in. $A_{v,\sup} = 1.23$ in$^2$. Run stirrups until $V_u < \phi V_c/2 = 60.33$ kips.

7. Using the trial method, compute the area of flexural steel required to carry the maximum negative moment $M_u = 709.38$ ft · kips

$$M_u = \phi T\left(d - \frac{a}{2}\right) \qquad \text{guess } a = 3 \text{ in}$$

$$12(709.38) = 0.9T(18 - 1.5) \qquad T = 573.24 \text{ kips}$$

Since equating $T = C$ gives $a = 3.12$ in, the initial value of $a$ is satisfactory.

$$A_s = \frac{T}{f_y} = \frac{573.24 \text{ kips}}{60 \text{ kips/in}^2} = 9.55 \text{ in}^2$$

$$A_{s,\min} = \frac{200b_w d}{f_y} = \frac{200(72)(18)}{60,000} = 4.32 \text{ in}^2$$

Use ten no. 9; $A_{s,\sup} = 10$ in$^2$. Computations for the area of flexural steel required to carry the positive moment, $M_u = 467.34$ ft · kips at the exterior face of column $B$ indicate that $A_{s,\text{req}} = 6.11$ in$^2$. Use eight no. 8 bars to give $A_s = 6.28$ in$^2$.

8. Design 22-in-deep footing to carry positive moment of $M_u = 20.14$ ft · kips at the exterior face of column $A$. Since the moment is small (well below the cracking moment), the flexural capacity of the uncracked concrete will be checked. For evaluating the flexural capacity of plain concrete, the concrete will be treated as an elastic material and a reduction factor of $\phi = 0.65$ used, as required by ACI Code §9.3.2

$$\phi M_n = \frac{\phi f_r I_g}{y_t} = \frac{0.65(7.5)\sqrt{3000}(63,888)}{11(12,000)} = 129.24 \text{ ft · kips}$$

Although the flexural capacity of the unreinforced concrete is adequate, four no. 8 bars from the positive steel at the right end will be run continuously along the bottom to the left end to tie the footing together and to hold the stirrups in position (see Fig. 8.24*b*).

9. Check the development length of the top steel in step 7

$$l_{d,\,req} = \frac{0.04 A_b f_y (1.4)}{\sqrt{f'_c}}$$

$$= \frac{0.04(1)(60,000)(1.4)}{\sqrt{3000}} = 61.3 \text{ in} = 5.1 \text{ ft}$$

The anchorage is OK since 6.84 ft is available on the left end if the bars extended to the centerline of each column and a greater distance is available on the right side.

10. *Area of transverse steel required under columns* At column $A$, the critical section is at the face of the column. The effective width of the local strip is

$$b_e = \frac{d}{2}(2) + \text{column width} = \tfrac{18}{2}(2) + 16 = 34 \text{ in}$$

$M_u$ at the face of the column (see Fig. 8.23), assuming the entire column load is carried by the local transverse beam, is

$$45.3(\tfrac{28}{12})(\tfrac{14}{12}) = 123.3 \text{ ft} \cdot \text{kips}$$

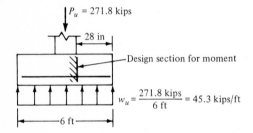

$P_u = 271.8$ kips

28 in

Design section for moment

$w_u = \dfrac{271.8 \text{ kips}}{6 \text{ ft}} = 45.3$ kips/ft

6 ft

**Figure 8.23** End view of footing showing design section for transverse moment.

For a value of $d = 18$ in, the required $A_s$ to carry 123.3 ft · kips is 1.58 in$^2$

$$A_{s,\,min} = \frac{200}{f_y} bd = \frac{200(34)(18)}{60,000} = 2.08 \text{ in}^2 \qquad \text{controls}$$

Use five no. 6 to give $A_s = 2.2$ in$^2$. At column $B$ a similar analysis for the transverse steel indicates that $A_{s,\,req} = 2.65$ in$^2$ controls; again use six no. 6 bars.

A check of development length indicates that 15.4 in are required. If bars terminated 2 in from edge of footing, 25 in are available; OK (see Fig. 8.24 for details of main reinforcement).

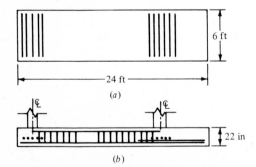

6 ft

24 ft

(a)

22 in

(b)

**Figure 8.24** Details of reinforcement: (a) plan view showing transverse reinforcement, (b) elevation showing longitudinal and shear reinforcement.

# PROBLEMS

**8.1** A combined footing 18 ft long by 6 ft wide supports two columns that carry the factored loads shown in Fig. P8.1. If the foot is assumed to be rigid, (*a*) determine the distribution of the soil pressure under the base produced by the column loads; the weight of the footing may be neglected. (*b*) Draw the shear and moment curves for the footing in the longitudinal direction.

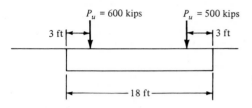

**Figure P8.1**

**8.2** A wall footing (Fig. P8.2) is founded 6 ft below grade on a soil with an allowable bearing capacity of 6 kips/ft$^2$ at the base of the footing. If the footing supports a 10-in-thick concrete wall that carries a uniform load of $D = 20$ kips/ft and $L = 8$ kips/ft, design the footing; $f_c' = 3$ kips/in$^2$, and $f_y = 60$ kips/in$^2$.

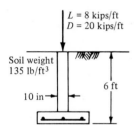

**Figure P8.2**

**8.3** Design a square footing to support an 18- by 18-in column (Fig. P8.3). The allowable soil pressure is 10 kips/ft$^2$; $f_c' = 3$ kips/in$^2$, and $f_y = 60$ kips/in$^2$.

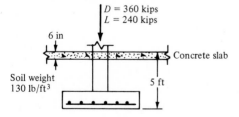

**Figure P8.3**

**8.4** Repeat Prob. 8.3 for service loads of $D = 600$ kips and $L = 360$ kips.

**8.5** Design a rectangular footing to support a 16- by 16-in column that carries service loads of $D = 300$ kips and $L = 240$ kips. The footing is 4 ft below grade, and the width is not to exceed 7 ft. The allowable soil pressure is 8 kips/ft$^2$; $f_c' = 3.5$ kips/in$^2$, and $f_y = 50$ kips/in$^2$. The soil weighs 130 lb/ft$^3$.

**8.6** The bottom of a centrally loaded square footing 6 by 6 ft in plan and 18 in deep (Fig. P8.6) is reinforced with eight no. 5 bars in each direction. Determine the maximum value of service load the footing can support if dead load makes up 50 percent of the service load. Check both the soil strength and the capacity of the footing for beam shear, punching shear, and moment. Allowable soil pressure is 6 kips/ft$^2$, the unit weight of soil is 125 lb/ft$^3$, $f'_c = 3$ kips/in$^2$, and $f_y = 60$ kips/in$^2$.

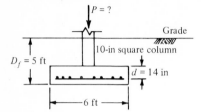

**Figure P8.6**

**8.7** A 10- by 10-in column, located at the center of a square two-way footing, carries a vertical load of 240 kips and a horizontal load of 10 kips (Fig. P8.7). The soil study indicates that a 7.5- by 7.5-ft footing is required to prevent overstress of the soil. Establish the depth and reinforcement required for the footing. The horizontal load can act in either direction. $f'_c = 3.5$ kips/in$^2$, and $f_y = 60$ kips/in$^2$.

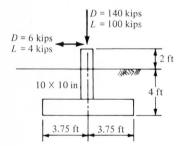

**Figure P8.7**

**8.8** Design a combined footing to support the columns shown in Fig. P8.8. The column footing cannot extend more than 6 in beyond the left face of column $A$. The allowable soil pressure is 5 kips/ft$^2$, the unit weight of soil is 130 lb/ft$^3$, $f'_c = 3$ kips/in$^2$, and $f_y = 60$ kips/in$^2$.

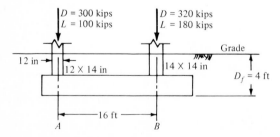

**Figure P8.8**

**8.9** Design a rectangular combined footing to support two 10-in walls spaced 10 ft apart (Fig. P8.9). Each wall carries service loads of $D = 22$ kips/ft and $L = 20$ kips/ft. The allowable soil pressure at the base of the footing is 6 kips/ft². The unit weight of the soil is 125 lb/ft³; $f'_c = 3$ kips/in², and $f_y = 60$ kips/in².

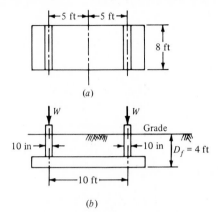

(a)

(b)

**Figure P8.9** (a) Plan, (b) elevation.

# REFERENCE

1. R. Peck, W. Hanson, and T. Thornburn: "Foundation Engineering," 2nd. ed., Wiley, New York, p. 309, 1974.

Guggenheim Museum, Fifth Avenue, New York City, designed by Frank Lloyd Wright, architect. The museum's collection is displayed on the curved walls that spiral around the building. (*Photograph by the Guggenheim Museum.*)

# NINE

## THE DESIGN AND ANALYSIS OF MULTISTORY BUILDING FRAMES

### 9.1 INTRODUCTION

The first seven chapters have covered the behavior and design of beams, one-way slabs, and beam-columns. For the most part, the members studied were either statically determinate beams or single-story lengths of columns. In every case, the members were simple to analyze or the end forces were given as known quantities. Since most reinforced concrete elements actually exist as components of highly indeterminate, continuous, three-dimensional frames, we now consider the analysis and design of large multistory rigid-jointed frames.

The basic objective of the frame analysis is to determine the moment at the ends of each member. After the member end moments have been established, the end shears can be computed by applying the equations of statics to a free body of each member. With the shears and moments known, the axial forces in members can be established from the equilibrium of the joints (see Fig. 7.2).

Although some modification of moments is permitted to account for inelastic behavior, the ACI Code specifies that the forces in a reinforced concrete frame, acted upon by factored loads, are to be determined by an *elastic* analysis. In practice, rigid frames of reinforced concrete are most commonly analyzed by moment distribution or by computer. Regardless of the method of analysis or the precision of the computations, the results of the analysis will not give the exact values of internal forces in members. The actual values of forces will deviate from the computed values for the following reasons:

The properties of the cross section are uncertain because the extent and position of cracking are indeterminate.

Most analyses neglect the influence of reinforcement on the flexural stiffness of the member.

Creep, shrinkage, and differential settlements produce deformations of unknown magnitude. These deformations create additional moments in indeterminate structures.

Certain highly stressed sections of reinforced concrete structures are loaded into the inelastic region. Once a section deforms plastically, an elastic analysis no longer predicts the distribution of forces.

The stiffening effects of nonstructural walls and partitions are neglected.

Concrete is not a linear-elastic material.

Nevertheless, practical experience with concrete structures over time indicates that structures designed by an elastic analysis are safe if the provisions of the ACI Code are followed and if an adequate value of live load is selected. In a well-designed, ductile, indeterminate structure, local overstress does not produce failure but results in a shift of force to less stressed sections.

Since the stiffness of the members of an indeterminate structure influences the distribution of forces in the structure, the size of the members must be known before the structure can be analyzed precisely. In other words, the size of columns and walls and the proportions of the floor system have to be established by a preliminary design so that the properties of the gross sections can be used to compute, for example, the distribution factors for analysis by moment distribution or the stiffness coefficients for a matrix analysis. If the behavior of members is understood, estimates of member size can often be made accurately by approximate computations or knowing when minimum depths control.

Once the frame has been analyzed and the forces—shear, moment, and axial load—have been determined at all sections, the member can be proportioned and the appropriate reinforcement selected. If the final proportions of members are the same as those estimated for the initial analysis, the design is complete, but if the analysis indicates that the initial estimates of the dimensions of certain members were inaccurate, the analysis must be repeated using revised cross-sectional properties, which can be estimated more accurately once the structure has been analyzed. For many indeterminate structures a small change in member size produces little change in the results of the analysis; therefore, building frames rarely have to be analyzed more than twice by experienced designers.

Although many of the computations required to analyze a highly indeterminate reinforced concrete structure are relatively routine once mastered, the establishment of a logical structural system that harmonizes with the architecture is a much more sophisticated process. Since each structure tends to be unique in certain of its architectural requirements: design loads, foundation conditions, and so forth, it is not possible to set up a standard procedure for automatically establishing an optimum structural system. The process of selecting an optimum structure is well stated by Siegel,[1] who writes of the great Italian engineer Pier Luigi Nervi

Nervi says that all his designs emerge spontaneously from a knowledge of the distribution of the forces and a search for forms that express, simply and economically, the way they interact. This is very much to the point. Such forms are not "computed," they are designed. Computations are simply means of checking whether the form is compatible with the stresses. Other forms might also be proposed and found capable of withstanding the loads. However, only one form is best, or as Nervi says, structurally "right," and this is the form in which the principle of economy is most clearly expressed. Searching for it and finding it is a creative act, which fundamentally has nothing to do with "computations." By the "principle of economy" we mean more than simply economical in price. It is a complex idea that permeates the entire design process. The greatest effects, even esthetic effects, are achieved with the most sparing means; this is a principle that applies to every area of creative activity ...

Before discussing the analysis of multistory structures we make some observations on the influence of column spacing on the configuration of the floor system in Sec. 9.2 to provide some understanding of the conditions under which a particular floor system is appropriate.

Examples in this chapter will deal with one-way floor systems, floors in which the loads are transmitted by flexure in one direction only. The design of two-way systems, which bend in double curvature, will be discussed in Chap. 10.

## 9.2 FACTORS THAT INFLUENCE THE CONFIGURATION OF FLOOR SYSTEMS

The structure of a building is basically a series of floors supported at more or less regular intervals by columns and walls. The floors, which are interconnected to the columns by continuous reinforcement to form rigid joints, are typically composed of slabs or slabs stiffened by beams. The spacing of columns is primarily determined by the architectural requirements of the structure. For example, if large open unobstructed floor areas are needed, most of the columns will be located around the perimeter of the building and the number of interior columns will be held to an absolute minimum. On the other hand, if the floors are to be partitioned into offices or rooms, closer spacing is possible.

As a general rule the weight of the building frame, the floors and the columns, decreases as the distance between columns decreases; i.e., the closer the columns, the lighter the building. This reduction in weight occurs because the depths of slabs and beams in the floor system reduce as their spans (and consequently the loads they must carry) reduce. As long as the capacity of most building columns is controlled by strength rather than buckling, the total area of all columns at a given level is approximately proportional to the total vertical load they support. Except for the columns in the upper two or three floors of a structure, where the loads are relatively small (and dimensions may be controlled by minimum clearances required for the concrete to flow into forms) a reduced floor weight permits a reduction in column areas at all floor levels. In general practice, the minimum spacing between columns is in the range of 8 to 10 ft (2.44 to 3.05 m) with an average spacing on the order of 15 to 16 ft (4.6 to 4.9 m). In unusual circumstances, columns might be placed as close as 4 or 5 ft (1.2 to 1.5 m) on the perimeter.

## Types of Floor Systems

The configuration of the floor system will be influenced primarily by the column spacing and the magnitude of the design loads. When a floor carries a uniformly distributed live load and the columns are closely spaced in straight lines forming a square or nearly square two-way grid (Fig. 9.1a), a simple slab (or flat plate) may have sufficient strength to transfer floor loads into the columns. Although a slab is simple and economical to form, produces a ceiling that requires little effort to finish, and provides a maximum of headroom, its ability to transfer shear into the columns is low. In addition slabs tend to be flexible and may deflect excessively. Therefore, the slab is most practical when floor loads are light and columns are closely spaced, say in the range of 10 to 15 ft (3.05 to 4.6 m).

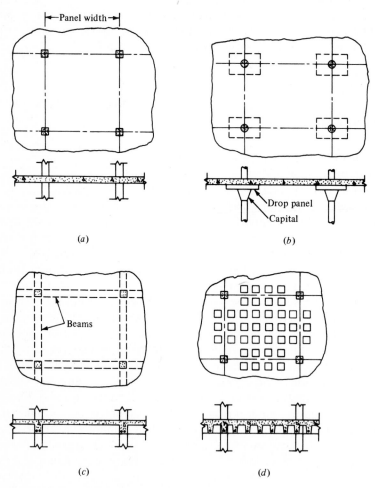

**Figure 9.1** Two-way slab systems: (a) flat plate, (b) flat slab, (c) flat slab with beams along column centerlines, (d) two-way grid or waffle slab.

As the distance between columns on a rectangular grid increases, say in the range of 18 to 24 ft (5.5 to 7.3 m), or if the magnitude of the floor loads is large, a slab of constant depth may not be able to carry the loads efficiently. To increase the capacity of the floor system, the underside of the slab may be thickened locally around columns with rectangular pads of concrete called *drop panels*, or the tops of columns may be flared outward to form an enlarged support (called a *capital*) (Fig. 9.1*b*), or beams may be added along the centerlines of columns (Fig. 9.1*c*).

If the spans are large, intermediate beams may be added to support the slab (Fig. 9.2*a*). These beams are often added at the midspan or third points of the panel and break the slab up into a series of narrow rectangular sections that span in the short direction between the intermediate beams. The intermediate beams are assumed to behave like continuous beams supported vertically by the girders; i.e., the girder is assumed to provide the equivalent of a simple vertical support at the points where it is intersected by the intermediate beams. In regions of positive moment, the floor slab acts with both the intermediate beams and the girders to

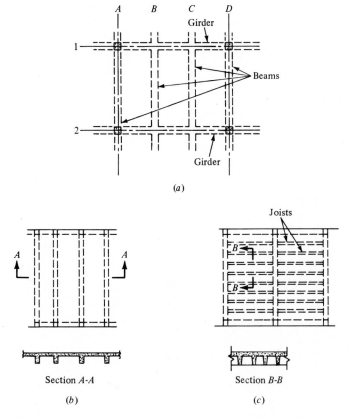

(*a*)

Section *A-A*

(*b*)

Section *B-B*

(*c*)

**Figure 9.2** One-way structural systems: (*a*) beams and girders, (*b*) T beam and slab, (*c*) one-way joist and girder.

form a T beam. To determine the best position for the intermediate beams, the designer should investigate alternative layouts and select the one that produces the most economical design.

If spans are large, say 25 to 30 ft (7.6 to 9.1 m), a two-way grid of beams may also provide an economical framing system. The use of beams spanning in two directions permits a shallower floor system than a floor composed of a smaller number of beams spanning in one direction. Two-way grids or *waffle* slabs are often formed by positioning heavy square metal pans with tapered sides in the bottom of a panel in the region of positive moment. The gaps between the metal pans, which are approximately 20 in (51 cm) on a side, form the beam webs in which the reinforcing steel is placed (see Fig. 9.1*d*). Since the slab acts with the stems to form a T beam in the regions of positive moment, the use of metal pans reduces the dead weight of the floor without any significant reduction in flexural capacity. To improve the transfer of shear into the columns, metal pans are omitted in the regions surrounding the column to increase the area through which shear is transferred.

If a column-free interior is required, the floor slab may be supported on deep girders spanning between columns located on the perimeter of the building. If the spacing between girders is large, say 15 ft (4.6 m) or more, shallow, closely spaced beams, which support a thin slab several inches thick, are often used to span between girders because of their greater strength and stiffness (Fig. 9.2*c*). These beams, called *joists*, are formed with heavy prefabricated metal pans.

For economy of construction, only a few sizes of beams are used in a typical floor system. By limiting the number of sizes, formwork is simplified and a desirable uniform surface appearance is obtained. Typically, the designer reviews the structural layout to determine where beams will be placed. From a group of similar-sized beams, the one that supports the greatest load will be designed with the maximum allowable area of steel to keep the beam size as small as possible. All other beams that are to be made the same dimensions will then be reinforced as required by the loads that they support.

## 9.3 BEHAVIOR OF RECTANGULAR SLABS SUPPORTED ON FOUR SIDES

In Figs. 9.1 and 9.2 several of the floor systems illustrated consist of rectangular slabs supported on four sides by beams. Since slabs of this type will be used in several examples in this chapter, their behavior, which is influenced by their proportions, will be reviewed.

At first glance, the problem of analyzing a continuous floor system composed of a multitude of slabs and beams appears to be a formidable task. Since a slab, which bends in two directions is a highly indeterminate structural element, an *exact* analysis of any slab is a complex mathematical problem regardless of the boundary conditions. Nevertheless, without going into the mathematics of slab analysis, it is often possible to gain insight into the distribution of moments in a slab by treating a slab as a two-way grid of interconnected beams that span between supports. By examining the loads carried by various beams the designer can often carry out a qualitative analysis of the slab. Also, since the moment in a beam is a direct function of curvature, $M = EI\, d^2y/dx^2$, the relative magnitude of moments on certain sections can be inferred from the deflected shape of beam strips crossing the section under study. If the variation of moments is known qualitatively, assumptions can often be made to simplify the analysis and design of a slab.

To illustrate the method, the moments in a uniformly loaded rectangular slab whose edges are assumed to be simply supported along all four sides will be investigated (see Fig. 9.3a). The slab is assumed to be approximately square and has a long- to short-side ratio $l_1/l_2$ slightly greater than 1. Since the strips of slab act as simply supported beams, the maximum moment in any strip will develop at midspan (Fig. 9.3b). By comparing the deflected shape of three beam strips in the $y$ direction (strips 1, 2, and 3) one can see that the bending moments are much larger in the center, where the deflections and the curvature of the strips are largest.

At the boundaries, the moment in the slab is zero both parallel and perpendicular to the edge support. The moment perpendicular to the edge is zero because of the simple support. The moment parallel to the edge is also zero since the edge support prevents bending (see strip 3). The approximate variation of midspan moment for strips of slab spanning in the short direction (the $y$ direction) is shown in Fig. 9.3d.

Since the slab is nearly square, significant curvature and consequently moment also develop in the long direction. If the curvature of strips 1 and 4 is compared, it is evident that strip 1 has a greater curvature than strip 4. Since both strips deflect the same amount at midspan, strip 1 must slope more steeply from the supports

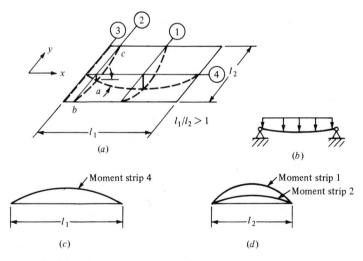

**Figure 9.3** Variation of moment in a uniformly loaded slab that is simply supported on all four sides: (a) deflections of slab, (b) cross section, (c) moment curve for strip 4, (d) moment curves for strips 1 and 2.

than strip 4 to deflect the same amount in a shorter distance. Figure 9.3c shows the variation of the slab moment in strip 4. The design of square or nearly square slabs will be covered in Chap. 10.

If a strip is not located on an axis of symmetry, it will twist as well as bend. For example, at points $b$ and $c$ the ends of strip 2 attached to the supports are horizontal in the direction parallel to the continuous edge support, but at point $a$ the strip has twisted through an angle that is equal to the slope of strip 4 at point $a$. The twist of the strip indicates that torsional moments have been created by the deformation of the slab.

The second slab to be considered is also simply supported on all four edges but has narrow rectangular proportions (see Fig. 9.4). The ratio of the long to short sides is assumed to be greater than 2. An examination of the deformations of slab strips 1, 2, and 3 in the short direction indicates that except for a short distance at either end, the curvature and consequently the moment are relatively constant throughout the length of the slab. The variation of moment in the short direction along strip 2 is shown in Fig. 9.4c. In the long direction (see strip 4) the deflections are nearly constant in the center region. The lack of curvature in the center region indicates that the moment in the longitudinal direction is nearly zero (see Fig. 9.4b). Therefore, a long rectangular slab supported on all four sides can be designed over most of its length for bending in the short direction only.

To compute the required slab reinforcement for a narrow rectangular slab, a strip of slab 1 ft (0.3 m) wide in the short direction is designed as a simple beam with a uniform load. The reinforcement for the strip is then used throughout the entire length of the slab, even at the ends where the moment in the short direction is small. Although no significant reinforcement is required in the long direction for flexure except near the ends of the slab, where the curvature in the long direction develops, some reinforcement must be provided throughout the entire length to

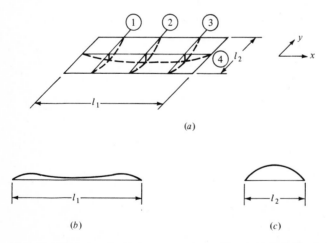

(a)

(b)                              (c)

**Figure 9.4** Variation of midspan moment in a uniformly loaded, simply supported, narrow rectangular slab: (a) deflected shape, (b) moment curve for strip 4, (c) moment curve for strip 2.

reduce the width of temperature and shrinkage cracks (see Sec. 3.14). At the ends of the slab in the long direction, the flexural strength produced by the temperature steel together with the steel in the short direction provides adequate strength to carry the design loads without excessive cracking.

If the slab is continuous over the supports, top steel must be added perpendicular to the support lines to carry the negative moments. In the design of one-way slabs all load is assumed to be carried by flexure in one direction. Torsion and bending in the long direction are neglected.

## 9.4 LATERAL BRACING OF BUILDINGS

To provide maximum rentable floor area, building columns are usually made as small as possible. While the axial capacity of a column with a small cross section can be adequate for gravity loads, its lateral bending strength and stiffness may be inadequate for lateral load. Therefore, if the area of the building columns is to remain small, a building frame acted upon by horizontal loads, e.g., those created by wind or earthquake, must be braced against lateral displacement to prevent excessive horizontal movement of the floors and overstress of the frame. Buildings without adequate lateral stiffness are susceptible to wind-induced vibrations that can create psychological stress and fatigue, and large lateral movements can also damage nonstructural elements attached to the building frame. Moreover, if the designer depends on the flexural strength of the frame to provide lateral resistance, the columns, which are then unbraced, will have a smaller capacity for axial load than identical columns in a frame braced laterally by attachment to a rigid structural element (see Fig. 7.10).

To brace a building frame laterally the engineer must attach the floors of the frame to walls or other rigid elements. If structural walls, called *shear walls*, cannot be added because of the architectural requirements, the elevator shaft or the stairwells can be designed as large hollow box beams to brace the frame. These elements, fixed at their base by the foundations, act like cantilever beams. Since the lateral bending (in-plane) stiffness of shear walls or the elevator core is many times greater than that of all the columns combined, lateral load is largely carried to the foundations by these elements. To prevent twisting of the structure when lateral loads act, the shear walls or other bracing elements should be positioned symmetrically in the structure. During the initial planning stages with the architect, it is important for the engineer to emphasize the need for incorporating shear walls of adequate stiffness into the structure.

## 9.5 PRELIMINARY DESIGN

After the structural system has been established, the size of all components—slabs, girders, columns and walls—must be estimated before the dead weight can be computed or the structure analyzed. If a designer has previously analyzed structures of the same type, information on member sizes will often be available from existing structural drawings. On the other hand, if a designer has no experience with a particular system, sizes can be estimated by simple computations. For example, in a uniformly loaded continuous beam of constant size, the maximum moments occur adjacent to interior supports and are somewhat larger than the value of the fixed-end moment, $wl^2/12$. To estimate the proportions of a beam's cross section, the designer might base the member's design on a moment of $wl^2/10$.

The required cross-sectional area of a column can often be closely approximated by neglecting the moment and dividing the factored axial load by a stress equal to a percentage of $f'_c$. For interior columns, in which moments are small, a stress of $0.55 f'_c$ is recommended, but it might be reduced to $0.4f'_c$ for exterior columns to account for the larger moments. With experience the structural engineer will develop a sense of the member size required for particular spans and loads.

## 9.6 ANALYSIS OF MULTISTORY BUILDING FRAMES

Because a three-dimensional reinforced concrete building frame has a large number of rigid joints, an indeterminate analysis of the entire frame for a particular set of design loads would be both time-consuming and expensive. Therefore, the designer typically divides the structure into simpler two-dimensional frames or continuous beams for analysis. Designers are given considerable latitude to make simplifications by ACI Code §8.6.1: "Any reasonable assumptions may be adopted for computing relative flexural and torsional stiffnesses of columns, walls, floors, and roof systems. Assumptions shall be consistent throughout the analysis."

Most structures require at least two distinct analyses. The first covers the design of the floors and roof for gravity loads. Since floor loads stress only the floor to which they are applied, the designer can analyze each floor independently without considering the loads on any other floor. This analysis is made by considering that the floor and its attached columns act as a rigid frame (Fig. 9.5b). The analysis is further simplified by assuming that the far ends of the columns are fixed and the floor does not sidesway. In the floor analysis load factors of 1.4 and 1.7 are applied to dead and live load respectively.

The second analysis considers the effect of lateral loads—wind and earthquake—in combination with dead and live loads. For this analysis the entire building frame must be considered (Fig. 9.5c) because the loads at all levels contribute to the lateral displacement of the building frame. For the wind analysis, the load factors of 1.4 for dead load and 1.7 for live and wind loads are reduced by an additional factor of 0.75.

If lateral loads might overturn the building or reverse the stress (tension) in certain columns on the windward side, an additional analysis for lateral load is required in which the live load is omitted. For this case the live load would tend to stabilize the structure and reduce overturning. Load factors of 0.9 and 1.3 are applied to the dead and wind loads respectively (see Example 1.1).

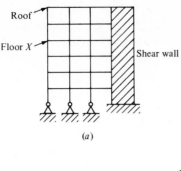

(a)

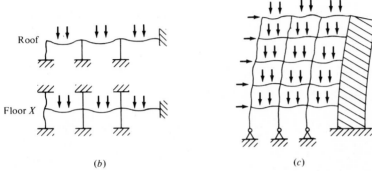

(b)                              (c)

**Figure 9.5** Simplified frames used to analyze a multistory building: (a) typical frame of a multistory building, (b) simplified frames used to determine the forces produced by vertical load in the roof and floor systems, (c) entire frame used for a lateral-load analysis.

If the analysis for lateral load is for earthquake forces rather than wind, the load factors for live and dead load remain the same but the load factors for earthquake induced forces are 10 percent greater than those for wind.

## Analysis of a One-Way Slab-and-Beam Floor

The analysis of the one-way floor system shown in Fig. 9.6 will be discussed to illustrate the simplifications normally made in analyzing the slabs and beams for gravity loads. This floor, constructed as a monolithic unit, is made up of a continuous slab supported by equally spaced beams that run in the $y$ direction. Continuous girders, located along column lines 1, 2, 3, and 4, support the beams and transfer their reactions into the columns. To simplify the discussion, shear walls, stairwells, elevator shafts, and duct openings are omitted.

The previous discussion of one-way slab behavior can be extended directly to the analysis of the uniformly loaded floor slabs in Fig. 9.6, where the intermediate beams divide the floor into a series of one-way slabs. If a typical 1-ft (0.3-m) width of floor is cut out as a free body in the longitudinal direction (see Fig. 9.6b), it is evident that the flexible slab will bend concavely upward between the beam stems and concavely downward over the stems.

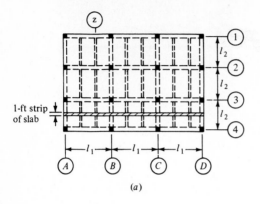

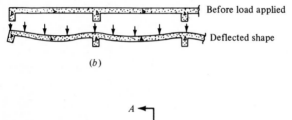

**Figure 9.6** One-way floor system: (*a*) floor plan, (*b*) cross section of floor showing slab and the intermediate beams, (*c*) slab modeled as a continuous beam on simple supports.

The deflected shape indicates that the slab in the short direction resembles a continuous beam spanning between the beam stems, which act as simple supports. The assumption of a simple support neglects the torsional stiffness of the beams that support the slab. If the slabs carry approximately the same load, and if the distance between beam stems is about the same, the slope of the slab over each interior support is nearly zero. If no significant rotation of the beam stems occurs, the torsional stiffness of the stems has little influence on the moments in the slab.

However, at the outside supports the beam stems, loaded from one side only, are twisted by the slab (Fig. 9.7). If the beam is small and its torsional stiffness low, the beam may not be able to offer any significant resistance to the end rotation of the slab, particularily to the slab located near the midspan of the panel away from the columns. For this condition of restraint the designer might assume that the exterior beam, which provides vertical support to the slab but no torsional resistance, is equivalent to a continuous pin support (Fig. 9.7*b*).

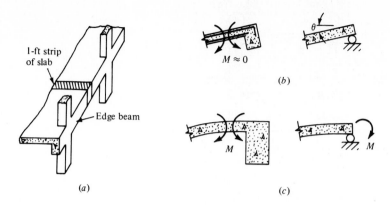

$M \approx 0$

(b)

(a)

(c)

**Figure 9.7** Restraint supplied by an edge beam: (a) slab supported on an edge beam, (b) edge beam with small torsional rigidity provides vertical restraint only, (c) edge beam with large torsional rigidity provides both vertical and rotational restraint.

On the other hand, if the exterior beam is large and has a high torsional rigidity, it will apply a restraining moment to the slab. The beam in turn will be subject to a torque applied by the slab (Fig. 9.7c). The design of the edge beams for torsion was illustrated in Example 5.5. The analysis of a continuous slab will be illustrated in Example 9.1.

To analyze a typical row of indeterminate beams that support the floor slab, e.g., the row of beams along line z in Fig. 9.6a, the designer will pass imaginary sections through the slab on either side of the beams and cut out a free body of the row of beams with the attached slabs (Fig. 9.8a). The intermediate beams are supported by the girders along column lines 1, 2, 3, and 4. As shown in Fig. 9.8, the intermediate beams together with the attached slabs form a T beam. Although the type of support supplied to the beams by the girders is left to the judgment of the designer, several assumptions are possible. At one extreme, if the girders are small and the columns flexible, the designer might assume that the girders supply vertical restraint only and provide the equivalent of simple supports (see Fig. 9.8b). At the other extreme, if the girders are large and have a high torsional rigidity and the columns are stiff, the designer might treat the floor system and the columns as a rigid frame (Fig. 9.8c). Since each column provides restraint to a half panel of floor on either side of the column centerline, each column resists the bending of three floor beams; although the floor beam framing into the column receives the greatest restraint, the designer might assume that all beams are equally restrained. Assuming that the far ends of the columns are fixed by the balance of the structure would be an appropriate asumption.

Finally, the girders together with the attached columns might be analyzed as a rigid frame loaded by the beam reactions, which might be treated as concentrated loads (Fig. 9.9). The girders will also carry a uniform load that represents both the weight of the girder and several feet of floor slab on either side of the girder centerline.

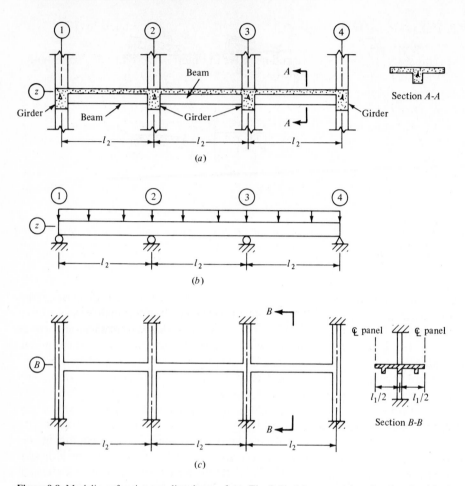

**Figure 9.8** Modeling of an intermediate beam from Fig. 9.6a: (a) cross section showing the side view of the intermediate beams, (b) intermediate beam modeled as a continuous beam, (c) columns with intermediate beams between panel centerlines modeled as a rigid frame.

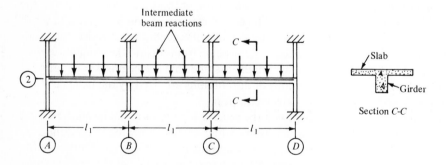

**Figure 9.9** Floor girders and columns from Fig. 9.6a modeled as a rigid frame.

# 9.7 POSITIONING LIVE LOAD FOR MAXIMUM MOMENT

The critical sections for positive and negative moment in the spans of a continuous girder are located at or near midspan and at the supports. These sections must be designed for the maximum moments created by both dead and live loads. Analysis for the dead load, which is fixed in position and magnitude, needs no discussion; however, the maximum value of positive and negative moment at critical sections in a given span are produced by different live-load patterns. Although influence lines can be used to determine the position of live load to maximize moment at a particular section, the same results can be developed for moment in a simpler fashion by applying the principle that *moment is a direct function of curvature*. In other words, the loading pattern that produces the greatest curvature at a particular section also produces the maximum moment at that section.

We shall first develop the rule for positioning live load to maximize the positive moment by establishing the pattern of live load that creates the greatest value of positive moment in span *CD* of the continuous beam in Fig. 9.10*a*. Since the greatest forces in a given span are created by loads applied directly to that span, it is evident that live load should be placed on span *CD*. Also because load applied anywhere on span *CD* contributes to its downward displacement and consequently the positive curvature in the center of the span, live load should be placed over the entire length of span *CD*. The downward displacement of span *CD* by live load causes the two adjacent spans *BC* and *DE* to deflect upward because of the continuity of girder slopes over the supports. The deflected shape produced by load on span *CD* only is shown by the dashed line in Fig. 9.10*a*.

In any span that tends to displace upward (spans *BC* or *DE*) the application of vertical load would push the beam back toward its original position and reduce the curvature in all spans; therefore, to maximize the curvature in span *CD*, the adjacent spans *BC* and *DE* and then the alternate spans (not shown in the figure) should remain unloaded. On the other hand, if spans *AB* and *EF* are deflected downward by the load on span *CD*, the application of live load to these spans will add to the deformations and curvature produced by the load on span *CD*.

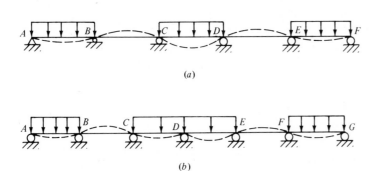

**Figure 9.10** Position of live load to maximize the moment in a continuous beam: (*a*) positive moment in span *CD*, (*b*) negative moment at support *D*.

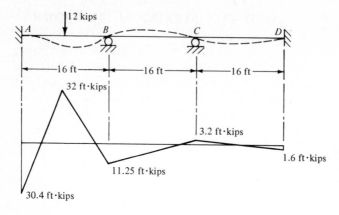

**Figure 9.11** Variation of moment with distance from the loaded span.

In summary, to maximize the positive moment due to live load in a particular span of a continuous structure, place the live load directly on the span being analyzed and then load alternate spans on each side of that span.

In Fig. 9.11 it can be seen that the deformations and moments produced by the load on span AB decrease rapidly with distance from the loaded span. Therefore, the designer can analyze a particular member of a continuous structure with reasonable accuracy by considering the influence of load on spans that are not further away than two spans on either side of the span being analyzed. Often to simplify the analysis of a member the designer will treat joints located a distance of two spans away from the span being analyzed as fixed-ended.

To establish the pattern of loading that will maximize the negative moment in a member at a support, the loading required to maximize the negative curvature at support D in the continuous beam of Fig. 9.10b will be discussed. Since load on both spans CD and DE contributes to the negative moment at D, live load should be placed on both these spans. Physically it can be seen that loads on both spans tend to bend the beam around the support at D. The deflected shape produced by the loads on spans CD and DE is shown by the dashed line in Fig. 9.10b. Using the same reasoning for this case as for the case of positive moment means that load should be placed on any span that deflects downward and omitted from any span that deflects upward. The resulting loading pattern indicates that to maximize the negative moment at a support the spans on either side of the support should be loaded as well as alternate spans.

## 9.8 ANALYSIS BY ACI COEFFICIENTS

ACI Code §8.3 permits the evaluation of moments and shears in continuous beams and one-way slabs by equations which give approximate values of the internal forces at critical sections. Analysis by this method is limited to structures in which the span lengths between supports are approximately equal (the larger of two

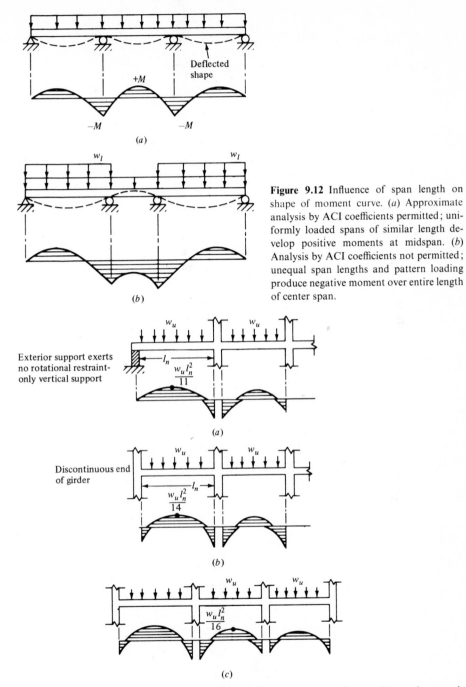

**Figure 9.12** Influence of span length on shape of moment curve. (*a*) Approximate analysis by ACI coefficients permitted; uniformly loaded spans of similar length develop positive moments at midspan. (*b*) Analysis by ACI coefficients not permitted; unequal span lengths and pattern loading produce negative moment over entire length of center span.

Exterior support exerts no rotational restraint—only vertical support

$$\frac{w_u l_n^2}{11}$$

(*a*)

Discontinuous end of girder

$$\frac{w_u l_n^2}{14}$$

(*b*)

$$\frac{w_u l_n^2}{16}$$

(*c*)

**Figure 9.13** ACI coefficients for *positive* moment in beams, slabs, and frames: (*a*) exterior span in which the discontinuous end of the girder is unrestrained, (*b*) exterior span in which the discontinuous end of the girder is integral with support, (*c*) interior span.

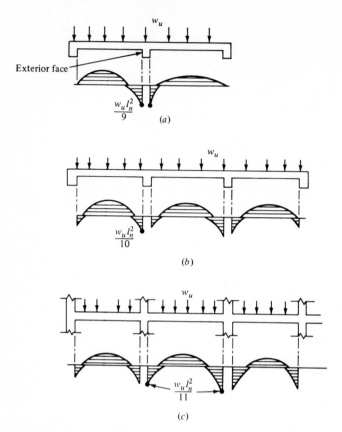

**Figure 9.14** ACI coefficients for negative moments in beams, slabs and frames: (a) at exterior face of the first interior support, two spans; (b) at exterior face of first interior support, more than two spans; (c) at other interior supports.

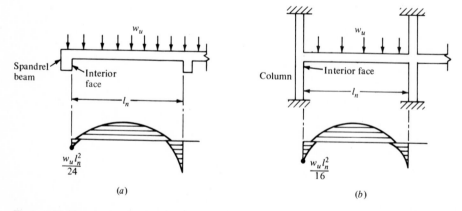

**Figure 9.15** Negative moment at the interior face of exterior supports for members built integrally with their supports: (a) spandrel beam, (b) column.

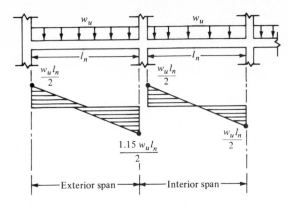

**Figure 9.16** Values of shear at face of supports.

adjacent spans does not exceed the shorter by more than 20 percent), the loads are uniformly distributed, and the live load does not exceed 3 times the dead load. These restrictions ensure that pattern loading (positioning live load in a pattern which maximizes moment at a given section) of spans will not create negative moment at sections which typically carry positive moment (Fig. 9.12).

ACI expressions used to establish values of positive and negative design moments in continuous beams and frames that satisfy the restrictions of ACI Code §8.3 are illustrated in Figs. 9.13 to 9.16. Although they show either beams or frames, the expressions apply to both types of structures. In all expressions, $l_n$ equals the clear span for positive moment or shear and the average of the adjacent clear spans for negative moment.

For slabs not exceeding 10 ft and for beams where the ratio of the sum of the column stiffnesses to beam stiffnesses exceeds 8 at each end of the span, the negative moment at the faces of all supports may be taken as $w_u l_n^2/12$ instead of the values given in Figs. 9.14 and 9.15.

**Example 9.1: Design of a slab and T-beam floor system** The T-beam–and–one-way–slab system in Fig. 9.17 carries a live load of 50 lb/ft² and a uniform dead load of 110 lb/ft². The dead load contains an allowance for the weight of the slab, beam stem, partitions, etc. The T beams are simply supported at each end and span 26 ft; $f'_c = 3$ kips/in², and $f_y = 60$ kips/in². Establish the dimensions of a typical interior beam and the slab and select the required reinforcement; interior exposure.

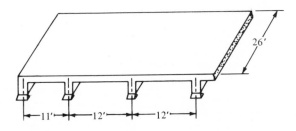

**Figure 9.17**

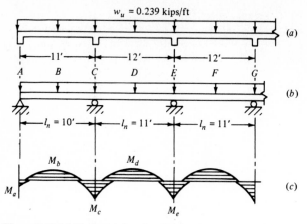

**Figure 9.18** (*a*) Typical 1-ft strip of slab, (*b*) slab modeled as a continuous beam, (*c*) design moments.

SOLUTION The minimum thickness of slab required to satisfy deflection limitations typically controls the thicknesses of slabs. Use Table 3.1

$$h_{min} = \frac{l}{28} = \frac{(12 \text{ ft})(12)}{28} = 5.14 \text{ in} \quad \text{use 5 in (127 mm) close enough}$$

The minimum depth of beam to control deflections is

$$h_{min} = \frac{l}{16} = \frac{(26 \text{ ft})(12)}{16} = 19.5 \text{ in} \quad \text{use } h = 20 \text{ in (508 mm)}$$

Try $b_w = 12$ in; the width must be adequate to carry shear and allow proper spacing between reinforcing bars.

Compute the design load per square foot of slab

$$w_u = 1.7(50 \text{ lb/ft}^2) + 1.4(110 \text{ lb/ft}^2) = 239 \text{ lb/ft}^2 \text{ (11.44 kPa)}$$

Analyze a 1-ft width of slab as a continuous beam using ACI coefficients to establish design moments for positive and negative steel (see Fig. 9.18)

$$M_a = \frac{w_u l_n^2}{24} = \frac{0.239(10^2)}{24} = 1.00 \text{ ft} \cdot \text{kips}$$

$$M_b = \frac{w_u l_n^2}{11} = \frac{0.239(10^2)}{11} = 2.17 \text{ ft} \cdot \text{kips}$$

$$M_c = \frac{w_u l_n^2}{10} = \frac{0.239(10.5^2)\dagger}{10} = 2.63 \text{ ft} \cdot \text{kips}$$

$$M_d = \frac{w_u l_n^2}{16} = \frac{0.239(11^2)}{16} = 1.81 \text{ ft} \cdot \text{kips}$$

$$M_e = \frac{w_u l_n^2}{11} = \frac{0.239(11^2)}{11} = 2.63 \text{ ft} \cdot \text{kips}$$

† For negative moment $l_n$ is the average of adjacent clear spans: $l_n = (10 + 11)/2 = 10.5$ ft.

Compute $A_s$ per foot of slab at the critical sections where the moments have been evaluated (Fig. 9.18). For example, at the first interior support, top steel must carry $M_c = 2.63$ ft · kips/ft (3.57 kN · m). Note that ACI Code §7.7.1 requires a minimum of $\frac{3}{4}$ in of cover for slab steel that is not exposed to weather or in contact with the ground.

$$d = h - \left(0.75 \text{ in} + \frac{d_b}{2}\right) = 5 \text{ in} - (0.75 + 0.25) = 4 \text{ in}$$

Use the trial method (Fig. 9.19)

$$M_u = \phi T\left(d - \frac{a}{2}\right) \qquad \text{guess } a = 0.4 \text{ in}$$

$$2.63(12) = 0.9T\left(4 - \frac{0.4}{2}\right)$$

$$T = 9.23 \text{ kips} \qquad A_s = \frac{T}{f_y} = \frac{9.23 \text{ kips}}{60 \text{ kips/in}^2} = 0.154 \text{ in}^2/\text{ft}$$

Equating $T = C$ gives $a = 0.3$ in    OK.

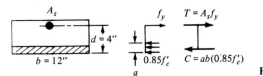

**Figure 9.19**

The minimum slab reinforcement $A_s$ is equal to temperature steel (see Sec. 3.13)

$$\text{Temperature steel} = 0.0018A_g = 0.0018(5 \text{ in})(12 \text{ in}) = 0.108 \text{ in}^2/\text{ft}$$

Use no. 3 bars at 12 in on center

$$A_{s,\text{sup}} = 0.11 \text{ in}^2/\text{ft}$$

Spacing of slab reinforcement to supply 0.154 in²/ft (use no. 3s). Reading in Table B.2

$$s = 8.0 \text{ in on center}$$

Also by ACI Code §7.6.5 maximum spacing of flexural steel not to exceed 18 in or 3 times the slab thickness

$$8 \text{ in} < 3(5 \text{ in}) \qquad \text{OK}$$

A sketch of the slab reinforcement is shown in Fig. 9.20.

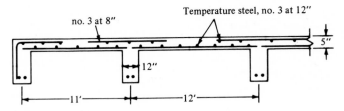

**Figure 9.20** Slab reinforcement.

Design reinforcement for a T beam.

SOLUTION The moment curve for the beam is shown in Fig. 9.21. The effective flange width of the T beam (see Fig. 9.22) is the smallest of

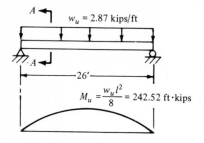

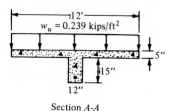

Section A-A

**Figure 9.21**

1. One-fourth of the beam's span

$$\frac{26 \text{ ft}}{4} = 6.5 \text{ ft} = 78 \text{ in} \qquad \text{controls}$$

2. Eight times the slab thickness on each side of stem plus the stem thickness

$$8(5 \text{ in})(2) + 12 \text{ in} = 92 \text{ in}$$

3. Center-to-center spacing of panel

$$(12 \text{ ft})(12) = 144 \text{ in}$$

Select the flexural steel $A_s$ ($M_u = 242.52$ ft · kips) using the trial method. Estimate $d$

$$d = h - 2.6 \text{ in} = 20 \text{ in} - 2.6 \text{ in} = 17.4 \text{ in}$$

$$M_u = \phi T \left( d - \frac{a}{2} \right) \qquad \text{guess } a = 0.8 \text{ in}$$

$$(242.52 \text{ ft} \cdot \text{kips})(12) = 0.9T \left( 17.4 \text{ in} - \frac{0.8}{2} \right)$$

$$T = 190.2 \text{ kips}$$

$$A_s = \frac{T}{f_y} = \frac{190.2 \text{ kips}}{60 \text{ kips/in}^2} = 3.17 \text{ in}^2 \ (2045.3 \text{ mm}^2)$$

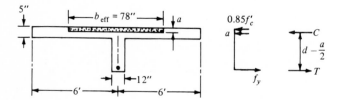

**Figure 9.22**

Check value of $a$

$$190.2 = T = C = ab_{\text{eff}}(0.85f'_c) = a(78 \text{ in})(0.85)(3 \text{ kips/in}^2)$$

$$a = 0.96 \text{ in} \qquad \text{initial guess OK}$$

$$A_{s,\text{min}} = \frac{200b_w d}{f_y} = \frac{200(12 \text{ in})(17.4 \text{ in})}{60,000 \text{ lb/in}^2} = 0.70 \text{ in}^2 \ (451.6 \text{ mm}^2)$$

Since 3.17 in$^2$ controls, use one no. 10 and two no. 9 bars

$$A_{s,\text{sup}} = 3.27 \text{ in}^2$$

Check $Z$

$$Z = 0.6f_y \sqrt[3]{d_c A} = 0.6(60) \sqrt[3]{2.6(24.23)} = 143.2 < 175 \qquad \text{OK}$$

## 9.9 MOMENT REDISTRIBUTION

As the load on an indeterminate ductile structure is increased beyond the service level, the most highly stressed sections are strained to their ultimate flexural capacity. Overstress at one section does not produce total failure of the structure if adjacent, less stressed sections can pick up additional load when the capacity of the overloaded section is exhausted. The ability to shift load to adjacent sections in an indeterminate structure, termed *moment redistribution,* is used in an ultimate-strength-design method called *plastic design.* Although plastic design has been used to size steel members for many years, its use is restricted by the current ACI Code because the ductility of underreinforced concrete members is limited. In this section the basic ideas of plastic design are briefly discussed and applied to the design of a continuous reinforced concrete beam.

### Concept of a Plastic Hinge

A simply supported reinforced concrete beam (Fig. 9.23) is considered to have failed when the steel yields and the nominal flexural capacity develops at the section of maximum moment. Although the beam can support the load producing failure, it has no capacity for additional load. Application of even the smallest force will cause the beam to deflect as the overstressed section rotates (Fig. 9.23b). In

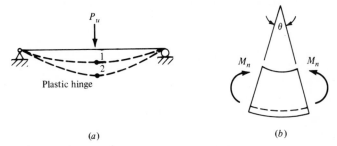

$(a)$ $\qquad\qquad\qquad\qquad\qquad$ $(b)$

**Figure 9.23** Formation of a plastic hinge: ($a$) deflected shape of a beam with a plastic hinge. ($b$) details of the plastic hinge.

Fig. 9.23a, for example, it is possible with the failure load in place to displace the beam from position 1 to position 2 with an infinitesimal force. If a section can rotate while sustaining its full flexural strength, a *plastic hinge* is said to have formed. A plastic hinge is denoted in the figures by a black dot at the point of maximum moment.

### Behavior of a Fixed-End Beam

The ACI Code specifies that an elastic analysis be used to determine the moments in an indeterminate structure. For the fixed-end beam carrying a uniform load of 2.5 kips/ft (Fig. 9.24) elastic theory indicates that the beam must be reinforced for 120 ft · kips at the supports and 60 ft · kips at midspan. If the beam is reinforced for the elastic moments, three plastic hinges will develop simultaneously at the supports and midspan when the design load is applied. These hinges convert the structure into a *linkage*, a structure that behaves as if it contained three hinges. Such a structure has no capacity for additional load.

As an alternate approach to design, let us assume that the ends of the beam are reinforced for a moment of 100 ft · kips and the center section is reinforced for 80 ft · kips. (The increment by which the moment has been reduced at the supports is added to the flexural capacity of the midspan section.) If a uniform load is applied that produces a moment of 100 ft · kips or less at the supports, the beam behaves elastically and the moment at the supports will be twice as large as the moment at midspan. The load that produces a moment of 100 ft · kips at the supports can be computed by using the elastic equation for the fixed-end moment

$$M = \frac{wl^2}{12}$$

$$w = \frac{12M}{l^2} = \frac{12(100)}{24^2} = 2.083 \text{ kips/ft}$$

If the uniform load $w = 2.083$ kips/ft is applied to the beam, plastic hinges form at each support (Fig. 9.25a). Since the beam is reinforced for 80 ft · kips at midspan, the moment of 50 ft · kips at that point is well below the flexural capacity of 80 ft · kips for which the beam has been reinforced and no plastic hinge forms. Because hinges develop only at the ends, the beam does not form a linkage and can

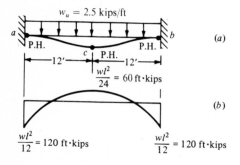

Figure 9.24 Fixed-end beam loaded to failure: (a) plastic hinges produce a linkage, (b) moments based on an elastic analysis.

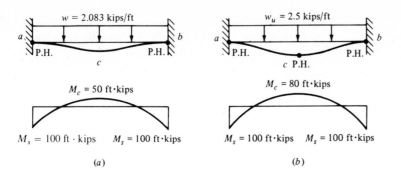

Figure 9.25 Moments in a fixed-end beam: (a) maximum load for which an elastic analysis is valid; (b) moments at failure, three plastic hinges form.

support additional load. If the load is increased, the beam will continue to deflect. Although rotation occurs at the supports, the moment remains constant because the section is only reinforced for 100 ft · kips. At midspan, however, the bending strains produced by the additional load increase the midspan moment. During this stage the beam acts as if it were simply supported. The load can be increased until the midspan moment reaches 80 ft · kips, and a third plastic hinge forms (Fig. 9.25b). After the third hinge develops, the beam, which is now converted into a linkage, fails.

The total load producing the linkage can be computed by considering the equilibrium of the left half of the beam between the support and midspan (Fig. 9.26). Because of symmetry, the shear is zero at midspan. Summing moments about the midspan section we can write

$$\Sigma M_c = 0$$

$$0 = 80 \text{ ft} \cdot \text{kips} + 100 \text{ ft} \cdot \text{kips} - 12w_u(12 \text{ ft}) + 12w_u(6 \text{ ft})$$

$$w_u = 2.5 \text{ kips/ft}$$

This analysis shows that the beam's capacity for load is the same in the two cases studied. Since the size of the beam is determined by the maximum design moment, the second case, in which the support moments are limited to 100 ft · kips, will permit a smaller member. Although the stresses in the steel at the supports under service loads will be higher in the second beam, the ultimate capacity of each will be the same because of the moment redistribution that takes place. In the ACI Code, the extent to which the moment can be reduced at a section is made a function

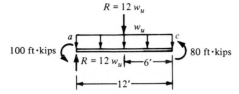

Figure 9.26 Free-body diagram of left half of beam from Fig. 9.25b.

of the area of flexural steel. The ACI expression states that reduction is permitted only if $\rho - \rho' \le 0.5\rho_b$ and that the percentage reduction of negative support moments in a beam equals

$$20\left(1 - \frac{[\rho - \rho']}{\rho_b}\right) \tag{9.1}$$

where $\rho$ = tension-steel–reinforcement ratio
   $\rho'$ = compression-steel–reinforcement ratio
   $\rho_b$ = balanced-steel–reinforcement ratio

Equation (9.1) ensures that the cross section has sufficient ductility to undergo the required rotation at a plastic hinge to allow moment redistribution to take place. Tests verify that beams designed by Eq. (9.1) do not crack or deflect excessively.

**Example 9.2: Redistribution of moment**   The reinforcement ratio in the continuous beam of Fig. 9.27 permits a 10 percent redistribution in support moment. Determine the corresponding increase in positive moment.

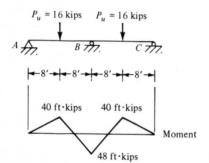

Figure 9.27 Moments from an elastic analysis.

SOLUTION With the elastic moment curve drawn, the designer can decide whether the support moment is to be adjusted up or down. Once the support moment has been modified, the shear and moment curves are completed by statics.
   Modify negative support moment

$$48 - 0.1(48) = 43.2 \text{ ft} \cdot \text{kips}$$

Using statics, compute the shear and moment curves for each span (Fig. 9.28).

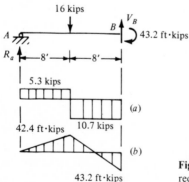

Figure 9.28 Shear and moment curves after moment redistribution.

## 9.10 STEEL VERSUS REINFORCED CONCRETE FOR BUILDING FRAMES

The choice between steel and concrete for building frames is often determined by the functional requirements of the structure. Certain span lengths, design loads, and architectural treatments lend themselves to a particular material. Although the relative costs of each depend on the availability of materials and the particular locale, concrete is often more economical than steel for structural frames with closely spaced columns that are less than 20 stories high. For taller buildings with larger spacing between columns, the greater strength and stiffness of steel permits smaller beams and columns. Smaller columns increase the rentable floor area, and shallower beams increase headroom.

The choice between steel and concrete may also be influenced by the time required for construction and how soon materials can be delivered to the construction site. While the frames of steel structures can be erected rapidly once steel has been delivered, a period of months (varying with the complexity of the job and the backlog at the rolling mill) must be allowed for the fabrication of the structural steel. On the other hand, construction of a concrete frame, which requires forms to be built, can be started within a matter of days after the construction contract is awarded. Also, structural steel must be protected from direct exposure to fire by insulation since its strength drops off rapidly as its temperature increases.

## PROBLEMS

**9.1** The one-way slab system in Fig. P9.1 supports service load composed of a uniform live load of 60 lb/ft$^2$ and a dead load including the weight of the slab of 160 lb/ft$^2$; $f'_c = 4$ kips/in$^2$, and $f_y = 60$ kips/in$^2$.

(a) Considering deflection requirements in Table 3.1, establish the minimum thickness of the slab.

(b) Determine the spacing of the no. 4 bars shown in spans $AB$ and $BC$.

(c) Determine the required spacing of temperature steel.

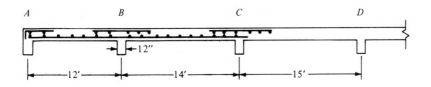

**Figure P9.1**

**9.2** Using ACI coefficients, select the dimensions and the reinforcement at critical sections for the three-span continuous beam shown in Fig. P9.2 if the live load is 80 lb/ft$^2$ and the dead load is 60 lb/ft$^2$ not including the weight of the beam. The width of the stem is not to exceed 14 in; $f'_c = 3$ kips/in$^2$, and $f_y = 60$ kips/in$^2$. Compression steel will be required at $B$ and $C$, where analysis shows that $M_u = 278$ ft · kips.

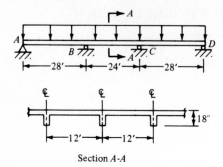

Section *A-A*                    **Figure P9.2**

# REFERENCE

1. Curt, Siegel: "Structure and Form in Modern Architecture," Reinhold, New York, 1962.

Detail of a waffle slab, a floor system composed of a shallow slab supported on a two-way grid of beams. (*Photograph by the Portland Cement Association.*)

# TEN

# DESIGN OF TWO-WAY SLABS

## 10.1 INTRODUCTION

Two-way slabs, i.e., slabs that bend in double curvature, require reinforcement in two directions to prevent excessive cracking and to limit deflections. The reinforcement is normally positioned parallel to the sides of the slab. Although mat foundations are occasionally designed as two-way slabs, most two-way slabs occur in reinforced concrete buildings in which floor slabs are supported on rows of columns. When the grid of columns is square or approximately square, two-way action develops (Fig. 10.1). Each rectangle of the grid formed by column centerlines is termed a *panel*. As shown to an exaggerated scale in Fig. 10.1, a uniform vertical load deforms the slab into a series of shallow hills at the columns and bowl-shaped valleys at the center of each panel. The position of flexural steel is determined by the curvature of the slab. Top steel is used in regions of negative curvature and bottom steel in regions of positive curvature.

The most common types of two-way slabs include the flat plate, the flat slab, the flat plate stiffened with beams along the column centerlines, and the two-way grid (Fig. 10.2). Under the provisions of the current ACI Code all these slab systems are analyzed in a similar manner.

Although an exact analysis of a two-way slab, a highly indeterminate element, is extremely complex, experimental studies indicate that it can be analyzed as a wide shallow beam. The designer assumes that the slab is cut into hypothetical strips, called *slab-beams*, along the panel centerlines. Although this approximate analysis of a slab as a beam gives reasonable values for moment, it does not provide information on the in-plane forces, often called the *membrane forces*, in the slab. By neglecting these forces the engineer produces a conservative design.

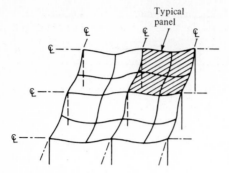

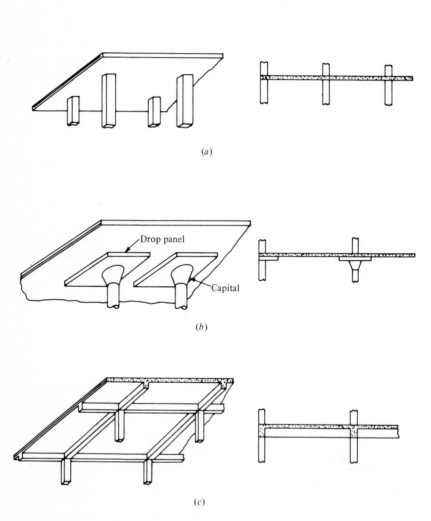

**Figure 10.1** Deflected shape of a uniformly loaded slab supported on a rectangular grid of columns.

**Figure 10.2** Two-way slab systems: (*a*) flat plate, (*b*) flat slab, (*c*) slab stiffened with beams on column centerlines.

and the summation sign indicates that the columns above and below are to be included

$$\Sigma K_c = \frac{4E(1728)}{15(12)}(2) = 76.8E \text{ in} \cdot \text{kips/rad}$$

Compute torsional stiffness of arm (see Fig. 10.15b for cross section).

$$K_t = \Sigma \frac{9E_{cs}C}{l_2(1 - c_2/l_2)^3} \tag{10.5}$$

where

$$C = \Sigma \left(1 - 0.63\frac{x}{y}\right)\frac{x^3 y}{3} \tag{10.6}$$

$$= (1 - 0.63|\tfrac{12}{24}|)\frac{12^3(24)}{3} + (1 - 0.63|\tfrac{8}{16}|)\frac{8^3(16)}{3}$$

$$= 11,340 \text{ in}^4$$

Since torsion arms of the same proportions exist on both sides of the column, two terms are included in the summation in Eq. (10.5)

$$K_t = \frac{9E(11,340)}{240(1 - \tfrac{12}{240})^3}(2) = 992E \quad \text{in} \cdot \text{kips/rad}$$

To compute $I_{sb}$ locate the centroid axis by summing moments of area about the base of beam (Fig. 10.16a)

$$A\bar{Y} = \Sigma A_n \bar{Y}_n$$

$$[240(8) + 12(16)]\bar{Y} = 1920(20) + 192(8)$$

$$\bar{Y} = 18.9 \text{ in}$$

$$I_{sb} = \Sigma I + Ad^2 = \frac{240(8^3)}{12} + 1920(1.1^2) + \frac{12(16^3)}{12} + 12(16)(10.9^2)$$

$$= 39,471 \text{ in}^4$$

Then compute $I_s$ (Fig. 10.16b)

$$I_s = \frac{240(8^3)}{12} = 10,240 \text{ in}^4$$

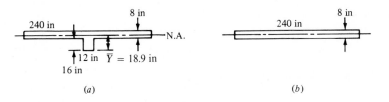

**Figure 10.16** (a) Slab-beam cross section, (b) slab.

Using Eq. (10.4a), we get

$$\frac{1}{K_{ec}} = \frac{1}{76.8E} + \frac{1}{992E(39,471/10,240)} \qquad \text{and} \qquad K_{ec} = 75.3E$$

## 10.6 MINIMUM THICKNESS OF TWO-WAY SLABS TO CONTROL DEFLECTIONS

To ensure that floors will be level within acceptable tolerances specified by the ACI Code, the deflections of two-way slabs either must be computed and compared with values specified in ACI Code §9.5 or limited by controlling the slab depth. Since the computation of slab deflections is complicated, most designers limit deflections by controlling the depth, the primary variable that influences the slab stiffness. Because the magnitude of the slab deflections depends on a large number of variables, three empirical equations, based on data from field tests of two-way slabs, are required to establish the minimum slab depth. These equations take into account the influence on deflections of the span length, the panel shape, continuity at the boundaries, the flexural stiffness of beams (if used), and the yield point of the steel. The minimum slab depth is given as

$$h = \frac{l_n(800 + 0.005f_y)}{36,000 + 5000\beta[\alpha_m - 0.5(1 - \beta_s)(1 + 1/\beta)]} \tag{10.7}$$

but not less than

$$h = \frac{l_n(800 + 0.005f_y)}{36,000 + 5000\beta(1 + \beta_s)} \tag{10.8}$$

Further, the thickness need not be more than

$$h = \frac{l_n(800 + 0.005f_y)}{36,000} \tag{10.9}$$

where $l_n$ = clear span in long direction
$\beta$ = ratio of clear spans in long to short directions
$\beta_s$ = ratio of length of continuous edges to total perimeter of panel under study
$\alpha_m$ = average value of $\alpha$ for all four sides of panel
$\alpha$ = ratio of flexural stiffness of effective beam (Fig. 10.8) to flexural stiffness of width of slab bounded laterally by the centerlines of adjacent panels (if any) on each side of beam

Specifically,

$$\alpha = \frac{E_{cb}I_b}{E_{cs}I_s} \tag{10.10}$$

where $E_{cb}$ and $E_{cs}$ refer to the modulus of elasticity of the concrete in the beam and slab, respectively. If no beams are used, the case of the flat plate, for example, $\alpha = 0$.

## Table 10.1  Minimum thickness of two-way slabs

| Type | in | mm |
|---|---|---|
| Slabs without beams or drop panels | 5 | 127 |
| No beams, but standard drop panels | 4 | 101.6 |
| Slabs with beams on all four sides whose $\alpha_m \geq 2$ | $3\frac{1}{2}$ | 88.9 |

If the discontinuous edge of a panel is not supported on an edge beam whose stiffness ratio $\alpha$ is at least 0.8, the minimum values of slab thickness given by Eqs. (10.7) to (10.9) must be increased by 10 percent.

If standard drop panels are provided, the minimum values of slab thickness given by Eqs. (10.7) to (10.9) may be reduced by 10 percent. A standard panel is defined as one which extends outward from the center of a support (the column) a distance equal to at least one-sixth of the center-to-center distance between supports and which also projects below the slab a distance equal to at least $\frac{1}{4}$ the slab thickness outside the drop panel.

Regardless of the values given by Eqs. (10.7) to (10.9), the thickness of two-way slabs must equal or exceed the values in Table 10.1.

**Example 10.4:** A flat plate floor (see Fig. 10.17) with panels 20 by 25 ft (6.1 by 7.6 m) is supported on 18-in (457.2-mm) square columns. No edge beams are used. From an investigation of a corner panel, what is the minimum required slab depth if deflection computations are not made? $f_y = 60$ kips/in$^2$ (413.7 MPa), and $f'_c = 3$ kips/in$^2$ (20.68 MPa).

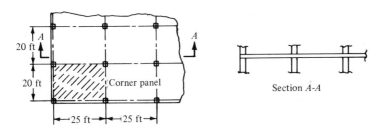

**Figure 10.17** Flat plate floor.

SOLUTION Since no edge beams are used, values of thickness given by Eqs. (10.7) to (10.9) must be increased by 10 percent

$$h_{min} = \frac{l_n(800 + 0.005 f_y)(1.1)}{36,000 + 5000\beta[\alpha_m - 0.5(1 - \beta_s)(1 + 1/\beta)]} = 10.29 \text{ in (261.4 mm)}$$

where $l_n = 23.5$ ft $= 282$ in,

$$\beta = \frac{l_{n,\text{long}}}{l_{n,\text{short}}} = \frac{23.5 \text{ ft}}{18.5 \text{ ft}} = 1.27$$

Since $\alpha = I_b/I_s = 0$ (no beams), $\alpha_m = 0$ and

$$\beta_s = \frac{25 + 20}{50 + 40} = 0.5$$

but not less than

$$h = \frac{l_n(800 + 0.005f_y)(1.1)}{36,000 + 5000\beta(1 + \beta_s)} = 7.5 \text{ in (190.5 mm)} \tag{10.8}$$

or more than

$$h = \frac{l_n(800 + 0.005f_y)(1.1)}{36,000} = 9.48 \text{ in (240.8 mm)} \quad \text{controls} \tag{10.9}$$

Use a $9\frac{1}{2}$-in-thick slab.

**Example 10.5:** For the floor system shown in Fig. 10.21 verify that a slab 8 in (203.2 mm) thick for an interior panel satisfies the ACI Code minimum-thickness requirements to limit deflections; $f_y = 60$ kips/in$^2$, and $f'_c = 3$ kips/in$^2$.

SOLUTION Properties of the floor system (see Example 10.1 for $I_b$) are as follows ($\alpha_1$ is the long and $\alpha_2$ the short direction):

$$\alpha_1 = \frac{EI_b}{EI_s} = \frac{23,864}{10,240} = 2.33$$

where $I_s$ for an 8-in-thick slab with a width of 20 ft equals

$$I_s = \frac{bh^3}{12} = 10,240 \text{ in}^4$$

$$\alpha_2 = \frac{EI_b}{EI_s} = \frac{23,864}{15,360} = 1.56$$

where $I_b$ is given above and $I_s$ for an 8-in-deep slab with a width of 30 ft is $bh^3/12 = 15,360$ in$^4$

$$\alpha_m = \frac{\alpha_1 + \alpha_2}{2} = \frac{2.33 + 1.56}{2} = 1.94$$

$$\beta_s = \frac{\text{length of continuous edges}}{\text{total perimeter}} = 1$$

$$\beta = \frac{l_{n,\text{long}}}{l_{n,\text{short}}} = \frac{29}{19} = 1.53$$

By Eq. (10.7)

$$h_{\text{min}} = \frac{348[800 + 0.005(60,000)]}{36,000 + 5000(1.53)[1.94 - 0.5(1 - 1)(1 + 1/1.53)]}$$

$$= 7.53 \text{ in (191.3 mm)}$$

But $h$ must not be less than the value specified by Eq. (10.8)

$$h = \frac{348[800 + 0.005(60,000)]}{36,000 + 5000(1.53)(1 + 1)} = 7.46 \text{ in (189.5 mm)}$$

and $h$ need not exceed the value given by Eq. (10.9)

$$h = \frac{348[800 + 0.005(60,000)]}{36,000} = 10.63 \text{ in (270 mm)}$$

Therefore 7.53 in controls, and an 8-in depth is satisfactory.

## 10.7 ANALYSIS OF TWO-WAY SLABS

Two-way slabs may be analyzed and designed by any method ensuring that all the strength and serviceability requirements of the ACI Code will be satisfied. When slabs are supported, for the most part, on a rectangular grid of columns (the usual condition), they are analyzed most conveniently either by the *direct design method* or the *equivalent-frame method*. These two methods, which are detailed in chapter 13 of the ACI Code, will be discussed fully in this section.

If slabs have an unusual geometric configuration, if supports are spaced at irregular intervals, or if the continuity of the slab is interrupted by large openings, neither of the code methods may be applicable. For these special cases, the designer may analyze the slab by using finite elements, by a model analysis, by a yield-line analysis (an ultimate-strength analysis based on the equilibrium of the cracked slab in its final deflected position), or by one of the classical methods of analysis that treats the slab as an elastic continuum.

### Analysis by the Direct Design Method

The direct design method is an empirical procedure for establishing the design moments at critical sections in uniformly loaded slab-beams of equivalent frames (see Fig. 10.7c, for example). In this method the simple beam moment in each span of the frame is distributed between the sections of maximum negative moment at the supports and the section of maximum positive moment at midspan. Since the properties of the members, except in the exterior panels, are not required to determine the moments in the longitudinal direction, computations are greatly simplified and can be carried out rapidly.

To ensure that the flexural strength of the slab at the critical sections of positive and negative moment will be adequate to carry the moments produced by the most unfavorable pattern of live load, the ACI Code requires that the following design conditions be satisfied:

1. At least three continuous spans must exist in each direction.
2. The spans must be similar in length. Specifically, adjacent spans in each direction must not differ in length by more than one-third of the longer span length.

3. The live load must be uniform and less than 3 times the dead load.
4. The panels must be rectangular, and the ratio of the long to short spans must not exceed 2.
5. If beams are present along column centerlines in each direction, their relative stiffness $\alpha_1 l_2^2/\alpha_2 l_1^2$ must not be less than 0.2 or greater than 5.
6. Columns must not be offset more than 10 percent of the distance between center-lines of successive columns in each direction.

The direct design method is based on the fact that the change in moment between the supports and midspan of a uniformly loaded beam equals $Wl_n^2/8$, the area under the shear curve between the two sections (see Fig. 10.18). If equal values of negative moment $M_s$ are assumed to develop at each support, the sum of the midspan moment $M_c$ and $M_s$ is

$$M_c + M_s = M_0 \tag{10.11}$$

For a uniformly loaded slab-beam with a width $l_2$, a clear span $l_n$, and a uniform load $w_u$, $M_0$ equals

$$M_0 = \frac{w_u l_2 l_n^2}{8} \tag{10.12}$$

where $l_n$ is measured between faces of supports (columns, capitals, or walls) but must not be less than $0.65l_1$, the center-to-center distance between supports.

Thus the determination of positive and negative design moments in the slab-beams of equivalent frames by the direct design method requires three steps:

1. The simple beam moment $M_0$ in each span is computed using Eq. (10.12).
2. $M_0$ is divided between the support and midspan sections of the slab beam as in Fig. 10.18; that is, $M_s + M_c = M_0$.

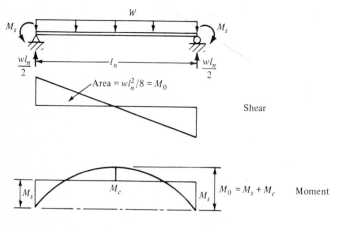

**Figure 10.18** Continuous beam.

3. $M_c$ and $M_s$ are distributed across the width of the slab-beam between the column and the middle strips. If the column strip contains a beam, the column strip moment is divided between the beam and the balance of the column strip.

Details of steps 2 and 3 are discussed in the next two sections.

**Positive and negative moments in a slab-beam**   The fraction of the simple beam moment $M_0$ that is assigned to the end and the midspan sections of a slab-beam is shown in Fig. 10.19 for both an interior and an exterior panel. For an interior panel, the positive moment $M_c$ at midspan equals $0.35\,M_0$, and the negative moments $M_s$ at each support equal $0.65\,M_0$. These values of moment, which are approximately the same as those in a uniformly loaded fixed-end beam, are based on the assumption that an interior joint undergoes no significant rotation. We can make this assumption because the end moments applied to each side of an interior joint are approximately equal and opposite in magnitude. In the direct design method, this balanced condition is assured by the restrictions that limit (1) the difference in length between adjacent spans and (2) the ratio of the live to the dead load.

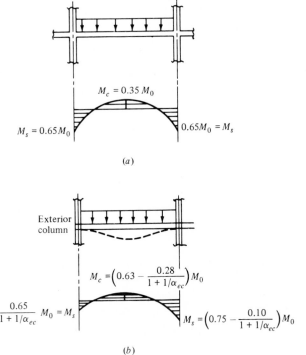

**Figure 10.19** Positive and negative design moments in a slab-beam expressed as a function of $M_0$: (a) interior panel, (b) exterior panel. $\alpha_{ec} = K_{ec}/\Sigma(K_s + K_b)$.

Since load is applied to the exterior column from one side only, the exterior joint is always subject to an unbalanced moment. Rotation of the exterior joint reduces the negative moment in the slab-beam at that point and increases the moments at midspan and at the first interior support. Since the magnitude of the rotation of the exterior joint depends on the stiffness of the exterior equivalent column relative to that of the slab-beam, the coefficients applied to $M_0$ in an exterior panel are given as a function of $\alpha_{ec}$ (see Fig. 10.19$b$), where

$$\alpha_{ec} = \frac{K_{ec}}{\Sigma(K_s + K_b)} \qquad (10.13)$$

If the stiffness of the exterior equivalent column is large compared with that of the slab-beam, the column will restrain the exterior end of the slab against rotation, producing a condition similar to that of a fixed support. For the limiting case of an equivalent column of infinite stiffness, that is, $\alpha_{ec}$ approaches infinity, the coefficients applied to $M_0$ in an exterior span reduce to the same values as those in an interior panel (0.65 at the ends and 0.35 at midspan). On the other hand, if the equivalent column is very flexible, that is, if $\alpha_{ec}$ tends to zero, the column, which now behaves like a pin support, applies almost no rotational restraint to the exterior end of the slab. For this case, the moment at the exterior end of the slab-beam approaches zero, the moment at midspan tends to $0.63M_0$, and the moment at the first interior support approaches $0.75M_0$.

If the magnitude of the negative moments on opposite sides of an interior support are unequal because the spans differ in length, the ACI Code specifies that the flexural reinforcement be designed for the larger moment unless an analysis (equivalent to one cycle of moment distribution) is carried out in which the unbalanced moment is distributed to the members framing into the joint in proportion to their flexural stiffness.

To account for the ability of an underreinforced slab to redistribute moment (typically the negative moment decreases and the positive moment increases) the ACI Code permits a design moment to be modified by 10 percent providing the total static moment for the panel in the direction considered is not less than $M_0$ as given by Eq. (10.12).

**Transverse distribution of moments at critical sections**   The final step in the direct design method is to distribute the positive and negative design moments $M_c$ and $M_s$ between the column strip and the half-middle strips of each slab-beam. The percentage of each design moment to be distributed to the column strip is listed in Tables 10.2 (positive moment), 10.3 (interior negative moment), and 10.4 (exterior negative moment). In Tables 10.2 and 10.3 the distribution factors are tabulated for three values (0.5, 1, and 2) of the panel dimensions $l_2/l_1$ and two values (0 and 1) of $\alpha_1 l_2/l_1$. If the proportions of the slab results in values not specifically given in the column headings the designer uses a linear interpolation between the rows and columns of the table. The portion of the design moment not assigned to the column strip is distributed uniformly across the half-middle strips.

**Table 10.2 Percentage of positive moment to column strip**

| $\alpha_1 \dfrac{l_2}{l_1}$ | $l_2/l_1$ | | |
|---|---|---|---|
| | 0.5 | 1.0 | 2.0 |
| 0 | 60 | 60 | 60 |
| $\geq 1$ | 90 | 75 | 45 |

**Table 10.3 Percentage of negative moment to the column strip at an interior support**

| $\alpha_1 \dfrac{l_2}{l_1}$ | $l_2/l_1$ | | |
|---|---|---|---|
| | 0.5 | 1.0 | 2.0 |
| 0 | 75 | 75 | 75 |
| $\geq 1$ | 90 | 75 | 45 |

For an exterior slab-beam, the designer uses Table 10.4 to divide the exterior design moment between the column and the middle strips. Since the distribution of moment between the column and middle strips is influenced by the torsional stiffness of the *spandrel beam* (the beam along the outside edge of the building), an additional parameter $\beta_t$, the ratio of the torsional stiffness of the spandrel beam to the flexural stiffness of the slab, is required. If the torsional stiffness of the spandrel beam is small ($\beta_t$ approaches zero), the column provides most of the flexural restraint to the exterior end of the slab. For this case, 100 percent of the design

**Table 10.4 Percentage of negative moment to the column strip at the exterior support**

| $\alpha_1 \dfrac{l_2}{l_1}$ | $\beta_t$ | $l_2/l_1$ | | |
|---|---|---|---|---|
| | | 0.5 | 1.0 | 2.0 |
| 0 | 0 | 100 | 100 | 100 |
| 0 | $\geq 2.5$ | 75 | 75 | 75 |
| $\geq 1$ | 0 | 100 | 100 | 100 |
| $\geq 1$ | $\geq 2.5$ | 90 | 75 | 45 |

*Notes*: $\beta_t$ is the ratio of torsional stiffness of the edge beam section to the flexural stiffness of a width of slab equal to the span length of the beam, center to center of supports

$$\beta_t = \frac{E_{cb} C}{2 E_{cs} I_s} \qquad (10.14)$$

If the exterior support is a column or wall extending for a distance equal to or greater than $\frac{3}{4}$ times the $l_2$ used to compute $M_0$, the exterior negative moment is considered to be uniformly distributed across $l_2$.

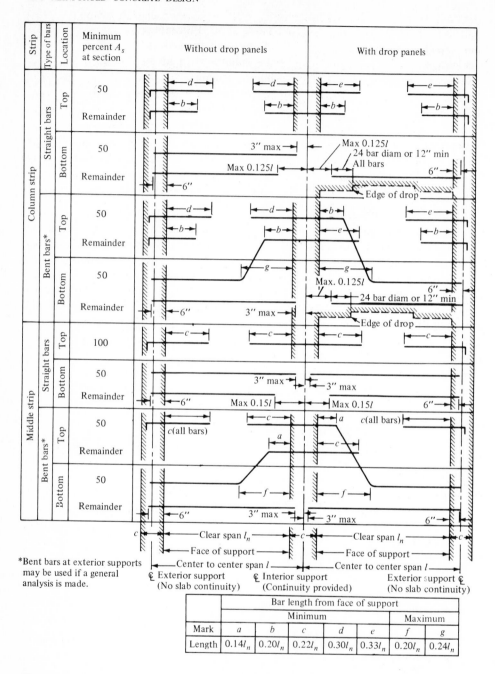

*Bent bars at exterior supports may be used if a general analysis is made.

| | Bar length from face of support | | | | | | |
|---|---|---|---|---|---|---|---|
| | Minimum | | | | | Maximum | |
| Mark | $a$ | $b$ | $c$ | $d$ | $e$ | $f$ | $g$ |
| Length | $0.14l_n$ | $0.20l_n$ | $0.22l_n$ | $0.30l_n$ | $0.33l_n$ | $0.20l_n$ | $0.24l_n$ |

**Figure 10.20** Minimum bend-point locations and extensions for reinforcement in slabs without beams.[5]

moment $M_s$ is concentrated in the column strip. Since the design moment in the middle strip is zero or close to zero, its area of flexural steel is controlled by minimum steel requirements.

When a beam is located along the column centerlines, it is proportioned for 85 percent of the moment in the column strip if $\alpha_1 l_2/l_1 \geq 1$. If the value of $\alpha_1 l_2/l_1$ falls between 1 and 0, the percentage of the column-strip moment assigned to the beam is determined by linear interpolation between 85 and 0 percent. For example, if $\alpha_1 l_2/l_1 = 0.8$, the beam would be designed to carry $0.8(0.85) = 68$ percent of the moment in the column strip. The portion of the column strip moment not assigned to the beam is distributed uniformly over the slab on either side of the beam.

After the steel is selected at the three critical sections, it is extended throughout the slab as detailed in Fig. 10.20. This procedure ensures that adequate reinforcement will be supplied at all other sections in the span.

**Example 10.6: Determination of the design moments by the direct design method in a two-way slab stiffened by beams** The two-way slab system in Fig. 10.21 is stiffened by beams 24 in (609.6 mm) deep along all column lines. Using the direct design method, determine the magnitude of the positive and negative design moments at all critical sections in the exterior and interior spans of the slab-beam along column line 2. $f'_c = 3$ kips/in² (20.68 MPa), $f_y = 60$ kips/in² (413.7 MPa), $w_d = 150$ lb/ft² (7.18 kPa), and $w_l = 50$ lb/ft³ (2.39 kPa).

SOLUTION Design moments in the interior slab-beam between column lines $B$ and $C$ (Fig. 10.21). Compute the simple beam moment using Eq. (10.12)

$$w_u = 0.15(1.4) + 0.05(1.7) = 0.295 \text{ kips/ft}^2 \ (14.12 \text{ kPa})$$

$$M_0 = \frac{w_u l_2 l_n^2}{8} = \frac{0.295(20)(29^2)}{8} = 620 \text{ ft} \cdot \text{kips} \ (840.7 \text{ kN} \cdot \text{m})$$

Divide $M_0$ between sections of positive and negative moment

At midspan:     $M_c = 0.35M_0 = 0.35(620 \text{ ft} \cdot \text{kips}) = 217 \text{ ft} \cdot \text{kips}$

At each end:    $M_s = 0.65M_0 = 0.65(620 \text{ ft kips}) = 403 \text{ ft} \cdot \text{kips}$

For the distribution of the midspan moment $M_c$ between column and middle strips use Table 10.2 to establish the percent of $M_c$ to the column strip. See Example 10.5 for the computation of $\alpha_1$.

$$\frac{l_2}{l_1} = \frac{20 \text{ ft}}{30 \text{ ft}} = 0.67$$

$$\alpha_1 \frac{l_2}{l_1} = \frac{EI_b}{EI_s} \frac{l_2}{l_1} = \frac{23,864}{10,240} \frac{20}{30} = 1.55$$

Using the above parameters, interpolate between the first two columns in the second row of Table 10.2 to get 85 percent

Moment to column strip $= 0.85(217) = 184 \text{ ft} \cdot \text{kips}$

Moment to middle strip $= 0.15 (217) = 33 \text{ ft} \cdot \text{kips}$

Design moment per foot of middle strip $= \frac{33}{10} = 3.3 \text{ ft} \cdot \text{kips/ft}$

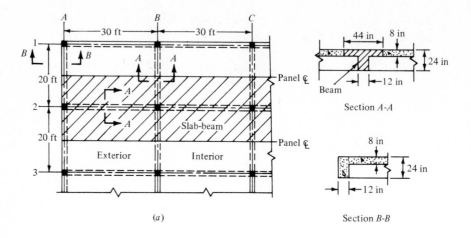

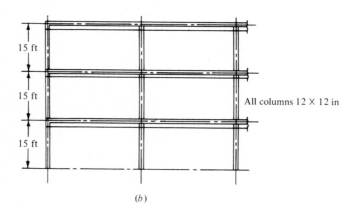

**Figure 10.21** Two-way slab with beams: (*a*) floor plan, (*b*) vertical cross section.

Since $\alpha_1 l_2/l_1 > 1$, 85 percent of the moment in the column strip [0.85 (184) or 156.4 ft · kips] is assigned to the beam. The balance of the column-strip moment, or 27.6 ft · kips, is distributed uniformly over the 3-ft 2-in widths of slab on either side of the beam to give a design moment per foot of slab of

$$\frac{27.6}{6.33} = 4.36 \text{ ft · kips/ft}$$

See Fig. 10.22*b* for summary.

Distribute the end moment $M_s = 403$ ft · kips between the column and middle strips. For $l_2/l_1 = 0.67$ and $\alpha_1 l_2/l_1 = 1.55$, Table 10.3 gives 85 percent

Moment to column strip = 0.85(403) = 343 ft · kips

Moment to middle strip = 0.15(403) = 60 ft · kips or 30 ft · kips to each $\frac{1}{2}$ middle strip

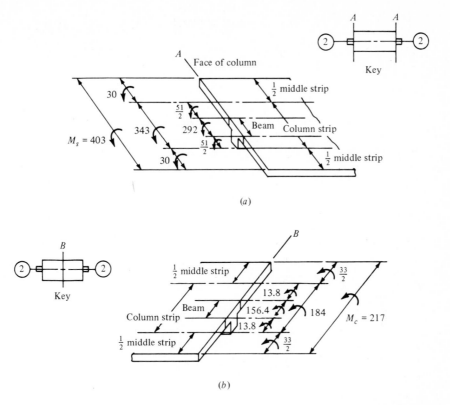

**Figure 10.22** Summary of design moments at sections of an interior slab-beam; numbers give moments in foot-kips: (a) negative slab moments on transverse section at face of columns, (b) positive slab moments at midspan.

Since $\alpha_1 l_2 / l_1 > 1$, 85 percent of the column strip moment, or 292 ft · kips, is assigned to the beam. The balance of the column-strip moment, $343 - 292 = 51$ ft · kips, is distributed to the slab on either side of the beam (Fig. 10.22a).

*Design moments at critical sections in the exterior slab beam*   The magnitude of the moments at critical sections in the exterior slab-beam is a function both of $M_0$, the simple beam moment, and $\alpha_{ec}$, the ratio of the stiffness of the exterior equivalent column to the sum of the stiffnesses of the slab and the beam framing into the exterior joint. Assuming that the beam and the slab are of constant cross section between the column center-lines in the direction of the span, $\alpha_{ec}$ as given by Eq. (10.13) equals

$$\alpha_{ec} = \frac{K_{ec}}{\Sigma(K_s + K_b)} = 0.199 = 0.2$$

where $K_{ec} = 75.3E$ (see Example 10.3) and

$$K_s = \frac{4EI_s}{l_1} = \frac{4E(10{,}240)}{30(12)} = 114E$$

$$K_b = \frac{4EI_b}{l_1} = \frac{4E(23{,}864)}{30(12)} = 265E$$

Evaluate the design moments at critical sections (Fig. 10.19$b$). At the exterior column face

$$M_s = \frac{0.65}{1 + 1/0.2} M_0 = 0.108(620) = 66.96 \text{ ft} \cdot \text{kips}$$

midspan

$$M_c = \left(0.63 - \frac{0.28}{1 + 1/0.2}\right) M_0 = 0.583(620) = 361.5 \text{ ft} \cdot \text{kips}$$

and on the interior column face

$$M_s = \left(0.75 - \frac{0.1}{1 + 1/0.2}\right) M_0 = 0.683 M_0 = 0.683(620) = 423.46 \text{ ft} \cdot \text{kips}$$

In the exterior slab-beam, the factors for distributing the design moments $M_c$ at midspan and $M_s$ at the face of the interior column between the column and the middle strips are identical to those for the interior slab-beam since the dimensions of the two members are the same. The distribution of moments on these two sections is computed below and summarized in Table 10.5.

At midspan the moment to column strip (see Table 10.2) is

$$0.85 M_c = 0.85(361.5) = 307.3 \text{ ft} \cdot \text{kips}$$

Of the total column-strip moment, 85 percent (261.2 ft · kips) is carried by the beam, and the balance (46.1 ft · kips) is distributed uniformly across the slab in the column strip not occupied by the beam. The moment to the middle strip is

$$0.15 M_c = 0.15(361.5) = 54.2 \text{ ft} \cdot \text{kips, or } 27.1 \text{ ft} \cdot \text{kips to each half middle strip}$$

At the face of the interior column the moment to the column strip (see Table 10.3) is

$$0.85 M_s = 0.85(423.46) = 359.94 \text{ ft} \cdot \text{kips}$$

Again 85 percent of the column strip moment (305.95 ft · kips) is carried by the beam. The moment to the middle strip is

$$0.15 M_s = 0.15(423.46) = 63.52 \text{ ft} \cdot \text{kips, or } 31.76 \text{ ft} \cdot \text{kips to each half middle strip}$$

At the exterior end of the slab-beam the lateral distribution of the design moment is given by Table 10.4. In addition to $l_2/l_1 = 0.67$ and $\alpha_1 l_2/l_1 = 1.55$, the parameter $\beta_t$ is required. Using values of $C = 11,340 \text{ in}^4$ and $I_s = 10,240 \text{ in}^4$ (see Example 10.3), and noting that $E_{cb} = E_{cs}$ since the slab and beams are constructed from the same strength concrete, we calculate $\beta_t$ by Eq. (10.14) as

$$\beta_t = \frac{E_{cb} C}{2 E_{cs} I_s} = 0.55$$

Interpolating in Table 10.4 gives 96 percent. The moment to the column strip is

$$0.96(66.96) = 64.3 \text{ ft} \cdot \text{kips}$$

of which 85 percent (54.7 ft · kips) is carried by the beam. The moment to be divided between the two half middle strips is

$$0.04(66.96) = 2.68 \text{ ft} \cdot \text{kips}$$

**Table 10.5 Summary of design moments in foot-kips at critical sections in the exterior slab-beam**

|  |  | Negative moment exterior end | Positive moment at midspan | Negative moment interior end |
|---|---|---|---|---|
| Half middle strip |  | 1.34 | 27.1 | 31.76 |
|  |  | 4.8 | 23.05 | 26.99 |
| Column strip | Beam | 54.7 | 261.2 | 305.95 |
|  |  | 4.8 | 23.05 | 26.99 |
| Half middle strip |  | 1.34 | 27.1 | 31.76 |

**Pattern loading**   In the direct design method, the distribution of the simple beam moment between the sections of maximum positive and negative moment is based on the assumption that the total live and dead loads are distributed uniformly over all spans. When spans are loaded uniformly, the rotation of the interior joints is small because the member end moments on each side of the joint balance each other. As discussed in Chap. 9 (see Fig. 9.10), the maximum moment at a section is created by *pattern loading* (the live load acts on certain spans but not on others) rather than by uniform load on all spans. While the negative moment at a support produced by pattern load is only slightly greater, say 4 or 5 percent, than that produced by uniform load, the positive moment at a midspan section produced by pattern loading can be substantially greater (as much as 40 to 50 percent) than that produced by uniform load. As the live load increases relative to the dead load, the magnitude of the positive moment produced by pattern loading increases relative to that produced by uniform load. Of course, if the live load is small compared with the dead load, most of the moment is produced by dead load and the position of the live load is of less consequence. Also as the stiffness of the columns decreases and less restraint is applied to the ends of the slab, the curvature and correspondingly the moment at midspan increase rapidly (Fig. 10.23).

A moderate increase in positive moment due to pattern loading can be accommodated by the slab (the Code permits up to 33 percent) because of the ductile slab's ability to redistribute moments between the end and midspan sections; however, a larger increase in positive moment may lead to overstress or excessive sag and cracking of floors. If the unfactored live load is equal to or larger than half the dead load, the ACI Code requires that $\alpha_c$, the ratio of the sum of the stiffness of the columns above and below the slab to the sum of the flexural stiffnesses of the beams and slabs framing into a joint must be equal to or greater than the minimum values of the stiffness ratio $\alpha_{min}$ in Table 10.6. If the stiffness ratio $\alpha_c$ is less than that required by Table 10.6, and if the designer is prevented by architectural considerations from increasing the column size, the Code requires the positive

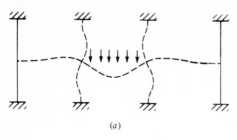

(a)

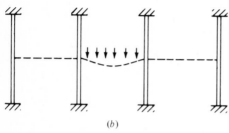

(b)

**Figure 10.23** Influence of column stiffness on midspan slab curvature, pattern loading: (a) flexible columns, rotation of joints increases midspan curvature; (b) rigid columns, large end restraint reduces midspan curvature.

moment in the slab from the direct design analysis to be increased by the factor $\delta_s$ as specified by

$$\delta_s = 1 + \frac{2 - \beta_a}{4 + \beta_a}\left(1 - \frac{\alpha_c}{\alpha_{min}}\right) \tag{10.15}$$

In this equation $\beta_a$ is the ratio of unfactored dead load per unit area to the unfactored live load per unit area. $\alpha_c$ is the ratio of the sum of the flexural stiffnesses of the columns above and below the slab to the sum of the beam and slab stiffnesses

$$\alpha_c = \Sigma K_c / \Sigma(K_b + K_s) \tag{10.16}$$

and $\alpha_{min}$ is the minimum required value of $\alpha_c$ as listed in Table 10.6.

**Example 10.7:** If the service loads on the floor system in Fig. 10.21 consist of 100 lb/ft² of live load and 150 lb/ft² of dead load, does the stiffness of the columns satisfy the requirements of Table 10.6, which limits the increase in positive moment at midspan in a typical panel?

SOLUTION The parameters required to enter Table 10.6 are

$$\beta_a = \frac{w_d}{w_l} = \frac{150 \text{ lb/ft}^2}{100 \text{ lb/ft}^2} = 1.5 \qquad \frac{l_2}{l_1} = \frac{20}{30} = 0.67 \qquad \alpha = \frac{E_{cb}I_b}{E_{cs}I_s}$$

Since the same strength concrete is used in the beams and the slabs, $E_{cs} = E_{cb}$.

$$\alpha = \frac{23,864}{10,240} = 2.33$$

See Example 10.1 for value of $I_b$ and Example 10.3 for value of $I_s$. From Table 10.6 read $\alpha_{min} = 0$. Compute $\alpha_c$

$$\alpha_c = \frac{\Sigma K_c}{\Sigma(K_s + K_b)} = \frac{76.8E}{2(113.78E + 265.2E)} = 0.1$$

**Table 10.6 Values of $\alpha_{min}$**

| $\beta_a$ | Aspect ratio $l_2/l_1$ | Relative beam stiffness $\alpha$ | | | | |
|-----|-------|------|------|------|------|------|
| | | 0 | 0.5 | 1.0 | 2.0 | 4.0 |
| 2.0 | 0.5–2.0 | 0 | 0 | 0 | 0 | 0 |
| 1.0 | 0.5 | 0.6 | 0 | 0 | 0 | 0 |
| | 0.8 | 0.7 | 0 | 0 | 0 | 0 |
| | 1.0 | 0.7 | 0.1 | 0 | 0 | 0 |
| | 1.25 | 0.8 | 0.4 | 0 | 0 | 0 |
| | 2.0 | 1.2 | 0.5 | 0.2 | 0 | 0 |
| 0.5 | 0.5 | 1.3 | 0.3 | 0 | 0 | 0 |
| | 0.8 | 1.5 | 0.5 | 0.2 | 0 | 0 |
| | 1.0 | 1.6 | 0.6 | 0.2 | 0 | 0 |
| | 1.25 | 1.9 | 1.0 | 0.5 | 0 | 0 |
| | 2.0 | 4.9 | 1.6 | 0.8 | 0.3 | 0 |
| 0.33 | 0.5 | 1.8 | 0.5 | 0.1 | 0 | 0 |
| | 0.8 | 2.0 | 0.9 | 0.3 | 0 | 0 |
| | 1.0 | 2.3 | 0.9 | 0.4 | 0 | 0 |
| | 1.25 | 2.8 | 1.5 | 0.8 | 0.2 | 0 |
| | 2.0 | 13.0 | 2.6 | 1.2 | 0.5 | 0.3 |

where $\qquad\qquad \Sigma K_c = 76.8E \qquad$ see Example 10.3

$$K_s = \frac{4EI_s}{L_s} = \frac{4E(10{,}240)}{30(12)} = 113.78E$$

$$K_b = \frac{4EI_b}{L_b} = \frac{4E(23{,}864)}{30(12)} = 265.2E$$

Since $\alpha_c > \alpha_{min}$, the columns have adequate stiffness to limit the positive moment

## Analysis by the Equivalent-Frame Method

If two-way floor systems do not satisfy the loading and geometric conditions required by the direct design method (see Sec. 10.7), the ACI Code permits a building to be analyzed by subdividing the structure into *equivalent frames*, which are then analyzed elastically, often by moment distribution, for all conditions of load. To represent the interaction between the columns and the floor slab as closely as possible, the designer must use equivalent columns of the type shown in Fig. 10.10 to carry out the analysis.

When the analysis includes lateral loads, e.g., wind or earthquake forces, frames running the full depth of the building must be used because at each level the forces in members are a function of the lateral forces on all floors above the level under consideration. If the analysis is limited to the action of gravity loads, computations can be simplified by analyzing each floor and its attached columns separately. For this case, the far ends of the columns are assumed to be fixed at the point where they connect to the floors above and below the floor being analyzed (see Fig. 10.7).

When an individual floor consists of many panels, the moment in the slab-beam at a particular support can be adequately determined if the end of the slab-beam is assumed to be fixed at joints two panels away from the particular support. This simplification is possible because load on a particular panel produces significant forces only in that panel and the adjacent panels. Vertical load, two or more panels away from a particular support, contributes very little to the forces (shear, moment, axial force) at that support.

**Properties of members** For the analysis of an equivalent frame, the Code permits the properties of the frame members to be based on the area of the gross cross section. If the designer were to attempt to account for the variation in cross section produced by flexural cracking, computations would become long and difficult even for rather simple structures.

The Code also specifies that the designer account for the variation in moment of inertia along the longitudinal axis of both the columns and the slab-beams. This variation is most complex within the joints formed by the intersection of the slabs and the columns. Within the joint, one must bear in mind a complex situation: the column acts to stiffen the slab, and the slab acts to stiffen the column. The significant region of the joint starts at the column centerline and runs to the *face of the support*, defined as the point of intersection of the face of the column, its bracket or its capital, or a wall equivalent to a wide column with the underside of the slab or drop panel. As illustrated in Fig. 10.24, between the column centerline and the

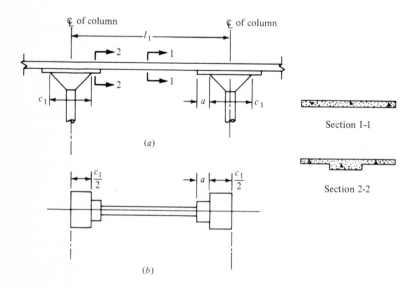

**Figure 10.24** Properties of slab-beams: (*a*) slab with drop panels and capitals, (*b*) assumed variation of moment of inertia.

face of the support, the column with its capital or bracket, if any, may be considered as acting like a thickened section of the floor slab. Because the length of contact between the slab-beam and the column face is small (only the width of the column or the column and its capital), only a limited width of slab at each end of the slab-beam acts effectively in flexure.

Two opposing factors affect the flexural stiffness: the large increase in depth provided by the column and its bracket, if any, versus reduced effective width, i.e., the column in limited contact with the slab-beam. To approximate the effect of both these factors, the Code specifies that between the column centerline and the face of the support the designer use a constant value of moment of inertia equal to the moment of inertia of the slab-beam's cross section at the face of the supports divided by the factor $(1 - c_2/l_2)^2$, where $c_2$ equals the width of the column side perpendicular to the direction of the span and $l_2$ equals the width of the slab-beam. Figure 10.24b shows the idealized variation of the moment of inertia for a typical slab-beam.

Similarly, the slab acts to stiffen the column: between the centerline of the slab and its surfaces, the half depth of the surrounding floor slab may be considered as a thickened section of the column with a moment of inertia assumed equal to infinity (an assumption made to facilitate computations). If beams, capitals, or brackets are present, the region in which the moment of inertia of the column is assumed to be infinite extends to the bottom edge of these elements (Fig. 10.25).

Since the computations of the properties of the columns and the slab-beams for an elastic analysis are complex when the variation of the moment of inertia, as specified by the Code, is considered, various design aides have been prepared to simplify the computations of member properties. If an equivalent frame is to be analyzed by moment distribution, stiffness factors, coefficients for fixed-end moments and carryover factors can be obtained from Tables 10.7 to 10.10.

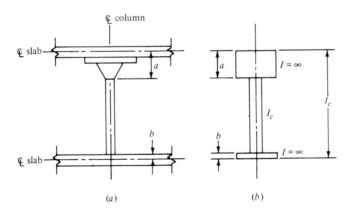

**Figure 10.25** Properties of columns: (a) column with a drop panel and capital, (b) assumed variation of moment of inertia.

## Table 10.7  Column stiffness coefficients[3] $k_c$

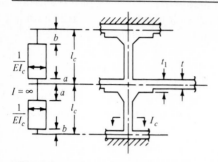

| b/l_c[†] | | | | | | | | | | |
|---|---|---|---|---|---|---|---|---|---|---|
| a/l_c[†] | 0.00 | 0.02 | 0.04 | 0.06 | 0.08 | 0.10 | 0.12 | 0.14 | 0.16 | 0.18 | 0.20 |
| 0.00 | 4.000 | 4.082 | 4.167 | 4.255 | 4.348 | 4.444 | 4.545 | 4.651 | 4.762 | 4.878 | 5.000 |
| 0.02 | 4.337 | 4.433 | 4.533 | 4.638 | 4.747 | 4.862 | 4.983 | 5.110 | 5.244 | 5.384 | 5.533 |
| 0.04 | 4.709 | 4.882 | 4.940 | 5.063 | 5.193 | 5.330 | 5.475 | 5.627 | 5.787 | 5.958 | 6.138 |
| 0.06 | 5.122 | 5.252 | 5.393 | 5.539 | 5.693 | 5.855 | 6.027 | 6.209 | 6.403 | 6.608 | 6.827 |
| 0.08 | 5.581 | 5.735 | 5.898 | 6.070 | 6.252 | 6.445 | 6.650 | 6.868 | 7.100 | 7.348 | 7.613 |
| 0.10 | 6.091 | 6.271 | 6.462 | 6.665 | 6.880 | 7.109 | 7.353 | 7.614 | 7.893 | 8.192 | 8.513 |
| 0.12 | 6.659 | 6.870 | 7.094 | 7.333 | 7.587 | 7.859 | 8.150 | 8.461 | 8.796 | 9.157 | 9.546 |
| 0.14 | 7.292 | 7.540 | 7.803 | 8.084 | 8.385 | 8.708 | 9.054 | 9.426 | 9.829 | 10.260 | 10.740 |
| 0.16 | 8.001 | 8.291 | 8.600 | 8.931 | 9.287 | 9.670 | 10.080 | 10.530 | 11.010 | 11.540 | 12.110 |
| 0.18 | 8.796 | 9.134 | 9.498 | 9.888 | 10.310 | 10.760 | 11.260 | 11.790 | 12.370 | 13.010 | 13.700 |
| 0.20 | 9.687 | 10.080 | 10.510 | 10.970 | 11.470 | 12.010 | 12.600 | 13.240 | 13.940 | 14.710 | 15.560 |

† $a$ = length of rigid column section at near end; $b$ = length at rigid column section at far end.

## Table 10.8  Values of torsion constant[3]  $C = \left(1 - \dfrac{0.63x}{y}\right)\dfrac{x^3 y}{3}$

| y \ x[†] | 4 | 5 | 6 | 7 | 8 | 9 | 10 | 12 | 14 | 16 |
|---|---|---|---|---|---|---|---|---|---|---|
| 12 | 202 | 369 | 592 | 868 | 1188 | 1,538 | 1,900 | 2,557 | | |
| 14 | 245 | 452 | 736 | 1096 | 1529 | 2,024 | 2,566 | 3,709 | 4,738 | |
| 16 | 388 | 534 | 880 | 1325 | 1871 | 2,510 | 3,233 | 4,861 | 6,567 | 8,083 |
| 18 | 330 | 619 | 1024 | 1554 | 2212 | 2,996 | 3,900 | 6,013 | 8,397 | 10,813 |
| 20 | 373 | 702 | 1167 | 1782 | 2553 | 3,482 | 4,567 | 7,615 | 10,226 | 13,544 |
| 22 | 416 | 785 | 1312 | 2011 | 2895 | 3,968 | 5,233 | 8.317 | 12,055 | 16,275 |
| 24 | 548 | 869 | 1456 | 2240 | 3236 | 4,454 | 5,900 | 9,469 | 13,895 | 19,005 |
| 27 | 522 | 994 | 1672 | 2583 | 3748 | 5,183 | 6,900 | 11,197 | 16,628 | 23,101 |
| 30 | 586 | 1119 | 1888 | 2926 | 4260 | 5,912 | 7,900 | 12,925 | 19,373 | 27,197 |
| 33 | 650 | 1243 | 2104 | 3269 | 4772 | 6,641 | 8,900 | 14,653 | 22,117 | 31,293 |
| 36 | 714 | 1369 | 2320 | 3612 | 5284 | 7,370 | 9,900 | 16,381 | 24,860 | 35,389 |
| 42 | 842 | 1619 | 2752 | 4298 | 6308 | 8,828 | 11,900 | 19,837 | 30,349 | 43,581 |
| 48 | 970 | 1869 | 3184 | 4984 | 7332 | 10,286 | 13,900 | 23,293 | 35,836 | 51,773 |

† $x$ is smaller dimension of rectangular cross section.

## Table 10.9  Moment distribution factors for slab-beam elements (flat plate)[3]

FEM (uniform load $w$) = $M w l_2 l_1^2$
$\quad$ $K$ (stiffness) = $k E I_2 t^3 / 12 l_1$
$\quad$ Carryover factor = $C$

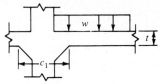

| $c_1/l_1$ | | $c_2/l_2$ 0.00 | 0.05 | 0.10 | 0.15 | 0.20 | 0.25 | 0.30 | 0.35 | 0.40 |
|---|---|---|---|---|---|---|---|---|---|---|
| 0.00 | $M$ | 0.083 | 0.063 | 0.083 | 0.083 | 0.083 | 0.083 | 0.083 | 0.083 | 0.083 |
|  | $k$ | 4.000 | 4.000 | 4.000 | 4.000 | 4.000 | 4.000 | 4.000 | 4.000 | 4.000 |
|  | $C$ | 0.500 | 0.500 | 0.500 | 0.500 | 0.500 | 0.500 | 0.500 | 0.500 | 0.500 |
| 0.05 | $M$ | 0.083 | 0.084 | 0.084 | 0.084 | 0.085 | 0.085 | 0.805 | 0.086 | 0.086 |
|  | $k$ | 4.000 | 4.047 | 4.093 | 4.138 | 4.181 | 4.222 | 4.261 | 4.299 | 4.334 |
|  | $C$ | 0.500 | 0.503 | 0.507 | 0.510 | 0.513 | 0.516 | 0.518 | 0.521 | 0.523 |
| 0.10 | $M$ | 0.083 | 0.084 | 0.085 | 0.085 | 0.086 | 0.087 | 0.087 | 0.088 | 0.088 |
|  | $k$ | 4.000 | 4.091 | 4.182 | 4.272 | 4.362 | 4.449 | 4.535 | 4.618 | 4.698 |
|  | $C$ | 0.500 | 0.506 | 0.513 | 0.519 | 0.524 | 0.530 | 0.535 | 0.540 | 0.545 |
| 0.15 | $M$ | 0.083 | 0.084 | 0.085 | 0.086 | 0.087 | 0.088 | 0.089 | 0.090 | 0.090 |
|  | $k$ | 4.000 | 4.132 | 4.267 | 4.403 | 4.541 | 4.680 | 4.818 | 4.955 | 5.090 |
|  | $C$ | 0.500 | 0.509 | 0.517 | 0.526 | 0.534 | 0.543 | 0.550 | 0.558 | 0.565 |
| 0.20 | $M$ | 0.083 | 0.085 | 0.086 | 0.087 | 0.088 | 0.089 | 0.090 | 0.091 | 0.092 |
|  | $k$ | 4.000 | 4.170 | 4.346 | 4.529 | 4.717 | 4.910 | 5.108 | 5.308 | 5.509 |
|  | $C$ | 0.500 | 0.511 | 0.522 | 0.532 | 0.543 | 0.554 | 0.564 | 0.574 | 0.584 |
| 0.25 | $M$ | 0.083 | 0.085 | 0.086 | 0.087 | 0.089 | 0.000 | 0.091 | 0.093 | 0.094 |
|  | $k$ | 4.000 | 4.204 | 4.420 | 4.648 | 4.887 | 5.138 | 5.401 | 5.672 | 5.952 |
|  | $C$ | 0.500 | 0.512 | 0.525 | 0.538 | 0.550 | 0.563 | 0.576 | 0.588 | 0.600 |
| 0.30 | $M$ | 0.083 | 0.085 | 0.086 | 0.088 | 0.089 | 0.091 | 0.092 | 0.094 | 0.095 |
|  | $k$ | 4.000 | 4.235 | 4.488 | 4.760 | 5.050 | 5.361 | 5.692 | 6.044 | 6.414 |
|  | $C$ | 0.500 | 0.514 | 0.527 | 0.542 | 0.556 | 0.571 | 0.585 | 0.600 | 0.614 |
| 0.35 | $M$ | 0.083 | 0.085 | 0.087 | 0.088 | 0.090 | 0.091 | 0.093 | 0.095 | 0.096 |
|  | $k$ | 4.000 | 4.264 | 4.551 | 4.864 | 5.204 | 5.576 | 5.979 | 6.416 | 6.888 |
|  | $C$ | 0.500 | 0.514 | 0.529 | 0.545 | 0.560 | 0.576 | 0.593 | 0.609 | 0.626 |
| 0.40 | $M$ | 0.083 | 0.085 | 0.087 | 0.088 | 0.090 | 0.092 | 0.094 | 0.095 | 0.097 |
|  | $k$ | 4.000 | 4.289 | 4.607 | 4.959 | 5.348 | 5.778 | 6.255 | 6.782 | 7.365 |
|  | $C$ | 0.500 | 0.515 | 0.530 | 0.546 | 0.563 | 0.580 | 0.598 | 0.617 | 0.635 |

**Design loads for equivalent frames**  Equivalent frames are normally analyzed for pattern loading to establish the design moments at the end and midspan sections of each panel. If the service live load is less than 75 percent of the service dead load, the Code permits frames to be analyzed for the single case of total factored load on all spans, eliminating the need for multiple analyses. Although a particular loading pattern may produce a larger moment at one section than a uniform load on all spans would produce, it will not create as large a moment at adjacent sections as a uniform load. Therefore, if a certain load pattern produces a small overstress at one section, the slab will deform and redistribute load to the adjacent understressed sections.

**Table 10.10  Moment distribution factors for slab-beam elements with drop panels**

FEM (uniform load $w$) = $Mwl_2 l_1^2$
$K$ (stiffness) = $kEl_2 t^3/12l_1$
Carry over factor = $C$

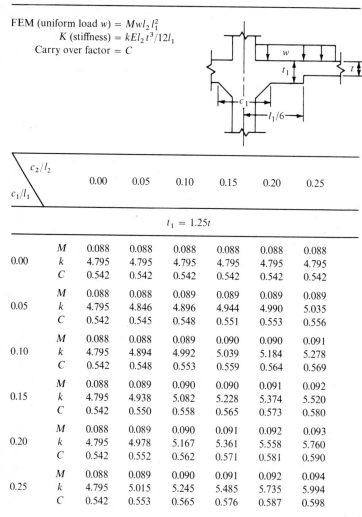

| $c_1/l_1$ | | $c_2/l_2$ | | | | | |
|---|---|---|---|---|---|---|---|
| | | 0.00 | 0.05 | 0.10 | 0.15 | 0.20 | 0.25 |
| | | | | $t_1 = 1.25t$ | | | |
| 0.00 | $M$ | 0.088 | 0.088 | 0.088 | 0.088 | 0.088 | 0.088 |
| | $k$ | 4.795 | 4.795 | 4.795 | 4.795 | 4.795 | 4.795 |
| | $C$ | 0.542 | 0.542 | 0.542 | 0.542 | 0.542 | 0.542 |
| 0.05 | $M$ | 0.088 | 0.088 | 0.089 | 0.089 | 0.089 | 0.089 |
| | $k$ | 4.795 | 4.846 | 4.896 | 4.944 | 4.990 | 5.035 |
| | $C$ | 0.542 | 0.545 | 0.548 | 0.551 | 0.553 | 0.556 |
| 0.10 | $M$ | 0.088 | 0.088 | 0.089 | 0.090 | 0.090 | 0.091 |
| | $k$ | 4.795 | 4.894 | 4.992 | 5.039 | 5.184 | 5.278 |
| | $C$ | 0.542 | 0.548 | 0.553 | 0.559 | 0.564 | 0.569 |
| 0.15 | $M$ | 0.088 | 0.089 | 0.090 | 0.090 | 0.091 | 0.092 |
| | $k$ | 4.795 | 4.938 | 5.082 | 5.228 | 5.374 | 5.520 |
| | $C$ | 0.542 | 0.550 | 0.558 | 0.565 | 0.573 | 0.580 |
| 0.20 | $M$ | 0.088 | 0.089 | 0.090 | 0.091 | 0.092 | 0.093 |
| | $k$ | 4.795 | 4.978 | 5.167 | 5.361 | 5.558 | 5.760 |
| | $C$ | 0.542 | 0.552 | 0.562 | 0.571 | 0.581 | 0.590 |
| 0.25 | $M$ | 0.088 | 0.089 | 0.090 | 0.091 | 0.092 | 0.094 |
| | $k$ | 4.795 | 5.015 | 5.245 | 5.485 | 5.735 | 5.994 |
| | $C$ | 0.542 | 0.553 | 0.565 | 0.576 | 0.587 | 0.598 |

| | | $t_1 = 1.5t$ | | | | | |
|------|------|------|------|------|------|------|------|
| | $M$ | 0.093 | 0.093 | 0.093 | 0.093 | 0.093 | 0.093 |
| 0.00 | $k$ | 5.837 | 5.837 | 5.837 | 5.837 | 5.837 | 5.837 |
| | $C$ | 0.589 | 0.589 | 0.589 | 0.589 | 0.589 | 0.589 |
| | $M$ | 0.093 | 0.093 | 0.093 | 0.093 | 0.094 | 0.094 |
| 0.05 | $k$ | 5.837 | 5.890 | 5.942 | 5.993 | 6.041 | 6.087 |
| | $C$ | 0.589 | 0.591 | 0.594 | 0.596 | 0.598 | 0.600 |
| | $M$ | 0.093 | 0.093 | 0.094 | 0.094 | 0.094 | 0.095 |
| 0.10 | $k$ | 5.837 | 5.940 | 6.042 | 6.142 | 6.240 | 6.335 |
| | $C$ | 0.589 | 0.593 | 0.598 | 0.602 | 0.607 | 0.611 |
| | $M$ | 0.093 | 0.093 | 0.094 | 0.095 | 0.095 | 0.096 |
| 0.15 | $k$ | 5.837 | 5.986 | 6.135 | 6.284 | 6.432 | 6.579 |
| | $C$ | 0.589 | 0.595 | 0.602 | 0.608 | 0.614 | 0.620 |
| | $M$ | 0.093 | 0.093 | 0.094 | 0.095 | 0.006 | 0.096 |
| 0.20 | $k$ | 5.837 | 6.027 | 6.221 | 6.418 | 6.616 | 6.816 |
| | $C$ | 0.589 | 0.597 | 0.605 | 0.613 | 0.621 | 0.628 |
| | $M$ | 0.093 | 0.094 | 0.094 | 0.095 | 0.096 | 0.097 |
| 0.25 | $k$ | 5.837 | 6.065 | 6.300 | 6.543 | 6.790 | 7.043 |
| | $C$ | 0.589 | 0.598 | 0.608 | 0.617 | 0.626 | 0.635 |
| | $M$ | 0.093 | 0.094 | 0.095 | 0.096 | 0.097 | 0.098 |
| 0.30 | $k$ | 5.837 | 6.099 | 6.372 | 6.657 | 6.953 | 7.258 |
| | $C$ | 0.589 | 0.599 | 0.610 | 0.620 | 0.631 | 0.641 |

If the live load exceeds 75 percent of the dead load, pattern loading using three-fourths of the full factored live load may be used; however, the moments must not be less than those produced by the full factored loads on all spans. The first requirement recognizes that a ductile underreinforced slab, which has the ability to redistribute moments between sections of maximum positive and negative moments, would have sufficient strength to carry the forces produced by pattern loading using the full live load.

**Design moments**  Since the analysis of a structure is based on centerline dimensions, the end moments in a slab-beam produced by the analysis of an equivalent frame are referred to the centerline of supports. Because the depth of the slab-beam's cross section increases sharply within the joint, the flexural capacity of the slab-beam for a given area of top steel is smaller at the face of a support than within the support (Fig. 10.26a); the Code therefore specifies that the reinforcement for negative moment may be designed for a value of moment smaller than that at the centerline of the support. For an *interior* column, the reinforcement may be designed for the value of moment at the face of the support (column or capital) but not at a distance greater than $0.175l_1$ from the center of the column, where $l_1$ is the distance between column centers in the direction of the span.

If capitals or brackets are used at *exterior* supports, the critical section for negative moment is taken at a distance from the face of the column equal to half the distance between the face of the column and the edge of the support (Fig. 10.26b).

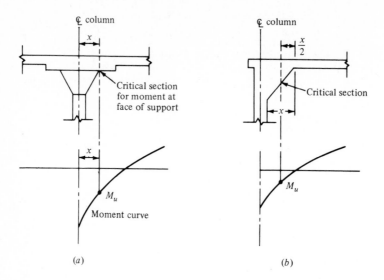

**Figure 10.26** Critical section for moment: (a) interior column, (b) exterior column.

If a floor system that satisfies the loading and geometrical conditions required by the direct design method is analyzed instead by the equivalent-frame method, the Code permits the resulting moments to be reduced by a constant factor so that the absolute sum of the positive and the average negative moments used in design does not exceed $M_0 = w_u l_2 l_n^2 / 8$ (see Example 10.8). This provision ensures that both the direct design method and the equivalent-frame method will produce similar results.

**Example 10.8:** Analysis by the equivalent-frame method gives the positive and negative moments shown in Fig. 10.27. If the slab system also satisfies the design conditions for the direct design method, by what fraction can the moments from the equivalent-frame analysis be reduced if the simple beam moment $M_0 = 240$ ft · kips for the span?

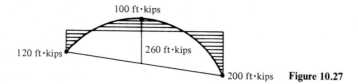

SOLUTION Compute the sum of the average negative moment and the positive moments

$$\frac{120 + 200}{2} + 100 = 260 \text{ ft · kips}$$

Since 260 ft · kips exceeds $M_0 = 240$ ft · kips, all moments from the equivalent frame analysis can be reduced by $\frac{240}{260} = 0.923$.

**Example 10.9:** Analyze a typical interior frame in the east-west (25-ft) direction of the flat-plate floor system shown in Fig. 10.28. The floor system consists of a slab 9-in (229-mm) thick supported on 18-in (457-mm) square columns and spans four bays in each direction. Analyze by moment distribution and consider the variation in cross section of the columns and slabs (see the frame in Fig. 10.29). $f_c' = 3$ kips/in² (20.68 MPa), and $f_y = 60$ kips/in² (413.7 MPa).

SOLUTION Compute the stiffness of the columns

$$l_c = 12 \text{ ft} = 144 \text{ in}$$

$$I_c = \frac{bh^3}{12} = \frac{(18)(18^3)}{12} = 8748 \text{ in}^4$$

$$k_c = 4.71$$

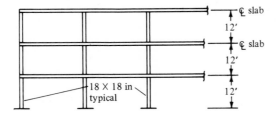

Section A-A

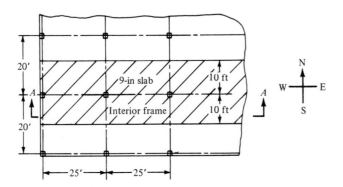

**Figure 10.28** Floor plan.

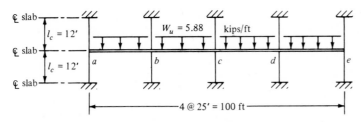

**Figure 10.29** Typical interior frame in east-west direction.

See Table 10.7 for $a/l_c = b/l_c = 4.5/[12(12)] = 0.031$, where $a$, $b$, and $l_c$ are shown in Fig. 10.30

$$\Sigma K_c = 2\frac{k_c EI_c}{l_c} = 2\frac{4.71(E8748)}{144} = 572E$$

In the above equation the factor 2 accounts for columns above and below the slab.

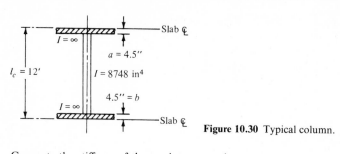

Figure 10.30 Typical column.

Compute the stiffness of the torsion arms whose cross section is a 9- by 18-in rectangle

$$l_2 = 20 \text{ ft} = 240 \text{ in}$$

$C = 2996 \text{ in}^4$ is evaluated by Eq. (10.6) or Table 10.8 for $x = 9$ in and $y = 18$ in; $c_2 = 18$ in

$$K_t = \frac{\Sigma 9 E_{cs} C}{l_2(1 - c_2/l_2)^3} = 2\frac{9E(2996)}{240(1 - \frac{18}{240})^3} = 284\,E$$

Compute the stiffness $K_{ec}$ of the equivalent column using Eq. (10.4)

$$\frac{1}{K_{ec}} = \frac{1}{\Sigma K_c} + \frac{1}{K_t} = \frac{1}{572E} + \frac{1}{559E}$$

$$K_{ec} = 190\,E$$

Compute the stiffness of the slab-beam (see Fig. 10.31)

$$I_s = \frac{l_2 h^3}{12} = \frac{(20)(12)(9^3)}{12} = 14,580 \text{ in}^4$$

$$l_1 = l_s = 25(12) = 300 \text{ in}$$

In Table 10.9 for

$$\frac{c_1}{l_1} = \frac{18}{25(12)} = 0.06 \qquad \frac{c_2}{l_2} = \frac{18}{20(12)} = 0.075$$

read $k = 4.084$

$$K_s = \frac{kEI_s}{l_s} = \frac{4.084E(14,580)}{300} = 198.48E \qquad \text{use } 198E$$

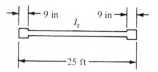

Figure 10.31 Typical slab-beam.

Distribution factors: for the exterior joint the sum of stiffnesses is

$$K_s + K_{ec} = 198E + 190E = 388E$$

$$\text{DF}_{slab} = \frac{K_s}{\Sigma K} = \frac{198E}{388E} = 0.51 \qquad \text{DF}_{ec} = \frac{K_{ec}}{\Sigma K} = 0.49$$

Since columns above and below the floor have the same stiffness, moment to the equivalent column is divided equally between top and bottom columns.

For the interior joint the sum of stiffnesses is

$$\Sigma K = K_s + K_s + K_{ec} = 2(198E) + 190E = 586E$$

$$\text{DF}_{slab} = \frac{K_s}{\Sigma K} = \frac{198E}{586E} = 0.34 \qquad \text{DF}_{ec} = \frac{K_{ec}}{\Sigma K} = \frac{283E}{586E} = 0.32$$

Compute the fixed-end moments. Assume that the full load acts on all spans (see Table 10.9 for coefficient $M$). For $c_1/l_1 = 0.06$ and $c_2/l_2 = 0.075$, read $M = 0.084$

$$FEM = Mw_u l_2 (l_1)^2 = 0.084(0.294)(20)(25^2) = 308.7 \text{ ft} \cdot \text{kips} \qquad \text{use } 309 \text{ ft} \cdot \text{kips}$$

Also read carry-over factor $= 0.506$ and round to 0.51.

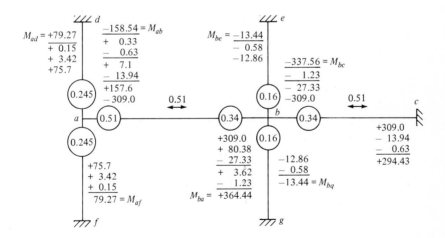

**Figure 10.32** Analysis of frame by moment distribution (moments at the far ends of columns not shown for clarity).

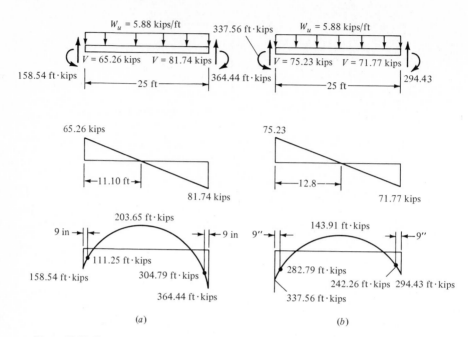

**Figure 10.33** Shear and moment curves for the slab-beams: (a) exterior span, (b) interior span.

*Moment distribution* Because of symmetry only half the structure need be analyzed. At the center, joint c can be treated as a fixed support (Fig. 10.32). Only the moments at the ends of the girder are evaluated.

The shear and moment curves for the exterior and the interior spans are computed (Fig. 10.33). Slabs can be designed for the negative moment at the face of supports.

## 10.8 DESIGN MOMENT FOR COLUMNS

If the analysis of the equivalent frames is carried out by the direct design method, the moment to be transferred to the columns above and below the slab is given by ACI equation 13.3 as

$$M = \frac{0.08[(w_d + 0.5w_l)l_2 l_n^2 - w_d' l_2'(l_n')^2]}{1 + 1/\alpha_{ec}} \tag{10.17}$$

where $w_d$, $w_l$ = factored dead and live load, respectively, per unit area on longer span

$w_d'$, $w_l'$ = factored dead and live load, respectively, per unit area on shorter span

$l_n'$ = length of shorter clear span

$l_n$ = length of longer clear span

$\alpha_{ec}$ = ratio of equivalent column stiffness to sum of stiffnesses of slabs and beams in direction in which moment is being computed = $K_{ec}/\Sigma(K_s + K_b)$

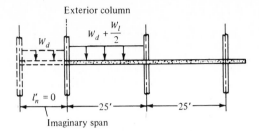

Exterior column

$W_d + \dfrac{W_l}{2}$

$W_d$

$l'_n = 0$

25'  25'

Imaginary span

**Figure 10.34** Loading specified by the ACI Code for column moments.

The total moment given by Eq. (10.17) is divided between the top and bottom columns at the joint in proportion to their flexural stiffness.

Equation (10.17) represents the column moment from a one-cycle moment distribution. In this computation 50 percent of the live load is assumed to act on the longer span only. Dead load acts on both spans. Although it is not specifically authorized by the ACI Code, engineers frequently apply Eq. (10.17) to an exterior joint by assuming that an imaginary span of zero length extends outward from the exterior column (see Fig. 10.34).

If the equivalent-frame method is used to determine the design moments to the columns, ACI Code §13.7.7.6 specifies that the results from the frame analysis be used to establish the design moments. Since the ACI Code calls for the floors to be designed for either full live load or three-fourths live load, the design moments to the columns from an equivalent-frame analysis will be much larger than those from the direct design method. Since the columns are vital to the structural integrity of the building, a conservative procedure has merit.

## 10.9 TRANSFER OF THE FLOOR LOAD INTO COLUMNS

The maximum load a two-way floor slab can support is often controlled by the strength of the joint between the slab and the column. Although a shallow two-way slab may have adequate flexural strength to carry the moments produced by the design loads, it may lack the shear strength required to transfer the design loads into the columns. Transfer of load into the columns occurs on the area around the perimeter of the column. If the slab is shallow, this area is small and the stresses on it are large.

Under certain conditions, moment must also be transferred between the slab and the column. The transfer of moment also creates shear stresses that must be combined with those produced by vertical load. These stresses are greatest at the exterior columns, where moments are applied to the joint from one side only.

Although slabs reinforced with beams along the column centerlines will be discussed briefly, the major emphasis in this section is on the transfer of load from unstiffened slabs (see Fig. 10.2a) into columns.

## Slabs with Beams

The shear capacity of beams, which are used along the column centerlines to stiffen two-way slabs, must be adequate to transmit the tributary floor loads into the columns. If the beam stiffness $\alpha_1 l_2 / l_1$ is equal to or greater than 1, the beams are assumed to transmit the entire floor load into the columns. If no beams are used, that is, $\alpha_1$ equals zero, the entire floor load is transmitted into the columns through the slab. When $\alpha_1 l_2 / l_1$ lies between 0 and 1, the distribution of load between the beams and slab is determined by linear interpolation.

In addition to supporting all loads applied directly to the beam, beams are also assumed to support the area of floor that extends half a panel width on each side of the beam centerline and lies between 45° lines extending outward from the corners of the panel (Fig. 10.35). As a result of this assumption, beams will carry loads with either a triangular or trapezoidal distribution. When the shear curve for a beam is computed, the designer must also account for any difference in moment at the ends of the member.

## Slabs without Beams

As previously discussed, vertical loads from the slab are transmitted by shear stresses into the supporting columns. A heavily loaded slab will fail if the shear stresses on the area around the column perimeter exceed the shear strength of the concrete, which is also heavily cracked because of the large moments that develop at the supports. For the special case in which the moments are equal on each side of the column, only vertical load is transmitted into the column. This load tends to displace the slab uniformly downward (Fig. 10.36), producing vertical shear stresses on the critical section around the perimeter of the column (Fig. 10.36b). In turn these shear stresses induce diagonal tension stresses that can lead to a punching-shear failure if the shear strength of the slab is exceeded (Fig. 10.36c). Although a punching-shear failure occurs on a diagonal plane, the analysis specified

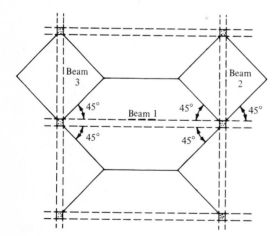

**Figure 10.35** Tributary areas supported by beams in a two-way floor.

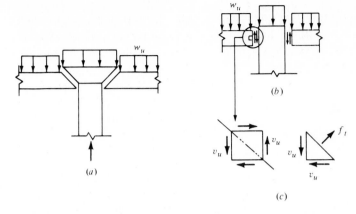

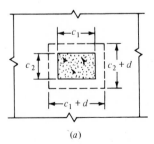

Figure 10.36 Transfer of vertical load into a column: (a) punching-shear failure, (b) shear stresses on vertical planes at the face of the column, (c) diagonal tension stresses created by shear.

in the ACI Code assumes that failure occurs on a fictitious surface lying a distance out from the face of the slab equal to half the effective depth of the slab. This critical section runs around the perimeter of the column, as shown in Fig. 10.37. The shear capacity of the concrete for punching shear is

$$V_c = \left(2 + \frac{4}{\beta_c}\right)\sqrt{f_c'}\,b_0 d \tag{4.24}$$

but not to exceed

$$4\sqrt{f_c'}\,b_0 d \tag{4.25}$$

where $d$ = effective depth of slab
$b_0$ = perimeter of the critical section
$\beta_c$ = ratio of long side of column or loaded area to short side

As long as $\beta_c$ is less than 2,

$$V_c = 4\sqrt{f_c'}\,b_0 d$$

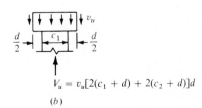

Figure 10.37 Critical section for punching shear: (a) plan, (b) shear stresses produced by vertical load assumed to be uniform on critical section.

**Shear stresses produced by unbalanced moment**  If unequal moments develop in the slab on opposite sides of a column, a portion of the unbalanced moment will be transmitted into the column by vertical shear stresses on the critical plane for punching shear (Fig. 10.38). The balance of the moment will create flexural stresses on a vertical surface directly at the column face (Fig. 10.42). Approximately 60 percent of the moment will be transmitted by conventional bending stresses and the balance by the vertical shear stresses shown in Fig. 10.39. When vertical load is transferred, the shear stress produced by direct stress and moment must be combined. A major contribution to the ACI design procedure, which follows, was made by Moe.[4]

The basic equation for computing maximum shear stress $v_t$ is

$$v_t = \frac{V}{A_c} + \gamma_v \frac{Mc}{J_c} \tag{10.18}$$

where $A_c$ = area of critical section around column

$J_c$ = polar moment of inertia of the areas parallel to applied moment plus moments of inertia of end area about centroidal axis of critical section

$M$ = total moment transferred into column

$\gamma_v$ = portion of moment producing shear stresses

$\gamma_v$ is evaluated by

$$\gamma_v = 1 - \frac{1}{1 + \frac{2}{3}\sqrt{(c_1 + d)/(c_2 + d)}} \tag{10.19}$$

For an interior column[5]

$$A_c = 2d(c_1 + c_2 + 2d) \tag{10.20}$$

$$J_c = \frac{d(c_1 + d)^3}{6} + \frac{(c_1 + d)d^3}{6} + \frac{d(c_2 + d)(c_1 + d)^2}{2} \tag{10.21}$$

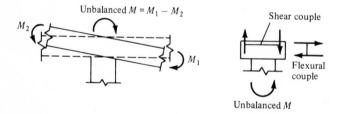

**Figure 10.38** Shear stresses produced by unbalanced moment: (*a*) rotation of slab by unbalanced moment, (*b*) transmission of unbalanced moment into column by shear and flexural couples.

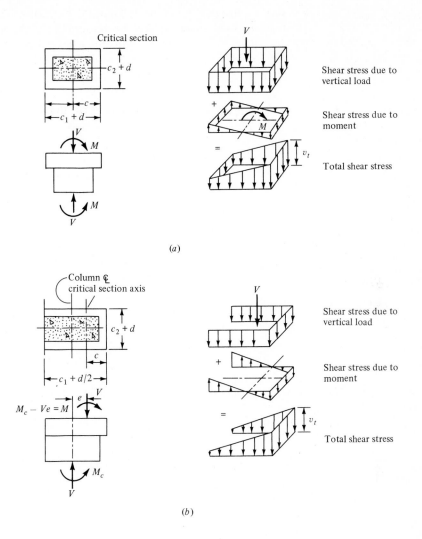

**Figure 10.39** Shear stresses on the critical surface produced by direct load and moment: (*a*) interior column, (*b*) exterior column.

**Example 10.10:** For the 9-in (229-mm)-thick slab in Fig. 10.28, compute the maximum shear stresses in the slab if a shear $V_u = 112.5$ kips (500.4 kN) and an unbalanced moment $M_u = 19$ ft·kips (25.76 kN·m) must be transferred through the critical section of the slab (Fig. 10.40). Assume $d = 7.75$ in (197 mm). The column is 18 in (457 mm) square so $c_1 = c_2 = 18$ in, $f'_c = 3000$ lb/in².

Critical section for shear

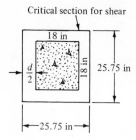

**Figure 10.40** Critical section for shear.

SOLUTION

$$v_t = \frac{V}{A_c} + \frac{\gamma_v Mc}{J_c}$$

$$\gamma_v = 1 - \frac{1}{1 + \frac{2}{3}\sqrt{(c_1 + d)/(c_2 + d)}} = 0.4 \tag{10.19}$$

$$J_c = \frac{d(c+d)^3}{6} + \frac{(c+d)d^3}{6} + \frac{d(c+d)(c+d)^2}{2} \tag{10.21}$$

$$= \frac{7.75(25.75^3)}{6} + \frac{25.75(7.75^3)}{6} + \frac{7.75(25.75^3)}{2} = 90{,}213 \text{ in}^4$$

$$v_t = \frac{112{,}500}{4(25.75)(7.75)} + \frac{0.4(19)(12{,}000)(12.87)}{90{,}213} \tag{10.18}$$

$$= 141 \text{ lb/in}^2 + 13 \text{ lb/in}^2 = 154 \text{ lb/in}^2 < \phi 4\sqrt{f'_c} = 186.2 \text{ lb/in}^2 \qquad \text{OK}$$

**Example 10.11:** The analysis of a flat plate floor indicates that a shear $V_u = 69.82$ kips and a moment $M_u = 110$ ft · kips must be transferred from a 9-in-thick slab into an 18-in-square exterior column (Fig. 10.41). Determine whether the shear capacity of the joint is adequate. The shear and moment above act at the centerline of the column. $f'_c = 4$ kips/in², $f_y = 60$ kips/in², and $d = 7.8$ in.

SOLUTION The critical section for shear is located a distance $d/2$ from the face of the column (see the dashed line in Fig. 10.41). Locate the centroid of the critical section by summing moments about axis 1-1

$$[21.9(2) + 25.8]\bar{X} = 21.9(2)\frac{21.9}{2}$$

$$\bar{X} = 6.9 \text{ in}$$

The area of the critical section $A_c$ is

$$A_c = 21.9(2)(7.8) + 25.8(7.8) = 542.9 \text{ in}^2$$

Next compute $J_c$. The contribution of sides parallel to the direction of moment is

$$I_{yy} + I_{xx} = \left[\left(\frac{bh^3}{12} + Ad^2\right) + \frac{bh^3}{12}\right]2$$

$$= \left[\frac{7.8(21.9^3)}{12} + 7.8(21.9)\left(\frac{21.9}{2} - 6.9\right)^2 + \frac{21.9(7.8^3)}{12}\right](2) = 20{,}990.36 \text{ in}^4$$

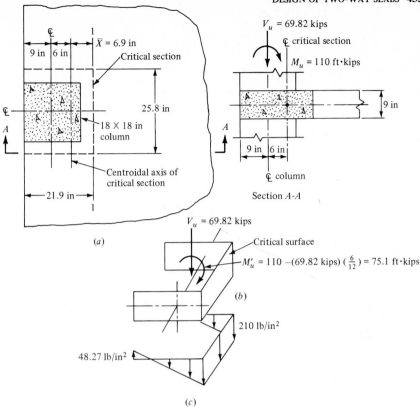

**Figure 10.41** Transfer of shear and moment through the slab: (a) plan at exterior column, (b) forces shifted from the centerline of the column to the centroid of the critical section, (c) shear stresses.

The contribution of side perpendicular to the direction of moment is

$$I = Ad^2 = 25.8(7.8)(6.9^2) = 9581.04 \text{ in}^4$$

$$J_c = 20,990.36 + 9581.04 = 30,571.4 \text{ in}^4$$

Compute the fraction of the moment $\alpha$ that is transferred by shear

$$\gamma_v = 1 - \frac{1}{1 + \frac{2}{3}\sqrt{(c_1 + d)/(c_2 + d)}} = 0.4$$

Since the analysis for the shear stresses is based on the forces acting at the centroid of the critical section, replace the shear and moment acting at the center of the column (from the frame analysis, Fig. 10.41a) by a statically equivalent force system acting at the centroid of the critical section (Fig. 10.41b).

Compute the maximum values of nominal shear stress. For the interior face

$$v_u = \frac{V_u}{A_c} + \frac{\gamma_v M_u' c}{J_c} = \frac{69,820}{542.9} + \frac{0.4(75.1 \text{ ft} \cdot \text{kips})(12,000)(6.9)}{30,571.4}$$

$$= 128.6 + 81.41 = 210 \text{ lb/in}^2 < \phi 4\sqrt{f_c'} = 215 \text{ lb/in}^2 \qquad \text{where } \phi = 0.85$$

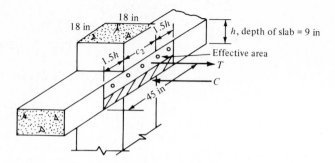

**Figure 10.42** Area of slab through which moment is transferred directly into column.

For the exterior edge (see Fig. 10.41b for $M'_u$)

$$v_u = \frac{V_u}{A_c} - \frac{\gamma_v M'_u c}{J_c} = \frac{69,820}{542.9} - \frac{0.4(75.1\ \text{ft}\cdot\text{kips})(12,000)(15)}{30,571.4}$$

$$= 128.6 - 176.87 = -48.27\ \text{lb/in}^2 < \phi 4\sqrt{f'_c} = 215\ \text{lb/in}^2$$

The distribution of shear stress is shown in Fig. 10.41c.

Verify that the flexural capacity of the critical area of the slab is adequate to transfer moment into the exterior column. As defined by ACI Code §13.3.4.2, the critical area of the slab extends a distance 1.5 times the depth of the slab on either side of the column (Fig. 10.42). The moment to be transferred by flexure is

$$0.6M'_u = 0.6(75.1) = 45.1\ \text{ft}\cdot\text{kips}$$

Using the trial method, determine $A_s$

$$M_u = 0.9T\left(d - \frac{a}{2}\right) \qquad \text{guess } a = 0.6\ \text{in}$$

$$45.1(12) = 0.9A_s(60)\left(7.8 - \frac{0.6}{2}\right)$$

$$A_s = 1.34\ \text{in}^2$$

Equating $T = C$ to check the initial value gives $a = 0.5$ in, which is OK. Use five no. 5 bars uniformly spaced over the 45-in width of the critical area, $A_{s,\,\text{sup}} = 1.53\ \text{in}^2$. Check bond

$$l_d = \frac{0.04A_b f_y}{\sqrt{f'_c}} = \frac{0.04(0.31)60,000}{\sqrt{4000}} = 11.8\ \text{in}$$

$$l_d = 0.0004d_b f_y = 0.0004(\tfrac{5}{8})(60,000) = 15\ \text{in} \qquad \text{controls}$$

The anchorage is OK since the column is 18 in deep.

## 10.10 SHEARHEAD REINFORCEMENT

In flat-slab construction the transfer of load into the column from the slab must be effected by shear stresses in the limited contact area between the intersection of the sides of the column and the slab. The area is small because of the shallowness of

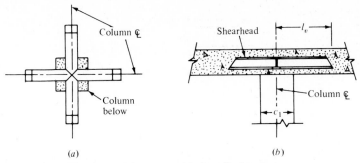

**Figure 10.43** Shearhead reinforcement: (*a*) plan, (*b*) elevation.

the slab, and shear stresses are therefore high. In flat-slab construction, the minimum dimensions of the column or the minimum thickness of the slab are governed typically by punching-shear requirements. As an alternative to increasing the column or the slab dimensions just to increase the perimeter of the critical section to carry shear, reinforcement: shearheads, for example, can be added.

Shearheads are cross-shaped elements constructed by welding structural-steel members, such as channels or I beams, into a rigid unit. Located above the column, the shearhead is used to increase the effective area through which shear can be transferred into an interior column from the surrounding slab (Fig. 10.43). Criteria have not yet been established to permit the design of shearheads at exterior columns where large torsional and bending moments in addition to shear must be transferred between slab and column. Shearhead design is not covered in this text.

## PROBLEMS

**10.1** The two-way floor system (Fig. P10.1) is composed of a 10-in slab supported on 14-in-wide beams. The overall depth of the floor from top of slab to underside of beam is 26 in. The distance between

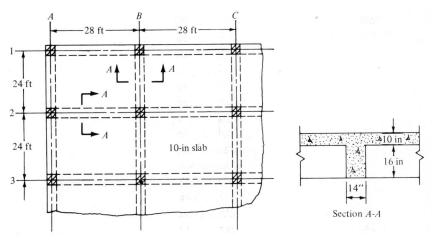

**Figure P10.1**

floors, which are of the identical construction, measures 14 ft. All columns are 14 by 14 in; $f'_c = 3$ kips/in$^2$, and $f_y = 60$ kips/in$^2$. The floor is to be designed for a live load of 80 lb/ft$^2$ in addition to the weight of the slabs and beams.

(*a*) Using the direct design method, determine the moments at critical sections in the exterior and interior slab-beams along column lines 2 and *B*.

(*b*) Repeat the analysis by the equivalent-frame method, taking into account the variation in the moment of inertia of the members.

(*c*) Determine whether the 10-in thickness satisfies the ACI equations to limit deflections.

(*d*) Design the exterior beam along column line 2 for shear and moment.

(*e*) Check the stiffness of the column to determine whether the positive moment in the slab beams must be increased for pattern loading.

(*f*) Determine the moments for which the exterior and interior columns must be designed when the analysis is made using the direct design method.

**10.2** Analysis of a flat-plate floor system (Fig. P10.2) indicates that a shear $V_u = 80$ kips and a moment $M_u = 72$ ft · kips must be transferred from the floor slab into an exterior column. Specified forces located at column centerline. Determine whether the shear capacity of the joint is adequate. $f'_c = 4$ kips/in$^2$.

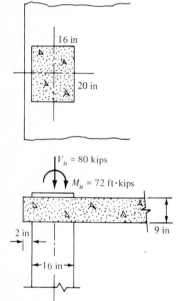

**Figure P10.2**

# REFERENCES

1. W. G. Corley and J. O. Jirsa: Equivalent Frame Analysis for Slab Design, *ACI Proc.*, vol. 67, no. 11, pp. 875–884, November 1970.
2. R. Park and W. Gamble: "Reinforced Concrete Slabs," Wiley, New York, pp. 562–612, 1980.
3. S. Simmonds and J. Misic: Design Factors for the Equivalent Frame Method, *ACI Proc.*, vol. 68, no. 11, pp. 68–71, November 1971.
4. Johannes Moe: Shear Strength of Reinforced Concrete Slabs and Footings under Concentrated Loads, *PCA Bull.* D47, p. 92, April 1961.
5. Commentary of Building Code Requirements for Reinforced Concrete (ACI 318-77), Detroit, 1977.
6. ASCE-ACI Task Committee 426: The Shear Strength of Reinforced Concrete Members—Slabs, *J. Struc. Div. ASCE*, vol. 100, no. ST8, pp. 1543–1591, August 1974.

Prestressing of high-strength steel in a precast beam with a hydraulic jack. (*Photograph by the Portland Cement Association.*)

# ELEVEN

## PRESTRESSED CONCRETE

### 11.1 INTRODUCTION

Although the low tensile strength of reinforced concrete means that it has certain weaknesses as a construction material, some aspects of its behavior can be improved, to eliminate cracking, reduce deflections, and increase shear strength, etc., by prestressing the concrete. *Prestressing* is a technique of introducing stresses of a predetermined magnitude into a structural member to improve behavior. Since the behavior of reinforced concrete can be substantially upgraded by the elimination of cracking, prestressing is used to introduce compression into regions where tension stresses will be created by applied loads. As discussed in Chap. 3, excessive cracking of concrete exposes the reinforcement to corrosion, decreases bending stiffness, reduces shear strength, increases deflection, reduces watertightness, and may detract from the appearance of the structure.

Typically, prestress is introduced into a member by high-strength steel tendons, in the form of bars, wires, or cables, that are first tensioned and then anchored to the member. The tendons are frequently passed through continuous channels formed by metal or plastic ducts, which are positioned securely in the forms before the concrete is cast.

Although prestressing produces many benefits, its use requires the designer to pay close attention to specific design and construction details. Because of the large forces carried by the tendons, the concrete is vulnerable to crushing at points where tendons are anchored or change direction. In addition, since the high stresses that are always present in prestressed concrete increase the axial and bending deformations caused by creep, the designer must detail the structure to ensure that the displacements of prestressed members can occur without damaging attached construction or creating unacceptable gaps between members.

Except for cylindrical concrete storage tanks, which have been prestressed since the middle 1930s by having their concrete walls wrapped with coils of highly tensioned wire, no major prestressed concrete structure was built in the United States until 1949. In that year the Walnut Lane Bridge, which focused the attention of the engineering profession on the possibilities of prestressed concrete, was erected in Philadelphia. The bridge was designed by Gustave Magnel, a Belgian professor and structural engineer, and was adopted for the Walnut Lane site when the low bid on the original design, a stone-faced concrete arch, was rejected because it exceeded the engineer's estimate by over $100,000.[1]

Since 1949, the number of buildings and bridges constructed of prestressed concrete has grown enormously; the reduced construction costs possible under certain design conditions, e.g., a structure with many repeating elements, make prestressed concrete an attractive option as a construction material. Prestressing also eliminates the need for on-the-site custom formwork for many members; it reduces dead weight by approximately 50 percent compared with reinforced concrete; and it shortens the construction time in the field by a considerable factor. Figure 11.1, which shows the increase in dollar volume of precast prestressed concrete products in the 30 years since the inception of the industry in 1950, illustrates the remarkable growth of prestressed construction in this country.

Although some structures are constructed and prestressed at the construction site, most prestressed members (about 80 percent) in the United States are produced in over 300 precasting plants. These plants manufacture a large variety of standard building and bridge components (beams, planks, wall panels, and piles), which are subsequently assembled in the field to form completed structures. Prestressing has also made it possible to extend the use of concrete to the construction of a wide range of structures not possible in reinforced concrete technology, including ocean-going ships, nuclear containment buildings, pressure vessels, and giant offshore drilling platforms.

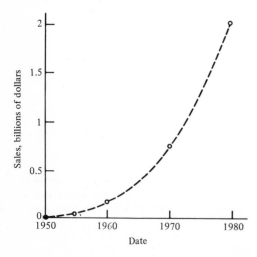

**Figure 11.1** Growth in sales of prestressed concrete in North America. (*Adapted from Ref. 2.*)

The fundamentals of prestressed concrete design are introduced in this chapter and applied primarily to the design of beams. Once the fundamentals of prestressing are understood, the application of prestressing to any structural element will be limited only by the imagination and creativity of the designer.

## 11.2 DESCRIPTION OF THE PRESTRESSING PROCEDURE

The sequence of steps required to prestress a member is illustrated in Fig. 11.2. The procedure outlined has been selected to show as clearly as possible the change in stress and the deformations that occur in both the steel and the concrete at each stage of prestressing. Figure 11.2$a$ shows a concrete member of constant cross section constructed with a narrow duct located at the centroid of the section and extending the full length of the member. At the time of prestressing the concrete

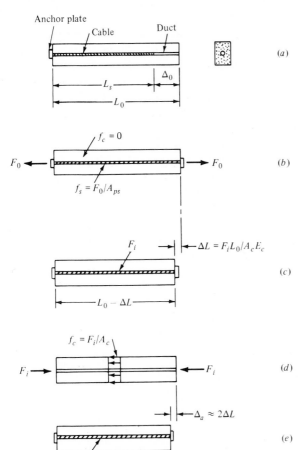

**Figure 11.2** The prestressing operation: ($a$) unstressed beam, ($b$) tensioning of the reinforcement, ($c$) transfer stage, ($d$) free-body diagram of concrete, ($e$) after creep and shrinkage.

has reached a compressive strength of $f'_{ci}$. A high-strength steel cable, which is shorter than the length of the beam by an amount $\Delta_0$, is inserted into the duct and connected to an anchor plate that bears against the left end of the member. At this stage both the steel and the concrete are unstressed.

In Fig. 11.2b forces of magnitude $F_o$ applied to each end of the cable by hydraulic jacks stretch the cable a distance $\Delta_0$. Now the length of the tensioned cable is equal to that of the concrete member. The force required to stretch the cable the distance $\Delta_0$ can be expressed in terms of the properties of the cable by

$$F_o = \frac{A_{ps} E_s}{L_s} \Delta_0$$

where $A_{ps}$ = cross-sectional area of cable
$E_s$ = modulus of elasticity of the tendon
$L_s$ = unstressed length of cable

After a second anchor plate has been connected to the right end of the cable, the jacking forces stretching the cable are released. As the cable tries to contract to its original length, it compresses the concrete member. The stage in which the force from the jack is transferred to the concrete is termed the *transfer stage*. As the member shortens elastically due to the compression of the concrete, the elongation of the cable reduces and the cable force decreases to a value of $F_i$ (Fig. 11.2c). To minimize loss of the cable force brought about by deformation of the concrete, a well-cured high-strength concrete is required.

With time, the compressed concrete will continue to shorten due to shrinkage and creep. This additional shortening can reduce the stress in the steel tendon by as much as 25 to 35 kips/in$^2$ (172.4 to 241.3 MPa). If a low-yield-point steel, e.g., a steel stressed to 20 kips/in$^2$ (137.9 MPa), were used as reinforcement, the shortening of the concrete after the tendon is anchored might eliminate all or a large part of the prestress. Therefore, to ensure that a significant portion of the prestress will be retained, high-strength steels, which require large elongations to stress, must be used to prestress concrete members. When supplied in the form of bars, wires, or cables, these steels can be tensioned to an initial stress of approximately 175 kips/in$^2$ (1206.6 MPa). If a loss in stress of 25 to 35 kips/in$^2$ (172.4 to 241.3 MPa) occurs with these steels, a large prestress still remains. After all losses have occurred, the effective force in the tendon reduces to a value of $F_e$.

**Example 11.1:** For the concentrically loaded member in Fig. 11.2, compute the deformations and the change in force that occur in the tendon and in the concrete before and after transfer. Given $A_{ps} = 0.5$ in$^2$, $A_c = 100$ in$^2$, $L_0 = 720$ in, $E_c = 4400$ kips/in$^2$, $E_s = 25,000$ kips/in$^2$, and $f_{si} = 180$ kips/in$^2$. The reduction in the initial tendon force due to elastic shortening of the concrete may be neglected.

SOLUTION Compute

$$F_o = A_{ps} f_{si} = 0.5(180 \text{ kips/in}^2) = 90 \text{ kips}$$

Compute the elongation of the cable as it is stressed to 180 kips/in² (base computations on $L_0$)

$$\Delta L = \frac{F_o L_0}{A_{ps} E_s} = \frac{90(720)}{0.5(25,000)} = 5.2 \text{ in}$$

Determine the change in length of the concrete due to $F_o$. Neglect the small change in force due to the deformation of the concrete

$$\Delta L = \epsilon L_0 = \frac{f_c L_0}{E_c} = \frac{\frac{90}{100}(720)}{4400} = 0.15 \text{ in}$$

Determine $F_i$. Steel and concrete are both shortened 0.15 in. Compute the change in force due to this deformation

$$\Delta F_s = \frac{\Delta L A_{ps} E_s}{L_0} = \frac{0.15(0.5)(25,000)}{720} = 2.6 \text{ kips}$$

$$F_i = F_o - \Delta F_s = 90 - 2.6 = 87.4 \text{ kips}$$

## 11.3 PRESTRESSING METHODS

Prestressed members are often classified by how the steel is stressed and anchored to the concrete. The member is said to be *pretensioned* if the steel is positioned in the form and stressed *before* the concrete is cast. A member is said to be *post-tensioned* when the steel is stressed *after* the concrete has hardened to a specific design strength.

### Pretensioning

Pretensioning is used primarily in precasting plants to mass-produce members whose size and weight are small enough to permit shipment to the site by truck. If pretensioning is carried out in the plant, the contractor is not required to supply equipment and trained personnel to prestress members in the field. In precast plants, members are commonly constructed on a long slab. These *casting beds*, which may be 400 to 500 ft (122 to 152 m) long, permit a number of members to be pre-tensioned simultaneously (Fig. 11.3). Large abutments, positioned at each end of the casting bed, are constructed with fittings to stress and anchor the tendons. After tendons have been tensioned and anchored to the abutments, forms are erected. Next, regular reinforcing steel required for carrying diagonal tension associated with shear, for controlling crack width produced by moment, or for

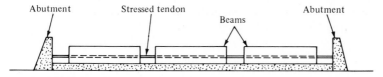

**Figure 11.3** Pretensioned beams constructed on a casting bed.

strengthening the anchorage zones is inserted into the form. Then concrete is cast and compacted. After the concrete reaches the required design strength, the tendons are cut. As the steel contracts, the force in the cable is transferred, primarily at the ends of the member, to the concrete by bond and by friction. As cables are elongated by tensioning, a lateral contraction of the tendon, due to the Poisson's ratio effect, takes place. After the tendons are cut and the reinforcement tries to return to its original unstressed dimensions, lateral expansion occurs. Wherever the reinforcement is encased in concrete, the lateral expansion creates high radial pressures between the concrete and the reinforcement, termed the *Hoyer effect*. The radial pressures allow large values of friction to develop between the concrete and the tendon, thereby permitting effective anchorage of cable strand and small-diameter-wire tendons (Fig. 11.4). This method cannot be used to anchor large-diameter [$\frac{3}{4}$ in (19 mm) and above] high-strength bars because the available friction is not adequate to anchor the large bar forces. Bearing plates must be used to anchor tendons with large forces.

Since steel forms and the capital costs of a prestressed plant are high, high-early-strength cement and steam curing are often used to accelerate the development of the concrete's strength in order to permit forms to be removed and reused as rapidly as possible. Under plant conditions, concretes with a compressive strength of 3 to 4 kips/in$^2$ (20.68 to 27.58 MPa) can be routinely produced in approximately half a day.

### Posttensioning

The other method of stressing the steel, posttensioning, is most logical when

1. Structures are too large to be pretensioned and shipped to the site.
2. The required cable shape (often called the cable *profile*) cannot be produced, e.g., a curved cable, if the cable is heavily tensioned since a tensioned cable tends to straighten between the points at which the tension is applied.
3. The design requires that the tendons be stressed in stages.
4. A structure is fabricated in sections to limit the weight of the element and then joined to other components by posttensioning to form a unit.

To ensure that tendons are free to elongate when the steel is tensioned, posttensioned construction requires the cable to remain unbonded until the concrete hardens. To prevent bond of the tendons to the concrete, cables may be enclosed in ducts that extend through the concrete or the tendons may be coated with grease or mastic and wrapped with paper. Ducts used to position tendons are often filled with cement grout after the tendons have been stressed and anchored. Grouting provides protection against corrosion and also raises the ultimate strength of the tendon at the section of maximum moment.

A large variety of mechanical devices to anchor tendons to concrete have been developed for posttensioned construction by manufacturers of prestressing systems. Fittings using wedges that lock the tendons to anchor plates by friction are frequently used to anchor tendons made of wire or strand. To minimize cable slipping

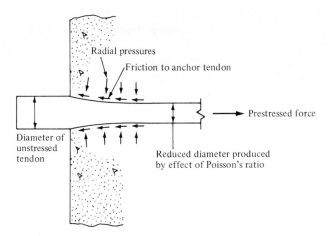

**Figure 11.4** The Hoyer effect. Lateral expansion of the tendon after the jack is released induces high radial pressures.

when the jacking force is released and the wedges forced into position, the surface of the wedge in contact with the tendon is grooved to produce sharp projections that dig into the cable surface.

Large-diameter high-strength bars may be anchored by wedges or threaded connections. Several examples of typical anchorage hardware are shown in Fig. 11.5. To prevent threading from lowering the strength of a bar by reducing the area of the end sections, the ends of bars to be threaded are often enlarged by forging (termed *upsetting*) to ensure that the cross section through the roots of the threads will be equal to or greater than the cross section of the unthreaded sections of the bar.

While engineers should be aware of the characteristics of the various types of prestressing systems so that their designs will provide adequate clearances for tendons and sufficient width for the end anchors, the designer typically specifies only the position of the centerline of the tendon and the magnitude of the prestress force. The contractor is then free to select the simplest and least expensive system supplying the required prestress.

Under certain design conditions members are both pretensioned and post-tensioned. For example, if many identical members are required in a structure, economy may be achieved by using a prestressing plant to produce the members. Pretensioning would be designed to carry all forces applied to the member during shipping and erection. After the members have been assembled in the field, additional tendons can be posttensioned to produce continuity or create additional strength.

## Full and Partial Prestressing

When a prestressed beam supports a live load that is large compared with the dead load and acts only intermittently, it may not be possible to select a prestress force and a cable profile that will produce the desired behavior in both the loaded and the

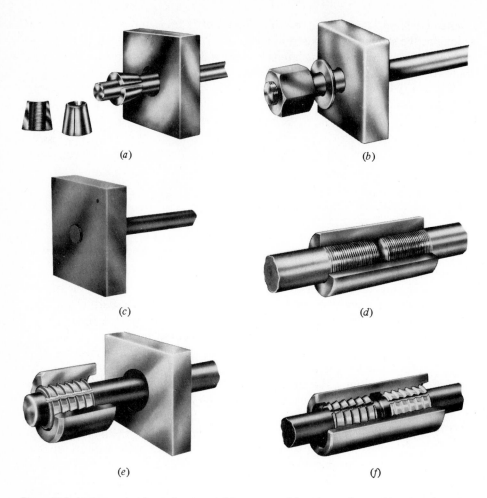

**Figure 11.5** Anchorage hardware for stressteel bar system: (*a*) wedge anchor and bearing plate, (*b*) nut and thread anchorage, (*c*) threaded plate (for use on end opposite jack), (*d*) threaded coupler, (*e*) Howlett grip nut, (*f*) Howlett grip coupler. (*Post-Tensioning Institute.*)

unloaded beam. A tendon that applies a prestress force large enough to neutralize the total tensile bending stresses created by both the dead and the live load may produce excessive camber (upward deflection) when the live load is absent. With time, creep, which can double or triple the magnitude of the initial camber, may produce an uneven floor or roof surface or damage nonstructural elements, e.g., tile or masonry partition walls, connected to, or supported by, the prestressed member. A large prestress force may also produce sufficient axial shortening in a member to damage other members or walls into which it frames.

Under these conditions, decreasing the intensity of the prestress force will lead to a better design (smaller camber under dead load and reduced axial shortening) even though tensile stresses and cracks develop when the full live load acts.

Limited cracking, which closes up when the full live load is removed, is acceptable as long as the steel is not exposed to corrosive conditions and the deflections are not excessive. Members in which tensile stresses develop under maximum service loads are termed *partially prestressed* beams.

When the service loads act, the ACI Code permits allowable tensile stresses as large as $6\sqrt{f_c'}$ lb/in$^2$ ($0.5\sqrt{f_c'}$ MPa) to develop in the precompressed tension zone of a partially prestressed member. The precompressed tension zone is the region of the member in which the applied loads create flexural tension stresses. To reduce these stresses, the prestressing tendon is positioned to introduce compression stresses into this zone. Since a tensile stress of $6\sqrt{f_c'}$ lb/in$^2$ ($0.5\sqrt{f_c'}$ MPa) is less than the tensile strength of the concrete, taken as $7.5\sqrt{f_c'}$ lb/in$^2$ ($0.62\sqrt{f_c'}$ MPa) but subject to a large variation, cracking is unlikely to occur.

If deflections (considering the influence of cracking) satisfy the provisions of ACI Code §9.5, and if the reinforcement is protected by the depth of cover specified by ACI Code §7.7.3.2, tensile stresses, computed on the basis of the gross uncracked section, may go as high as $12\sqrt{f_c'}$ lb/in$^2$ ($1.0\sqrt{f_c'}$ MPa).

For structures which are exposed to corrosive environments or which must be impermeable to fluids, cross sections must always be in compression to prevent cracks from developing. Members in which no tensile stresses are permitted under maximum service loads are said to be *fully prestressed*.

## 11.4 MATERIALS AND ALLOWABLE STRESSES

Because of the large axial shortening that occurs in prestressed concrete members as a result of such factors as creep, shrinkage, and elastic shortening, the initial prestress force in the tendons will gradually diminish. If a significant percentage of the initial prestress is to remain after these deformations have occurred, the strain required to stress the steel initially must be large compared with the strains produced by the shortening. Since the strains due to shortening correspond to a loss in steel stress that can range from approximately 20 to 35 kips/in$^2$ (137.9 to 241.3 MPa), steels are initially stressed to 125 to 175 kips/in$^2$ (862 to 1206.6 MPa). The lower stress applies to bars; the upper stress to strand and wire.

### Prestressing Steel

Prestressing steel is available in the form of cold-drawn wire, stranded cable (seven-wire strand is the most common), and alloyed steel bars. Wire and strand are produced with ultimate tensile strengths of 250 and 270 kips/in$^2$ (1724 and 1862 MPa). Smooth-surfaced high-strength bars with diameters between $\frac{3}{4}$ and $1\frac{3}{8}$ in (19 and 35 mm) are manufactured with ultimate tensile strengths of 145 and 160 kips/in$^2$ (1000 and 1103 MPa).

High-strength steels are produced by using alloying elements (manganese, silicon, carbon, etc.), by cold working, and by heat treating and tempering. Since the manufacturing techniques that produce high strength also reduce ductility and toughness, steels with yield points above 270 kips/in$^2$ (1862 MPa), which would be extremely brittle, are not used as tendons.

Typical stress-strain curves for high-strength bars and wires (Fig. 11.6) indicate that prestress steels lack a sharply defined yield point. To establish the beginning of the inelastic range, a yield strength $f_{py}$ for wire and strand is often defined as the stress associated with a 1 percent strain. For high-strength bars, the yield strength is frequently specified as the stress associated with the intersection of the stress-strain curve and a line parallel to the initial slope of the stress-strain curve that extends upward from a strain of 0.002 at zero stress. For most prestressed steels the yield strength $f_{py}$ will be equal to approximately 85 percent of the ultimate tensile strength $f_{pu}$. Table 11.1 lists allowable stresses for prestressing steels at various stages of construction.

For bars and wires, the modulus of elasticity $E_s$ can be taken as 29,000 kips/in$^2$ (200 GPa). Strand, whose individual wires spiral around a central wire core, has an effective modulus of elasticity of approximately 26,000 kips/in$^2$ (200 GPa). The reduced modulus results from both the stretching and uncurling of the wires, which tend to straighten as well as deform when the strand is loaded.

Highly tensioned steels are more vulnerable to corrosion than lightly stressed steels. This *stress corrosion* is most likely to occur in oil-tempered wire tendons exposed directly to a combination of air and moisture. Since calcium chloride, which accelerates the rate at which concrete gains strength, also increases the susceptibility of prestressed tendons to stress corrosion, concrete specifications typically specify that calcium chloride must not be added to cements or mortars in contact with prestressed reinforcement.

### Concrete for Prestressed Members

Prestressed concrete members are typically constructed from higher-strength concretes than regular reinforced concrete members. The use of high-strength concrete is dictated either by the requirements of the construction procedure in

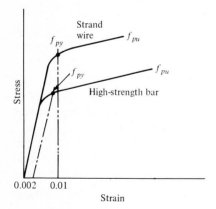

**Figure 11.6** Stress-strain curves for prestress steel.

**Table 11.1  Allowable stresses in tendons†**

| | |
|---|---|
| Maximum stress produced at the time of initial tensioning not to exceed the smaller of | $0.8f_{pu}$ or $0.94f_{py}$ |
| In addition stress must not exceed that recommended by the manufacturer of the prestressing tendons and anchorages | |
| Maximum stress in pretensioned tendons immediately after transfer or in post-tensioned tendons after tendon is anchored | $0.7f_{pu}$ |

† Based on ACI Code §18.5.

precasting plants or by the design conditions (long spans, heavy loads, or limitations on depth) under which the use of prestressed concrete is most advantageous.

In the precast plant, high-strength concrete provides the following advantages:

1. Speeds the gain in strength, permitting concrete to be prestressed the day after members are cast. Reduces overhead by allowing frequent reuse of expensive metal forms.
2. Produces the high bond strength required to anchor the wire and the strand used as reinforcement in pretensioned construction.
3. Increases the allowable bearing stresses, thus reducing the size of bearing plates required to anchor posttensioned tendons.
4. Minimizes losses in prestressed tendons by reducing elastic shortening, creep, and shrinkage of concrete.
5. Reduces the size and the weight of members, thereby saving materials and permitting the use of lighter equipment to handle precast elements.
6. Reduces the required area of shear reinforcement.

The structural system selected for a particular site and function is strongly influenced by the span length of the structure and the magnitude of the design loads. When spans are short [15 to 20 ft (4.6 to 6.1 m)], poured-in-place concrete with an $f'_c$ of 2.5 to 3 kips/in² (17.24 to 24.13 MPa) typically has adequate strength to support the design loads. Use of a stronger concrete would add to the cost of construction without any significant improvement in safety or any savings in reinforcement. However, when spans are long [over 75 ft (23 m)], the member's own weight typically constitutes the major load to be supported. By using high-strength concrete, the size of members can be significantly reduced. The weight reduction achieved results in reduced internal forces and leads to a further reduction in member size. By saving large quantities of materials and simplifying construction procedures the use of smaller members can produce a large reduction in construction costs. Lighter members further reduce costs by reducing the size of foundations on which the superstructure is supported.

Prestressed members are frequently constructed of concrete with a 28-day compressive strength of 4 to 6 kips/in² (27.58 to 41.37 MPa). Although concrete of this strength can be produced with careful supervision in the plant and in the field, the likelihood of producing concrete with higher compressive strengths is

## Table 11.2 Allowable stresses for concrete†

| At the time of initial tensioning before time dependent losses produced by creep, shrinkage, or relaxation have occurred | |
| --- | --- |
| 1. Maximum compressive stress | $0.6f'_{ci}$ |
| 2. Maximum tensile stress except as permitted in item 3 | $3\sqrt{f'_{ci}}$ |
| 3. Maximum tensile stress at the ends of simply supported members | $6\sqrt{f'_{ci}}$ |

If tensile stresses exceed the values in 2 and 3, reinforcement must be provided to carry the total tensile force in the concrete; ACI Code §18.4.1 permits cracking to be neglected and the computations for stress based on the properties of the gross section

| Service-load stresses, assuming all prestress losses have occurred | |
| --- | --- |
| 1. Maximum compressive stress | $0.45f'_c$ |
| 2. Maximum tensile stress in precompressed tensile zone | $6\sqrt{f'_c}$ |
| 3. Maximum tensile stress in precompressed tensile zone if deflections are within specified limits and concrete cover satisfies the provisions of ACI §7.7.3.2 for members subject to earth, weather, or corrosive conditions | $12\sqrt{f'_c}$ |

† The allowable stresses above may be exceeded if experimental or analytical studies indicate that behavior will be satisfactory.

less certain. Higher strengths are harder to achieve because the mixes, which require a low water-cement ratio, are less workable and more difficult to place and compact in forms. Table 11.2 summarizes the stress in concrete permitted by the ACI Code at various stages of prestressing.

## 11.5 FLEXURAL DESIGN REQUIREMENTS FOR PRESTRESSED BEAMS

Unlike a reinforced concrete beam, in which stresses vary with the intensity of load, a prestressed beam (loaded or unloaded) is always heavily stressed. Once the tendons in a prestressed concrete beam have been stressed and anchored, high longitudinal compressive stresses are induced throughout the member. In addition, if a tendon is curved or offset from the centroidal axis, moments are created that produce tensile and compressive bending stresses. To ensure that a prestressed beam will not be overstressed at transfer or under service loads, it must be analyzed elastically. Also, to provide an adequate factor of safety against failure, the nominal flexural strength must exceed the moment produced by factored loads. A complete design requires that a beam satisfy the following criteria:

1. The concrete must not be overstressed at the time of initial prestress (the transfer stage), when the prestress force is highest and the applied loads smallest (ACI Code §18.2.2).

2. The concrete must not be overstressed by service loads or any other loads that may be applied during manufacture in the plant, during transportation to the site, or during erection in the field.
3. The flexural design strength, the ultimate moment capacity, of the cross section $\phi M_n$ must be at least 20 percent greater than $M_{cr}$, the cracking moment of the cross section (ACI Code §18.8.3). This provision ensures that the steel tendon will not rupture with the formation of the first flexural crack. When a prestressed beam cracks, the steel tendon must absorb the tensile force previously carried by the concrete for the beam to remain stable.
4. The flexural design strength $\phi M_n$ must equal or exceed the ultimate moment $M_u$ produced by factored service loads (ACI Code §18.2.1).
5. The area of reinforcement must be small enough to ensure that failure will be initiated by overstress of the steel rather than by crushing of the concrete. In recognition of the satisfactory behavior of certain types of heavily reinforced prestressed systems, the ACI Code does allow the design of beams that are over-reinforced by a small amount; however, in keeping with the intention of the ACI Code to produce ductile members, the designer should avoid overreinforced beams wherever possible.

## 11.6 FORCES EXERTED BY STRESSED TENDONS

Before discussing the complete elastic analysis of a prestressed beam, we shall review the forces and stresses induced in the concrete of a beam by the prestressed tendon for several common tendon profiles. A knowledge of the forces exerted on the concrete by various tendon shapes provides the designer with the insight required to select an effective cable profile for a given set of design loads.

In addition to applying axial compression to the concrete, tendons can be shaped and positioned to apply both distributed and concentrated loads. At the exterior ends of members, tendons can also be used to create moment by placing the anchorage above the centroid of the concrete section. In general, the cable profile should be selected so that it creates loads with the same distribution as the design loads but in the opposite sense.

### A Straight Tendon at the Centroid

To begin with the simplest case, we shall consider a straight member with a constant cross section that is prestressed by a tendon located at the centroid of the concrete cross section (see Fig. 11.7). Through bearing plates, the straight cable exerts two equal and opposite compressive forces of magnitude $F$ on the ends of the concrete member (Fig. 11.7b and c). This force is then transmitted to all internal sections (Fig. 11.7d). Since the concrete is axially loaded, a state of uniform compression is produced on sections normal to the longitudinal axis except for a short distance at each end. At the ends, the tendon reaction produces high local stresses in the concrete (St. Venant's principle).

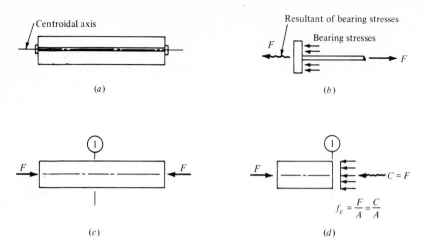

**Figure 11.7** Concentrically loaded concrete member: (a) tendon at the centroid, (b) end anchorage, (c) reactions of tendon on concrete, (d) stress in concrete due to prestress.

The uniform distribution of compressive stresses induced by a tendon at the centroid of an area is not an effective stress distribution for a section subject to high bending moment. Although the compression reduces the tensile bending stresses, it adds to the compressive bending stresses. If the tendon is positioned below the centroid of the section on the side in which tensile bending stresses develop, the prestress will produce a moment (opposite in sense to that of the applied loads) as well as compression. With proper choice of eccentricity, a triangular or nearly triangular stress distribution can be created (Fig. 11.22a). This distribution neutralizes the tensile bending stresses without adding to the compressive bending stresses at the top surface.

## A Straight Tendon with Constant Eccentricity

If a straight tendon is located a constant distance $e$ from the centroidal axis (Fig. 11.8a), the concrete member is stressed at each end by two eccentric loads $F$ that produce compression and bending. As shown in Fig. 11.8b, the eccentric loads are equivalent to an axial force $F$ and a moment $Fe$. Since the moment is constant along the length of the member, no shear develops from prestressing. To satisfy equilibrium in the horizontal direction, an internal compressive force equal to the force in the tendon must develop in the concrete on all vertical sections a distance $e$ below the centroidal axis (Fig. 11.8c). This internal force produces both direct and bending stresses. The shape of the resultant longitudinal stress distribution will depend on the relative magnitude of the axial force and its eccentricity. Wherever the tensile bending stresses produced by moment exceed the axial compressive stresses, tensile stresses develop on the cross section (Fig. 11.8d). If the axial stresses are larger than the maximum bending stresses, a trapezoidal variation of compressive stresses develops (Fig. 11.8e).

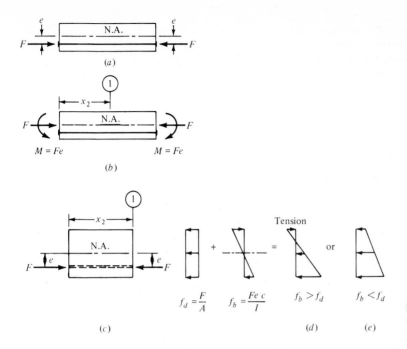

**Figure 11.8** Forces exerted on a concrete member by a tendon with constant eccentricity: (a) tendon with constant eccentricity, (b) equivalent-force system, (c) longitudinal stresses on a vertical section, (d) bending stresses greater than direct stresses, (e) direct stresses greater than bending stresses.

Additional insight into the influence of a tendon with constant eccentricity on the state of stress in a short-span, determinate beam can be secured by considering the variation of the resultant moment produced by the prestress force and the applied loads at both the transfer and the service-load stages (Figs. 11.9 and 11.10). Immediately after transfer, when the tendon force $F_i$ is highest, the maximum moment in the beam from loads and prestress occurs at the ends of the member where the dead-load moment is zero (Fig. 11.9). At midspan, where the dead-load moment is maximum, the net moment, i.e., the algebraic sum of the dead-load and the prestressed moments, is smallest. Since the dead-load moment is small compared with the moment due to prestress, negative moment exists over the entire span and produces an upward camber. If the live load acts infrequently, the upward camber will increase with time as a result of creep and shrinkage. Although prestressing will improve the distribution of moments when live load acts, under dead load it has created much larger moments than would exist in a nonprestressed beam.

When the total load acts (Fig. 11.10), the moment created by the prestressing, now reduced to $F_e$ because of creep and shrinkage primarily, reduces the moment at midspan produced by the service loads by approximately 50 percent. Although prestressing has created moment at the ends of the member, where none would normally exist, the maximum moment for which the beam must be sized has been reduced from 1620 to 820 in · kips (183 to 92.7 kN · m). Although a portion of the

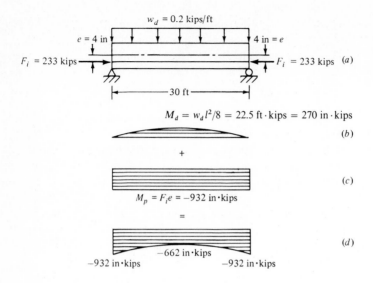

$$M_d = w_d l^2/8 = 22.5 \text{ ft} \cdot \text{kips} = 270 \text{ in} \cdot \text{kips}$$

$$M_p = F_i e = -932 \text{ in} \cdot \text{kips}$$

**Figure 11.9** Variation of moment produced by dead load and initial prestress: (*a*) beam, (*b*) dead-load moment, (*c*) prestress moment (due to eccentricity of tendon), (*d*) net moment, dead load plus prestress.

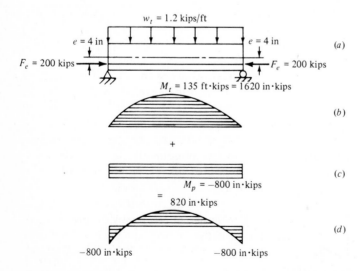

**Figure 11.10** Variation of moment produced by total load and effective prestress: (*a*) beam, (*b*) total load moment, (*c*) prestress moment, (*d*) net moment, total load plus prestress.

beam's strength must be used to carry the axial compression of 200 kips (889.6 kN), the large reduction in design moment permits the use of a much smaller cross section than would be possible in a nonprestressed beam.

## A Sloping Tendon

In a member stressed by a tendon with a constant eccentricity, the moment $Fe$ due to prestress is the same on all vertical sections along the length of the beam. Since the moment created by transverse load varies along the length of a member, a tendon with a constant eccentricity may produce an optimum state of stress on only one section for one particular loading condition. By sloping the tendon to vary the eccentricity, the moment created by prestress can be varied from section to section to improve the state of stress on additional sections. For example, by reducing the moment created by prestress near the ends of the beam the net moment can be reduced sharply, so that at transfer the end regions are lightly stressed. Reducing the tendon eccentricity will also reduce the long-term deflections due to creep.

The moments and stresses created by a tendon with a variable eccentricity will first be examined in a beam with a V-shaped tendon (Fig. 11.11a), in which the axis of the tendon intersects the longitudinal axis of the beam at a slope of $\theta$ at the left end and a slope of $\theta'$ at the right end. To establish the components of the tendon

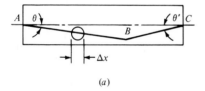

(a)

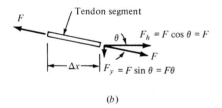

(b)

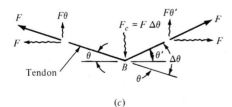

(c)

**Figure 11.11** Lateral forces produced by a change in tendon direction: (a) beam with a sloping tendon, (b) forces on a short segment of tendon, for small angles, $\cos \theta = 1$ and $\sin \theta = \theta$, (c) reaction of concrete on tendon at point where tendon changes direction, $F_c = F\theta + F\theta' = F(\theta + \theta') = F \Delta\theta$.

force, the state of stress in a short segment of tendon (shown circled in Fig. 11.1a) is examined in Fig. 11.11b. Since the tendon is assumed to be completely flexible, it transmits an axial force $F$ that is always tangent to the axis of the tendon. The horizontal and vertical components of the tendon force can be expressed as a function of the cable slope angle by

$$F_h = F \cos \theta \qquad F_y = F \sin \theta \tag{11.1}$$

For small slopes, typically those associated with prestressing tendons, $\cos \theta = 1$ and the $\sin \theta = \theta$ (in radians) can be substituted into Eq. (11.1) to give

$$F_h = F \qquad F_y = F\theta \tag{11.2}$$

At point $B$, where the tendon changes direction (Fig. 11.11c), a summation of forces on a short length of tendon shows that the tendon is in equilibrium in the horizontal direction because the horizontal components of the tendon force are equal in magnitude but oppositely directed. However, in the vertical direction equilibrium of the tendon segment requires that the concrete exert on the cable a lateral force $F_c$ that is equal to the sum of the vertical components of cable tension, $F\theta$ and $F\theta'$. Since no forces normal to the cable develop between points $A$ and $B$ or between points $B$ and $C$, the vertical force $F_c$ can develop only at point $B$, where the tendon profile undergoes a sharp change in angle of $\Delta\theta = \theta + \theta'$. Although the profile of a real tendon will be rounded at the point where the slope changes direction, the representation of the concrete reaction as a concentrated force introduces no significant error if the length of curvature is small.

At the point where the tendon changes direction, the designer should always estimate the magnitude of the bearing stresses created by the lateral force to ensure that the concrete does not crush locally. If the bearing stresses are excessive, some type of bearing plate or fitting should be used to reduce the intensity of the direct stresses created by the narrow tendon on the concrete.

In Example 11.2, which shows the forces created in the concrete of a prestressed beam by a sloping tendon, the tendon reactions are applied to a free body of the concrete only. If the dead load of the member is neglected, equilibrium of the concrete free body indicates that the vertical reaction created by prestress at supports $A$ and $B$ are zero. As shown in this example, the sloping cable creates both shear and moment. At each section along the length of the member, the moment produced by prestress is proportional to the eccentricity of the tendon at that point.

**Example 11.2: Shear and moment created by prestress**    Neglecting the weight of the beam in Fig. 11.12, (a) plot the variation of shear and moment created in the concrete by the sloping tendon, which is stressed to 240 kips (1068 kN), and (b) determine the distribution of longitudinal stresses in the concrete on a vertical section at point $C$.

SOLUTION (a) Compute the cable slope angles

$$\theta = \frac{6}{10(12)} = 0.05 \text{ rad} \qquad \theta' = \frac{6}{20(12)} = 0.025 \text{ rad}$$

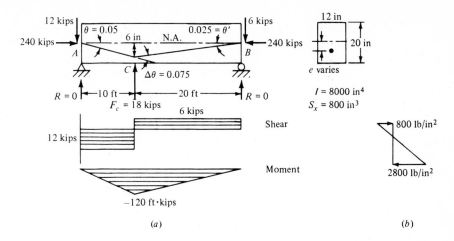

**Figure 11.12** Shear and moment created by prestress only: (*a*) shear and moment curves, (*b*) longitudinal stresses at point *C*.

The lateral forces applied by the cable to the concrete are

$$F_y = \begin{cases} F\theta = (240 \text{ kips})(0.05) = 12 \text{ kips} \downarrow (53.4 \text{ kN}) & \text{at } A \\ F\theta' = (240 \text{ kips})(0.025) = 6 \text{ kips} \downarrow (26.7 \text{ kN}) & \text{at } B \\ F \cdot \Delta\theta = (240 \text{ kips})(0.075) = 18 \text{ kips} \uparrow (80.1 \text{ kN}) & \text{at } C \end{cases}$$

The shear and moment curves produced by prestress only are shown in Fig. 11.12*a*.

(*b*) Longitudinal stresses at point *C* created by prestress only (Fig. 11.12*b*).

$$f = \frac{F}{A} \pm \frac{M}{S_x} = -\frac{240{,}000 \text{ lb}}{240 \text{ in}^2} + \frac{(120 \text{ ft} \cdot \text{kips})(12{,}000)}{800 \text{ in}^3}$$

$$= -1000 \text{ lb/in}^2 \pm 1800 \text{ lb/in}^2$$

$$f_{\text{top}} = -1000 + 1800 = +800 \text{ lb/in}^2 \text{ (5.52 MPa)}$$

$$f_{\text{bot}} = -1000 - 1800 = -2800 \text{ lb/in}^2 \text{ (19.31 MPa)}$$

**Example 11.3:** Neglecting the dead weight of the beam in Fig. 11.13, (*a*) plot the variation of shear and moment in the concrete of the beam in Example 11.2 due to the 240 kips prestress force in the tendon and an applied load of 24 kips at point *C*. The reactions of the tendon, computed in Example 11.2, are shown on the figure by dashed arrows. (*b*) Compute the distribution of the longitudinal stresses in the concrete on a vertical section at point *C*.

SOLUTION (*a*) The reactions of the tendon on the concrete, computed in Example 11.2, are shown with dashed arrows. The reactions at the supports are computed using the basic equations of statics. The shear and moment curves from all forces are shown in Fig. 11.13*a*.

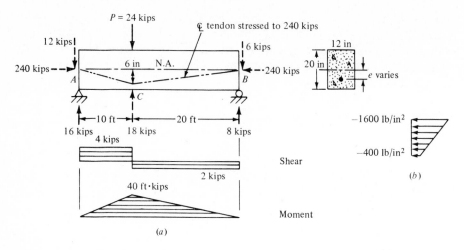

**Figure 11.13** Shear and moment curves produced by load and prestressing: (*a*) shear and moment curves, (*b*) longitudinal stresses at point *C*.

(*b*) Longitudinal stresses at *C* due to prestress and applied loads (Fig. 11.13*b*) are

$$\text{Top surface:} \quad f = \frac{F}{A} - \frac{M}{Z} = -\frac{240{,}000 \text{ lb}}{240 \text{ in}^2} - \frac{(40 \text{ ft} \cdot \text{kips})(12{,}000)}{800 \text{ in}^3} = -1600 \text{ lb/in}^2$$

$$(11 \text{ MPa})$$

$$\text{Bottom:} \quad f = -\frac{F}{A} + \frac{M}{Z} = -\frac{240{,}000 \text{ lb}}{240 \text{ in}^2} + \frac{(40 \text{ ft} \cdot \text{kips})/(12{,}000)}{800 \text{ in}^3} = -400 \text{ lb/in}^2$$

$$(2.76 \text{ MPa})$$

## A Curved Tendon

The lateral force exerted on the concrete by a curved tendon can be established by considering the equilibrium of a small segment of tendon of length *ds* and radius of curvature *r*; see Fig. 11.14, where the segment is oriented so that the vertical axis bisects the angle *dθ* between the radii drawn at each end of the segment. Although the load per unit length *w* exerted by the cable on the concrete acts

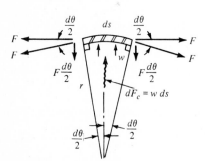

**Figure 11.14** Transverse pressure created by a curved tendon.

normal to the axis of the tendon, if the tendon slope is small, it can be assumed to act vertically without introducing any significant error. $w$ can be evaluated by considering the equilibrium of the tendon segment in the vertical direction. If we assume that the tendon force is constant over the length $ds$, a summation of forces in the $y$ direction yields

$$w \, ds = F \frac{d\theta}{2} 2$$

$$w = F \frac{d\theta}{ds} \tag{11.3}$$

where $d\theta/ds$ represents the curvature of the tendon segment. From the geometry of the segment in Fig. 11.14 it can be seen that $r \, d\theta = ds$, from which it follows that $d\theta/ds = 1/r$. Expressing the curvature as $1/r$, we can write Eq. (11.3) as

$$w = \frac{F}{r} \tag{11.4}$$

Since for small slopes the curvature can be expressed in terms of rectangular coordinates as $d\theta/ds = d^2y/dx^2$, Eq. (11.3) can also be written

$$w = F \frac{d^2 y}{dx^2} \tag{11.5}$$

Equation (11.5) can be used to show that parabolic tendons apply a uniformly distributed transverse load to a beam. Such a distribution of load produces an internal moment that is both proportional and opposite in sense to the moment created by a uniform design load.

A beam with a parabolic tendon is shown in Fig. 11.15. The symmetrical cable, which intersects the neutral axis of the beam at each end, carries a constant force $F$ throughout its length. The equation of a parabola in terms of a rectangular coordinate system with an origin located at the intersection of the tendon centerline and the vertical axis of the beam at midspan is given by

$$y = kx^2 \tag{11.6}$$

To evaluate $k$ in terms of the geometry of the cable, the values of $y = h$ when $x = L/2$ can be substituted into Eq. (11.6) to give

$$k = \frac{4h}{L^2} \tag{11.7}$$

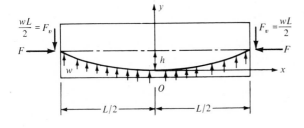

**Figure 11.15** Forces applied by a parabolic cable to the concrete; $w = 8hF/L^2$.

Substituting (11.7) into (11.6) gives the equation of the parabola in terms of the tendon geometry as

$$y = \frac{4h}{L^2} x^2 \tag{11.8}$$

Differentiating Eq. (11.8) twice with respect to $x$ leads to

$$\frac{d^2y}{dx^2} = \frac{8h}{L^2} \tag{11.9}$$

and substituting Eq. (11.9) into Eq. (11.5) gives

$$w = F \frac{8h}{L^2} \tag{11.10}$$

Since the curvature of a parabola is constant along its entire length, the parabolic tendon applies a uniform load $w$ normal to the tendon axis along its entire length. For shallow tendons the load may be assumed to act perpendicular to the longitudinal axis of the member.

**Example 11.4:** The beam in Fig. 11.16 carries a uniform service load of 1.5 kips/ft. If the beam is to be stressed by a parabolic tendon that produces a uniformly distributed upward load of 1.2 kips/ft, what tendon force is required? Determine the net moment at midspan.

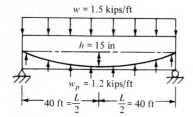

Figure 11.16

SOLUTION Using Eq. (11.10), set $w = 1.2$ kips/ft, $h = \frac{15}{12}$ ft, and solve for $F$

$$F = \frac{wL^2}{8h} = \frac{1.2(80^2)}{8(\frac{15}{12})} = 768 \text{ kips}$$

$$M = \frac{wL^2}{8} = \frac{(1.5 - 1.2)80^2}{8} = 240 \text{ ft} \cdot \text{kips}$$

## Forces Created by Prestress on a Vertical Section

Although this method of analysis (determining the effects of prestress by applying the cable reactions to the concrete) gives a clear picture of the structural action of the tendon (see Example 11.2), the designer is frequently concerned with the more limited problem of investigating the longitudinal stresses on a few critical sections. For this type of study, the resultant of the concrete stresses induced by the prestress only on a vertical section of a determinate beam can be represented by a vector. Equal in magnitude to the prestress force, the vector acts at the centerline of the tendon in a direction opposite to the cable tension. In other words, for any shape of tendon in a determinate member the resultant of the concrete stresses created by the prestress alone can be represented by a vector that is equal to, opposite in direction to, and collinear with, the tendon force. The validity of this representation of the prestress force on an internal section can be demonstrated if we consider the equilibrium of a free body, cut from the beam in Fig. 11.17a by section 1-1.

Since the tendon carries only direct stress, the tendon force $F$ is always directed tangent to the axis of the tendon. If we neglect all applied loads including the member's own weight, the free body in Fig. 11.17b can be in equilibrium only if the tendon force is balanced by an equal, oppositely directed, collinear force $C$ on the internal section. The force $C$, which equals the resultant of the concrete stresses on cross section 1-1, represents also the force applied to the free body by the concrete to the right of section 1-1.

If the resultant of the compressive stresses $C$ is broken into components, it can be seen that the horizontal component produces an internal moment $M_p$ of magnitude $Fe$ with respect to the neutral axis. The prestress also produces an axial compression $F$ and a shear of $F_v$ (Fig. 11.17c). Since $F$ is constant along the length of the member, the moment produced by the tendon force at each section is equal to the product of the tendon force and the eccentricity of the tendon. That is, in a determinate beam the moment created by prestress has the same variation as

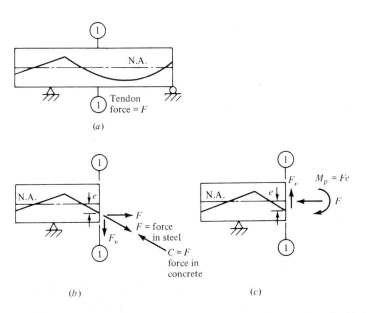

**Figure 11.17** Internal forces on a vertical section created by prestress only: (a) determinate prestressed beam, (b) free-body diagram of the concrete showing the resultant force $C$ on the concrete created by prestress, (c) alternate representation of the shear, axial force, and moment created by prestress.

the cable eccentricity. A reexamination of Example 11.2 confirms this result. For example, the moment at point $C$ created by prestress is equal to

$$M_p = Fe = \frac{(240 \text{ kips})(6 \text{ in})}{12} = 120 \text{ ft} \cdot \text{kips}$$

Example 11.5 illustrates how the shape of the moment curve, produced by the primary design loads, can be used to establish the shape of an effective tendon profile.

Although the tendon shape in this example has been selected to produce zero deflection under a particular set of loads, this is not meant to imply that the criterion of zero deflection under full service loads controls the shape of the tendon profile or the magnitude of the tendon force. Greater economy of design may result from increasing the tendon eccentricity or reducing the magnitude of the prestress force. Generally, in a well-designed beam, service loads and prestress create a triangular or nearly triangular stress distribution that varies from a small stress on the tension side to a maximum value of stress on the compression side (Fig. 11.22*b*).

**Example 11.5:** Neglecting the dead weight of the beam in Fig. 11.18, determine the required cable force and profile if no deflection of the beam is to occur when the two concentrated loads are applied. The cable eccentricity at point $B$ is specified as 1.25 ft (0.38 m). Determine the longitudinal stresses in the beam.

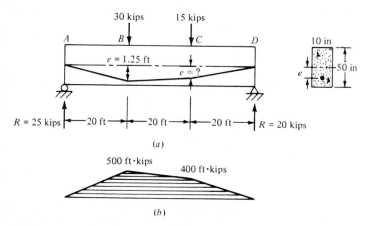

**Figure 11.18** (*a*) Prestressed beam, (*b*) moment due to transverse loads only.

SOLUTION If the beam is not to deflect, the moment created by prestress must be equal and opposite to the moment due to the applied loads at all sections. Compute the tendon force. At point $B$ equate the moment due to prestress to the moment due to load

$$Fe = M$$

$$F(1.25 \text{ ft}) = 500 \text{ ft} \cdot \text{kips}$$

$$F = 400 \text{ kips } (1779.2 \text{ kN})$$

The eccentricity at each section is $e = M/F$. Since $F$ is constant, the eccentricity must have the same variation as the moment curve. Under the 15-kip load the required eccentricity equals $M/F = (400 \text{ ft} \cdot \text{kips})/(400 \text{ kips}) = 1$ ft. Since the net moment is zero at all sections, only the axial load produces stress. The stress in beam at all sections is $P/A = (400 \text{ kips})/(500 \text{ in}^2) = 0.8 \text{ kips/in}^2$.

## Indeterminate Beams

Although the design of indeterminate prestressed beams will not be covered in this chapter, their behavior will be discussed briefly. When an indeterminate beam is prestressed, reactions develop at all supports if the beam has a tendency to displace laterally at any support. This aspect of behavior—the development of reactions—never occurs in a determinate beam since the beam is not restrained against deformation by the supports. For example, in Fig. 11.19a the simply supported beam will deflect upward when prestressed by an eccentric cable that produces a negative moment $Fe$ along the length of the member. If this same beam were supported on three supports, as shown in Fig. 11.19b, a downward reaction would develop at support $B$ to prevent the beam from lifting off the center support. To balance the reaction at $B$, vertical reactions must also develop at supports $A$ and $C$.

When reactions are created by prestress, the location of the resultant of the compression stress $C$ produced by prestress is no longer located at the level of the

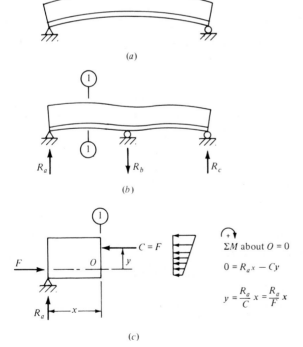

$$\Sigma M \text{ about } O = 0$$

$$0 = R_a x - Cy$$

$$y = \frac{R_a}{C} x = \frac{R_a}{F} x$$

**Figure 11.19** Forces in an indeterminate prestressed beam: (a) deformations of a simply supported beam, (b) reactions induced in an indeterminate beam by an eccentric tendon, (c) location of the resultant of the concrete stresses produced by prestress.

tendon. Now the position of $C$ is a function of both the tendon reactions and the support reactions. As illustrated in Fig. 11.19c, the position of $C$ on the cross section, measured by the distance $y$ from the centroid of the tendon, can be established by summing moments about the centroid of the tendon of all forces on a free-body diagram of the concrete to one side of a vertical section through the beam. Even though they may be substantial in magnitude, the additional moments produced by the support reactions are termed *secondary moments*. When reactions develop in an indeterminate beam because of prestress, the tendon is termed *nonconcordant*.

## 11.7 ELASTIC ANALYSIS OF PRESTRESSED BEAMS

To ensure that a prestressed beam will not be overstressed either at transfer or while service loads are in place, the ACI Code requires that beams be analyzed elastically to establish that the direct stresses due to axial load and bending at all critical sections are within a specified set of allowable stresses. The limit on the magnitude of stress permitted at transfer or with service loads in place prevents crushing of the concrete and controls cracking. Although a limit on stresses indirectly controls axial shortening and lateral deflections, a complete analysis also requires deflections to be computed to ensure that surfaces are level within specified tolerances.

Two methods of analysis, *superposition* and the *internal-couple* method, will be discussed in this section. Superposition is most commonly used in an investigation to establish the distribution of longitudinal stresses on cross sections whose proportions are given. The internal-coupled method is useful in a preliminary design to establish the proportions of the section and to locate the centroid of the prestress tendon. Since these methods assume that the cross section is uncracked, the maximum tensile stress must technically not exceed the modulus of rupture. Although the modulus of rupture is normally taken as $7.5\sqrt{f'_c}$ lb/in² ($0.62\sqrt{f'_c}$ MPa), the ACI Code considers an elastic analysis valid as long as the maximum tensile stresses do not exceed $12\sqrt{f'_c}$ lb/in² ($1.0\sqrt{f'_c}$ MPa).

Flexural stresses in the concrete for an elastic analysis are computed by the standard beam equation

$$f = \frac{My}{I} \tag{11.11}$$

where $M$ = bending moment
$y$ = distance from neutral axis to point where stress is to be evaluated
$I$ = moment of inertia of gross cross-sectional area about centroidal axis

Since Eq. (11.11) is primarily used to determine the maximum stresses on a cross section, $y$ is set equal to $c_t$ or $c_b$, the distances from the centroidal axis to the top and bottom fibers (Fig. 11.20). With $y$ set equal to $c_t$ or $c_b$, Eq. (3.1) can be written as

$$f_t = \frac{M}{I/c_t} = \frac{M}{Z_t} \qquad f_b = \frac{M}{I/c_b} = \frac{M}{Z_b} \tag{11.12}$$

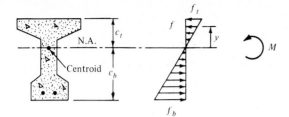

**Figure 11.20**

where $f_t$ and $f_b$ are the flexural stresses at the top and bottom of the section and $Z_t$ and $Z_b$ are the section moduli of the cross section. The section modulus, the ratio $I/c$, is an index of the flexural capacity of the cross section. A large value of section modulus is associated with a deep beam, which has a large bending stiffness. Values of $Z_b$ and $Z_t$, which are constants for a given cross section, are frequently tabulated in design handbooks that list section properties for standard beams (see Table 11.5).

## Superposition

The first method to be discussed, the method of superposition, is based on superimposing the stresses created by the applied loads with those created by the prestressing force. In this procedure the stresses produced by prestress and those produced by applied loads are computed independently and then combined. The forces induced in a concrete member by a prestressed tendon can often be determined most simply by analyzing the concrete for the loads applied by the tendon. For this portion of the analysis, the concrete is treated as an elastic material acted upon by the tendon reactions, which constitute a force system in equilibrium. Such a force system creates internal stresses in a *statically determinate* member but does not contribute to the reactions, which are functions only of the external loads.

Although sloping tendons create shear forces that can reduce the shear produced by the applied loads, the ACI Code does not apply superposition to shear forces.

The method of superposition, which has been previously used in stress computations, is illustrated in Example 11.6.

## Internal-Couple Method

A second method of elastic analysis, the internal-couple method, is often used to estimate the area of the prestressed steel and the dimensions of the concrete cross section in the initial stage of a prestressed beam design. The internal-couple method uses a free body that consists of both the concrete and the prestressed tendon (see Fig. 11.21). Analysis by this method is based on equating the moment at a section produced by external loads to the internal couple composed of the tendon force $F$ and the force $C$, the resultant of the longitudinal compressive stresses in the concrete. Equilibrium in the horizontal direction, of course, requires that

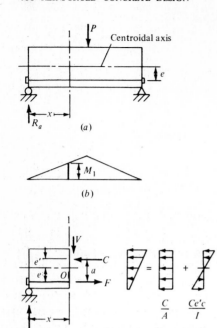

**Figure 11.21** Representation of the internal forces produced by applied load and prestress by the internal couple: (a) beam with service load, (b) moment curve due to service load, (c) internal couple to resist service load moment. $M_1 = Ca = Fa$, and $a = M_1/F$; $e' = a - e$.

$C$ equal $F$. Since the prestress force $F$ set by the designer is known, the distance of $C$ above the tendon, denoted by $a$, can be established by equating the internal couple $Fa$ to the moment $M_1$ produced by the applied loads. Solving for $a$, the arm of the couple, gives

$$a = \frac{M_1}{F} \tag{11.13}$$

Once the position of $C$ has been established, the stress distribution producing $C$ can be determined by combining the direct stress $(C/A = F/A)$ produced by $C$ with the bending stress created by the eccentricity $e'$ of $C$ from the centroid of the section (see Fig. 11.21c).

Equation (11.13) reveals an interesting aspect of the behavior of a prestressed beam. Since the prestress force is essentially constant as long as the concrete remains uncracked (the changes in tendon strain due to bending of the member are small compared with the large strains required to tension the tendon), the position of the resultant compression force $C$ on the cross section with respect to the position of the tendon varies directly with the magnitude of the moment. If, for example, the moment is zero, Eq. (11.13) indicates that the arm $a$ between $F$ and $C$ will be zero. Therefore, the resultant of the compressive stresses will be located at the level of the tendon (see Fig. 11.22a). When $C$ is located near the bottom of the section, the compressive stresses will be maximum at the bottom surface and decrease linearly toward the top surface.

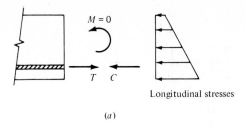

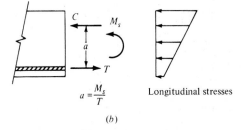

**Figure 11.22** Influence of the magnitude of the service-load moment on the position of $C$ and the variation of the longitudinal stresses in the concrete: ($a$) moment equals zero; $C$ and $T$ coincide, ($b$) moment due to service loads acts; $a = M_s/T$.

If applied loads now cause the moment on the cross section to increase, the internal moment must increase. Since the tendon force remains essentially constant, the internal moment can increase only if the arm of the couple increases. An increase in arm requires $C$ to rise. As $C$ moves upward, the compressive stresses decrease at the bottom of the section and increase at the top of the section (Fig. 11.22$b$).

**Example 11.6:** In Fig. 11.23 a simply supported beam is prestressed by a straight tendon that carries an effective force of 233 kips (1036.4 kN). If the beam carries a total service load of 1.2 kips/ft (17.5 kN/m), including a dead load of 0.21 kip/ft (3.06 kN/m), determine the distribution of compressive stresses on a vertical section at midspan produced by total load and prestress. Use the gross properties of the cross section to compute stresses. ($a$) Analyze by superposition and ($b$) verify results by the internal-couple method.

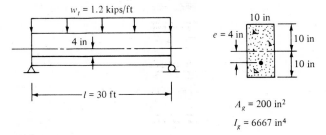

**Figure 11.23**

SOLUTION ($a$) The midspan moment due to uniform load is

$$M_s = \frac{wl^2}{8} = \frac{1.2(30^2)}{8} = 135 \text{ ft} \cdot \text{kips (183 kN} \cdot \text{m)}$$

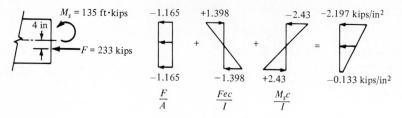

**Figure 11.24**

Stresses at midspan (Fig. 11.24) are as follows: prestress

$$f_a = \frac{F}{A} = \frac{233}{200} = -1.165 \text{ kips/in}^2$$

$$f_b = \frac{Mc}{I} = \frac{Fec}{I} = \frac{233(4)(10)}{6667} = \pm 1.398 \text{ kips/in}^2$$

applied load

$$f_b = \frac{M_s}{I} = \frac{135(12)(10)}{6667} = \pm 2.43 \text{ kips/in}^2$$

resultant stress at top

$$f_t = -1.165 + 1.398 - 2.43 = -2.197 \text{ kips/in}^2 \qquad \text{compression}$$

and resultant stress at bottom

$$f_b = -1.165 - 1.398 + 2.43 = -0.133 \text{ kips/in}^2 \qquad \text{compression}$$

(b) In Fig. 11.25 compute the internal couple arm with Eq. (11.13)

$$a = \frac{M_s}{T} = \frac{135(12)}{233} = 6.953 \text{ in}$$

$$e' = a - 4 = 2.953 \text{ in}$$

The stresses produced by $C = 233$ kips are

$$f_a = \frac{C}{A} = \frac{233}{200} = -1.165 \text{ kips/in}^2$$

$$f_b = \frac{Ce'c}{I} = \frac{233(2.953)(10)}{6667} = \pm 1.032 \text{ kips/in}^2$$

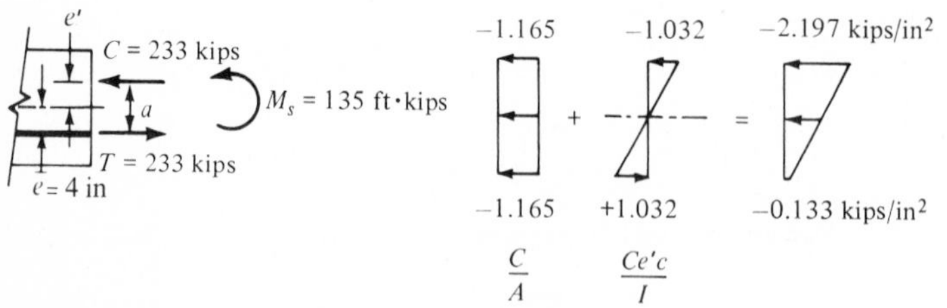

**Figure 11.25**

The resultant stress at the top is

$$f_t = -1.165 - 1.032 = -2.197 \text{ kips/in}^2$$

and the resultant stress at the bottom is

$$f_b = -1.165 + 1.032 = -0.133 \text{ kip/in}^2$$

## 11.8 KERN POINTS

The distribution of the longitudinal stresses on a vertical cross section of a pre-stressed beam will vary as the moment on the cross section changes. For example, if $C$, the resultant of the compressive stresses, is located at the centroid of the cross section, the stresses will be uniform. As the moment on the cross section changes, $C$ will move and produce a trapezoidal distribution of stress on the cross section.

If no tensile stresses are to be permitted on the cross section either at transfer or when the service loads act, $C$ must neither extend below nor rise above the kern points of the cross section. By definition the *kern* is that point on the section at which a resultant force will produce a triangular distribution of stress on the section. To produce a triangular distribution of stress, the tensile bending stresses on the outside edge produced by the eccentricity of a longitudinal force on the section must equal the direct stress at the same point produced by the same force. Of course, each section has both an upper and a lower kern point. Fig. 11.26 illustrates the stress distribution produced by a force $C$ at the upper kern point. Equating the direct and bending stresses at the bottom surface gives the distance to the upper kern point as

$$k_t = \frac{I}{Ac_b} \tag{11.14}$$

Similarily, the location of the bottom kern point is given as

$$k_b = \frac{I}{Ac_t} \tag{11.15}$$

All terms in Eqs. (11.14) and (11.15) are defined in Fig. 11.26.

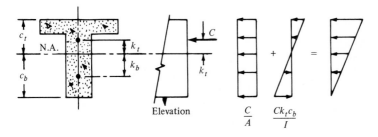

**Figure 11.26** Resultant of compressive stresses acts at top kern point.

## 11.9 THE CRACKING MOMENT

A prestressed beam is assumed to behave elastically and remain uncracked as long as the maximum tensile stress on the cross section does not exceed the modulus of rupture $f_r$. In combination with prestress the moment due to external loads that produces a tensile stress equal to the modulus of rupture is termed the *cracking moment* $M_{cr}$. As shown in Fig. 11.27, the cracking moment can be computed by summing the stresses on the bottom surface produced by prestress and by the cracking moment to give

$$f_r = -\frac{F}{A} - \frac{Fey_t}{I} + \frac{M_{cr}y_t}{I} \tag{11.16}$$

where $y_t$ = distance from neutral axis to tension surface
$M_{cr}$ = cracking moment
$F$ = effective prestress after all losses
$f_r$ = modulus of rupture

If $f_r$ is taken as $7.5\sqrt{f_c'}$, Eq. (11.16) can be solved for $M_{cr}$ to give

$$M_{cr} = \frac{I}{y_t}\left(7.5\sqrt{f_c'} + \frac{F}{A} + \frac{Fey_t}{I}\right) \tag{11.17}$$

Since most prestressed beams are designed to remain uncracked at service loads, the cracking moment is usually larger than the moment produced by service loads.

As long as the beam is uncracked, the effective force in the tendon remains nearly constant even though the applied loads may vary; however, once the cracking moment is exceeded, a crack forms and the tendon force increases sharply as it picks up the tension formerly carried by the concrete. To ensure that the tendon does not fail as its picks up the tension in the concrete, ACI Code §18.8.3 requires that the ultimate moment capacity of the member be at least 20 percent greater than the cracking moment. That is, $\phi M_n \geq 1.2M_{cr}$.

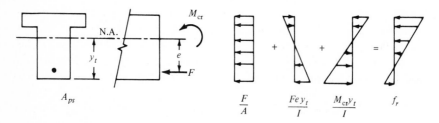

**Figure 11.27** State of stress produced by prestress and cracking moment.

**Example 11.7:** Determine the cracking moment for the cross section in Example 11.6; $f'_c = 6$ kips/in$^2$.

SOLUTION  Using Eq. (11.17), we compute

$$M_{cr} = \frac{I}{y_t}\left(7.5\sqrt{f'_c} + \frac{F}{A} + \frac{Fey_t}{I}\right)$$

$$= \frac{6667}{10}\left[\frac{7.5\sqrt{6000}}{1000} + \frac{233}{200} + \frac{233(4)(10)}{6667}\right] = 2096 \text{ in} \cdot \text{kips } (236.86 \text{ kN} \cdot \text{m})$$

# 11.10 BEHAVIOR OF A PRESTRESSED BEAM LOADED TO FAILURE

Before the details of prestressed concrete beam design are discussed, the behavior of a simply supported, fully prestressed, posttensioned concrete beam with a bonded tendon will be examined as it is loaded to failure by increasing increments of uniform load (Fig. 11.28). Particular attention will be paid to the variation of the tendon stress at midspan, the distribution of longitudinal stresses in the concrete at midspan, and the deflected shape of the beam. In Fig. 11.28 the longitudinal stresses at midspan for each stage of loading are plotted at the right end of each beam.

Immediately after the tensioned tendon has been anchored to the end of the beam and before any time-dependent losses have occurred, the tendon is stressed to $f_{si}$. During the transfer stage as the force in the cable is transferred to the concrete, the beam cambers upward under the negative moment created by the eccentricity of the prestress force (Fig. 11.28a). As the center of the beam rises from the forms, the weight of the girder $w_g$ produces a positive moment of $M_g$ at midspan.

As creep and shrinkage occur, the concrete progressively shortens and the stress in the tendon decreases further to a value of $f_{se}$. These time-dependent deformations, which occur most rapidly after tensioning, may continue for several years. When service loads are applied, the beam deflects downward and the tendon elongates. Since the bending strains at the level of the tendon are small compared with the strains in the tensioned steel, only a slight increase in tendon force occurs as the beam deflects. The increased internal moment required to equilibrate the moments produced by service loads is created by an increase in the arm between the components of the internal couple. The movement of the resultant of the compressive stresses causes the stresses in the concrete to change from the distribution in Fig. 11.28a to that shown in Fig. 11.28b. In a fully prestressed beam, the compressive stresses due to prestress are larger than the tensile bending stresses produced by service loads, and only compressive stresses exist on the cross section.

If the applied loads exceed the service loads, the tendon stress will continue to increase slowly as the beam deflection increases. When the tensile stress in the concrete at the bottom surface exceeds the modulus of rupture (the tensile

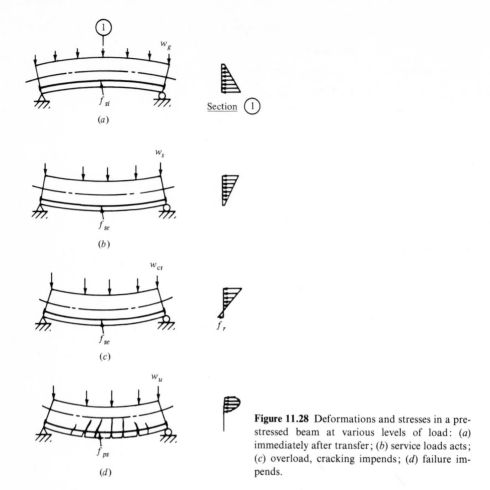

**Figure 11.28** Deformations and stresses in a prestressed beam at various levels of load: (*a*) immediately after transfer; (*b*) service loads acts; (*c*) overload, cracking impends; (*d*) failure impends.

strength of the concrete), vertical cracking occurs. If the beam is not to fail after cracking, the force in the tendon must increase by an amount equal to the tension formerly carried by the concrete. The tendon must have enough additional capacity to absorb the tension formerly carried by the concrete.

After a prestressed concrete beam cracks, it behaves like a regular reinforced concrete beam; i.e., additional moment capacity is created primarily by an increase in the magnitude of the components of the interval couple rather than by increase in the arm of the internal couple. Further increments of load applied to a cracked beam increase the deflection, cause cracks to extend and widen, and raise the level of stresses in the steel and in the concrete. In an underreinforced beam, final failure occurs after the tendon is stressed into the inelastic region to a value of $f_{ps}$ and the longitudinal strain at the top surface of the concrete reaches a value of 0.003 (Fig. 11.28*d*). Figure 11.29, which shows a plot of the variation of tendon stress with applied load from transfer to failure, summarizes the variation of the tendon stress at midspan for the beam in Fig. 11.28.

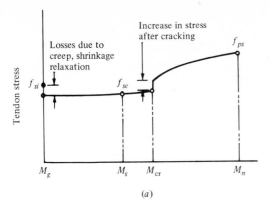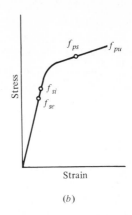

(a)  (b)

**Figure 11.29** Variation of tendon stress with level of load, bonded tendon: (a) tendon stress as a function of applied moment, (b) tendon stress plotted on a stress-strain curve.

## 11.11 STRENGTH DESIGN FOR FLEXURE

Although limiting the maximum values of tensile and compressive stress permitted on the concrete at various loading stages limits deflections, controls cracking, and prevents crushing of the concrete, an elastic analysis offers no control over the failure mode or the factor of safety of a prestressed concrete beam. To ensure that prestressed beams will be designed with an adequate factor of safety against failure, the ACI Code requires that $M_u$, the moment due to factored service loads, not exceed $\phi M_n$, the flexural design strength of the cross section.

The nominal bending strength $M_n$ of a prestressed beam is computed in nearly the same manner as that of a reinforced concrete beam. If a prestressed beam is underreinforced, failure initiates when the moment on a cross section strains the steel above the yield strength into the inelastic region of the stress-strain curve. Once the steel has been strained beyond the yield point, additional increments of load cause the neutral axis to shift toward the compression surface. As the shift occurs, the strains in the steel increase at an accelerated rate and the beam undergoes large deflections. Final failure occurs when the maximum compressive strain in the concrete reaches a value of 0.003 (Fig. 11.30). Since the stress-strain curves of

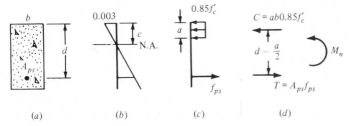

(a)  (b)  (c)  (d)

**Figure 11.30** State of stress in an underreinforced beam at failure: (a) cross section, (b) strains, (c) assumed stress distribution, (d) internal couple.

high-yield-point steels used as prestress tendons do not develop a horizontal yield range once the yield strength is reached but continue to slope upward at a reduced slope (see Fig. 11.6), the final stress in the steel at failure $f_{ps}$ must be predicted by empirical equations based on tests or by computing the strain in the steel at failure and reading the corresponding steel stress from a stress-strain curve of the material.

The expression for the nominal bending strength of a prestressed beam with a rectangular cross section can be established by summing moments of the internal forces about the tension steel (Fig. 11.30) to give

$$M_n = C\left(d - \frac{a}{2}\right) = ab(0.85f_c')\left(d - \frac{a}{2}\right) \tag{11.18}$$

or about the compression force to give

$$M_n = T\left(d - \frac{a}{2}\right) = A_{ps}f_{ps}\left(d - \frac{a}{2}\right) \tag{11.19}$$

where $a$ can be determined by equating $C = T$, to give

$$a = \frac{A_{ps}f_{ps}}{b(0.85f_c')} \tag{11.20}$$

Multiplying top and bottom of Eq. (11.20) by $d$ gives

$$a = \frac{A_{ps}}{bd}\frac{f_{ps}}{0.85f_c'}d = \rho_p\frac{f_{ps}}{0.85f_c'}d \tag{11.21}$$

where $\rho_p = A_{ps}/bd$. Letting $\omega_p = \rho_p(f_{ps}/f_c')$ permits Eq. (11.21) to be written

$$a = \omega_p d/0.85 \tag{11.22}$$

For tendons whose effective stress after losses $f_{se}$ is not less than $0.5f_{pu}$, ACI Code §18.7.2 permits $f_{ps}$, the stress in the steel at failure, to be evaluated by empirical equations

$$f_{ps} = \begin{cases} f_{pu}\left(1 - 0.5\rho_p\frac{f_{pu}}{f_c'}\right) & \text{bonded tendon} \tag{11.23} \\[2ex] f_{se} + 10,000 + \frac{f_c'}{100\rho_p} & \text{unbonded tendon} \tag{11.24} \end{cases}$$

where $\rho_p = A_{ps}/bd$ and $f_{pu}$ is the ultimate tensile strength of the tendon. In a T beam, $b$ in the equation for $\rho_p$ is taken as the width of the flange. $f_{ps}$ given by Eq. (11.24) is not to exceed either $f_{py}$ or $f_{se} + 60,000$, where all stresses are to be expressed in pounds per square inch.

In Equation (11.22) the quantity $\omega_p$ is a direct measure of the tension force in the tendon. To ensure that a beam is underreinforced by limiting the tension force that can develop in a tendon, ACI Code §18.8.1 specifies that $\omega_p \leq 0.3$. If the limiting value of $\omega_p = 0.3$ is substituted into Eq. (11.22), $a = 0.353d$. Substituting this value into Eq. (11.18), combining terms, and rounding off the constant gives the maximum nominal bending strength of a rectangular section as

$$M_n = 0.25bd^2f_c' \tag{11.25}$$

Applying the same criterion (that $a$ must not exceed $0.353d$) to a T beam gives the following expression for the maximum value of nominal bending strength

$$M_n = 0.25f_c'b_wd^2 + 0.85f_c'(b - b_w)h_f(d - 0.5h_f) \tag{11.26}$$

where $b_w$ = width of web
$\quad\;\, h_f$ = thickness of flange
$\quad\;\, b$ = width of flange

Examination of Eqs. (11.18) and (11.19) shows that $M_n$, the nominal bending strength associated with failure of an underreinforced prestressed beam, does not depend directly on the magnitude of the initial prestress force in the tendon. Although the size of the effective prestress force influences crack width and the magnitude of the deflections when service loads act, the nominal bending strength is primarily a function of the tendon area. Beams with lightly tensioned tendons require more deformation than a beam with a heavily tensioned tendon to strain the member to failure; however, the additional deformation, which produces a more ductile member, represents one of the advantages of reducing the magnitude of the initial effective prestress.

Figure 11.31 illustrates the influence of the initial level of prestress on the behavior of three beams of identical size reinforced with the same area of steel: only the level of the effective prestress is varied. Curve $A$ represents the load-deflection behavior of a hypothetical beam whose tendon has been prestressed to nearly $f_{ps}$. For this beam, in which the cracking moment equals $M_n$, failure occurs as soon as the beam cracks. Lacking ductility, beam $A$ would be undesirable for use as a structural member. The curve for beam $B$ represents the behavior of a fully prestressed beam in which the effective tendon stress has been sized to ensure that no tensile stresses develop in the concrete when the service loads act. Although its behavior is identical to that of beam $A$ as long as the service loads are not exceeded, it exhibits ductility; i.e., it undergoes significant deflections before failure occurs. Curve $C$ is associated with a partially prestressed beam; tensile stresses and

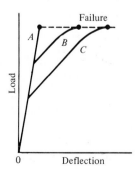

**Figure 11.31** Influence of initial level of prestress on ductility. (*Adapted from Ref. 3.*)

limited cracking develop when the full service loads are applied. Although this beam under service loads deflects more than the fully prestressed beam, it is the most ductile of the three beams compared.

**Example 11.8:** Compute the nominal moment capacity of the T beam in Fig. 11.32 if the section is reinforced with a bonded tendon $A_{ps} = 2.45$ in² (1581 mm²); $f'_c = 5$ kips/in² (34.47 MPa), $f_{pu} = 270$ kips/in² (1862 MPa), and the effective stress after losses is $f_{se} = 160$ kips/in² (1103 MPa).

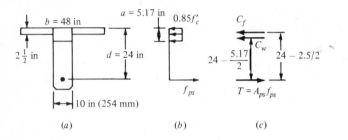

**Figure 11.32** (*a*) Cross section, (*b*) stresses, (*c*) internal couple.

SOLUTION Compute $f_{ps}$. Since $f_{se} > f_{pu}/2$, use Eq. (11.23)

$$f_{ps} = f_{pu}\left(1 - 0.5\rho_p \frac{f_{pu}}{f'_c}\right)$$

$$= 270 \text{ kips/in}^2\left[1 - 0.5 \frac{2.45}{48(24)} \frac{270}{5}\right] = 254.5 \text{ kips/in}^2$$

The tension force in the steel at failure is

$$T = A_{ps}f_{ps} = 2.45(254.5) = 623.5 \text{ kips}$$

To compute the area of concrete in the stress block, equate $T = C$

$$623.5 \text{ kips} = A_c(0.85f'_c)$$

$$A_c = 146.7 \text{ in}^2$$

Since the area of the flange is 120 in², the stress block extends into the web 2.67 in (146.7 − 120 = 26.7 in², and 26.7/10 = 2.67 in). Compute $M_n$ by breaking the compression zone into two areas and then summing moments of the forces about the tendon force $T$

$$M_n = C_{flg}\left(24 - \frac{2.5}{2}\right)\frac{1}{12} + C_{web}\left(24 - \frac{5.17}{2}\right)\frac{1}{12}$$

$$= 38(2.5)(0.85)(5)\frac{22.75}{12} + 5.17(10)(0.85)(5)\frac{21.42}{12}$$

$$= 1157.7 \text{ ft} \cdot \text{kips}$$

## 11.12 DESIGN OF A CROSS SECTION FOR MOMENT

Prestressed concrete beams are normally proportioned for moment and subsequently reinforced for shear in regions where the shear produced by factored loads exceeds the shear capacity of the concrete. Establishing the required cross section of a prestressed concrete beam for design moments can be carried out most simply by a trial procedure, in which the internal couple is equated to the maximum moment produced by service loads in order to establish expressions giving the approximate areas required for both the prestress steel and the cross section of the concrete. After a shape based on these areas has been established, the member is analyzed elastically to determine whether the longitudinal stresses produced by prestress and bending are within the range of allowable stresses specified by the ACI Code. If the initial elastic analysis indicates that the section is overstressed or significantly understressed (the section is oversized), the proportions are modified and the analysis repeated. Adjustments are made to the section until the longitudinal stresses indicate a well-proportioned section.

Even though the elastic analysis indicates that stresses are within acceptable limits, an ultimate-strength analysis is also necessary to ensure an adequate factor of safety against flexural failure. This analysis must verify that the flexural design strength of the section $\phi M_n$ is greater than the moment produced by factored loads.

To complete the design, both short- and long-term deflections should be checked to ensure that members have adequate bending stiffness to support the design loads without excessive deformations. Beams that lack adequate bending stiffness are often susceptible to live-load-induced vibrations. The procedure to be followed in establishing the trial cross section for the initial elastic analysis is detailed below.

### Establishing the Depth

Although many factors such as clearance, magnitude of the design loads, control of deflection, shape of the cross section, and span length influence the member's depth $h$, the suggested values of depth, expressed as the ratio of span to depth in Table 11.3, can be used as a guide to establish the depth of flexural members supporting normal building loads. If beams must support heavier than normal floor loads, the span-to-depth ratios in Table 11.3 may have to be reduced to values that range between 20 and 25 for an economical design.

Since prestressed concrete beams, which are usually uncracked at service loads and made of high-strength concrete, are much stiffer than nonprestressed concrete beams of the same cross section, and since the deflections created by the prestress forces are opposite in sense to those produced by load, the net deflections of prestressed beams with service loads in place are often small. Recognizing that the deflections of the stiffer prestressed beams are influenced by the tendon profile as well as by the depth, minimum depth requirements are not specified in the ACI Code to control the deflections of prestressed beams.

**Table 11.3 Typical span-to-depth ratios** $L/h$
**for flexural members in buildings†**

| Type of member | Continuous span | | Simple span | |
|---|---|---|---|---|
| | Roof | Floor | Roof | Floor |
| Beam | 35 | 30 | 30 | 26 |
| One-way solid slabs | 50 | 45 | 45 | 40 |

† Based on Ref. 4, table 4.2.1.

## Establishing the Shape

Most prestressed concrete beams are fabricated in precasting plants. To reduce fabrication costs, the industry has standardized the shapes of a large number of cross sections that are frequently used in precast construction. Properties of these cross sections together with tables that indicate the uniform design loads they can support as a function of span length, a recommended tendon profile, and tendon force are tabulated in Ref. 4. Tables 11.4 and 11.5 give the properties of several standard shapes—the double tee and the single tee, frequently used in the con-

**Table 11.4 Type B load table for double tee**[4]

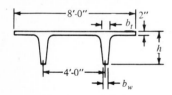

8 ft 0 in wide. Wide stem member. Normal-weight concrete

| | | | Section properties | | | | | |
|---|---|---|---|---|---|---|---|---|
| Designation | $h$, in | $b_t/b_w$, in | $A$, in$^2$ | $I$, in$^4$ | $y_b$, in | $Z_b$, in$^3$ | $Z_t$, in$^3$ | $wt$, lb/ft |
| 8DT16A | 16 | 8.00/6.00 | 388 | 8944 | 11.13 | 804 | 1837 | 404 |
| 8DT18A | 18 | 9.75/7.75 | 472 | 14,623 | 11.84 | 1235 | 2374 | 492 |
| 8DT20A | 20 | 8.00/5.50 | 435 | 16,117 | 13.72 | 1175 | 2566 | 453 |
| 8DT20B | 20 | 9.75/7.50 | 503 | 19,354 | 13.06 | 1482 | 2789 | 523 |
| 8DT24A | 24 | 8.00/5.00 | 478 | 25,686 | 16.33 | 1573 | 3349 | 498 |
| 8DT24B | 24 | 9.75/7.00 | 560 | 31,192 | 15.51 | 2011 | 3674 | 583 |
| 8DT32A | 32 | 8.00/4.00 | 549 | 51,286 | 21.71 | 2362 | 4984 | 572 |
| 8DT32B | 32 | 9.75/6.00 | 665 | 64,775 | 20.47 | 3164 | 5618 | 692 |

struction of roof and floor systems of precast prestressed concrete buildings. Table 11.6 shows the properties of standard AASHO girders produced for bridge construction. Typically, a cast-in-place floor slab is bonded to the upper flange of these I sections to produce a composite T beam that resists the live and the superimposed dead load.

When a standard section cannot be adapted for a particular design, the engineer must proportion the cross section for functional requirements as well as for strength. If spans are short and few beams are required, a rectangular section may be the most economical solution. Formwork for a rectangular cross section is least expensive to fabricate because of its simplicity. If spans are long and loads are heavy, a substantial saving in materials can be achieved by using box beams or I beams. The use of a flanged section with narrow webs instead of a solid section can significantly reduce dead weight, which may represent 80 to 90 percent of the total load to be supported by a long span structure. When cross sections are shaped, generous fillets should be provided at the junctions of the web and the flange to facilitate the flow of the plastic concrete into the flanges. Fillets also eliminate voids and honeycomb in the concrete along the line of intersection between the flange and the web.

If the dead load (usually just the weight of the girder) acting at the time of transfer is large, T sections are often economical. The tensile stresses created by a

**Table 11.5 Type B load table for single tee[4]**

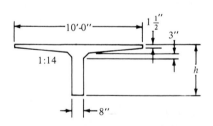

10 ft 0 in wide. Normal-weight concrete

| | | | Section properties | | | | |
|---|---|---|---|---|---|---|---|
| Designation | $h$, in | $A$, in$^2$ | $I$, in$^4$ | $y_b$, in | $Z_b$, in$^3$ | $Z_t$, in$^3$ | $wt$, lb/ft |
| 10ST24 | 24 | 590 | 22,914 | 18.73 | 1223 | 4348 | 615 |
| 10ST28 | 28 | 622 | 36,005 | 21.67 | 1662 | 5688 | 648 |
| 10ST32 | 32 | 654 | 53,095 | 24.51 | 2166 | 7089 | 681 |
| 10ST36 | 36 | 686 | 74,577 | 27.27 | 2735 | 8543 | 715 |
| 10ST40 | 40 | 718 | 100,819 | 29.97 | 3364 | 10,052 | 748 |
| 10ST44 | 44 | 750 | 132,171 | 32.61 | 4053 | 11,604 | 781 |
| 10ST48 | 48 | 782 | 168,968 | 35.19 | 4802 | 13,190 | 815 |

**Table 11.6 Type B load table for AASHO girders, normal-weight concrete[4]**

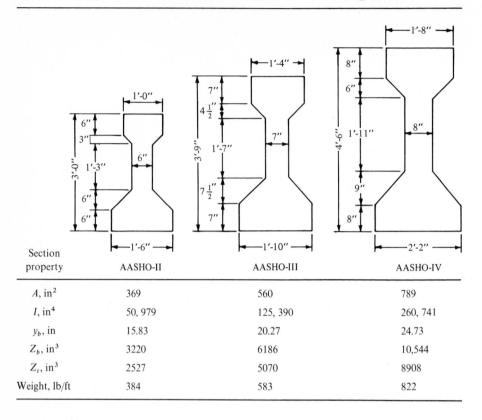

| Section property | AASHO-II | AASHO-III | AASHO-IV |
|---|---|---|---|
| $A$, in$^2$ | 369 | 560 | 789 |
| $I$, in$^4$ | 50, 979 | 125, 390 | 260, 741 |
| $y_b$, in | 15.83 | 20.27 | 24.73 |
| $Z_b$, in$^3$ | 3220 | 6186 | 10,544 |
| $Z_t$, in$^3$ | 2527 | 5070 | 8908 |
| Weight, lb/ft | 384 | 583 | 822 |

large dead-load moment balance the compression produced by the prestress force and eliminate the need for a bottom flange, which simplifies forming. If the dead load is small, a large lower flange may be required to carry the heavy compressive stresses induced by the prestress force at the transfer stage. When spans are long (say over 70 ft), hollow box beams (Fig. 11.33) with a deck slab are often economical. In addition to reducing dead weight with little loss of bending stiffness, compared with a solid section of the same outside dimensions, these sections also have a large torsional stiffness that resists lateral torsional buckling of the unsupported compression flange. By eliminating the need to provide temporary lateral bracing of the compression flange during erection of the box beam in the field, construction is speeded.

## Preliminary Design of the Cross Section

Once the depth and the shape of the cross section have been selected, approximate values for the area of the prestress steel $A_{ps}$ and the area of the concrete cross section $A_c$ can be established by equating the internal couple, consisting of the

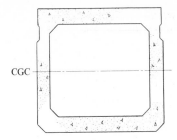

**Figure 11.33** Box beam.

tendon force and the resultant of the longitudinal compressive stresses, to the maximum moment produced by unfactored service loads (Fig. 11.34). Equating the internal couple expressed in terms of the tendon force and the arm $a_1$ to the service load moment $M_s$ gives

$$A_{ps} f_{se} a_1 = M_s$$

$$A_{ps} = \frac{M_s}{f_{se} a_1} \tag{11.27}$$

By estimating the value of $a_1$, the designer can use Eq. (11.27) to establish an approximate value of $A_{ps}$. A study of many computations of well-designed beams indicates that the arm of the internal couple with the service loads acting varies from $0.3h$ to $0.8h$.[5] The small arm applies to beams with a rectangular cross section that support a small moment due to girder weight. The larger arm is appropriate for an I beam or a box section that supports a large moment due to the girder dead weight. For many sections, Lin[5] indicates that an arm equal to $0.65h$ will produce a satisfactory estimate of $A_{ps}$.

If it is assumed that the service-load moment produces a triangular distribution of longitudinal stresses in the concrete, the resultant of the compressive stresses can be approximated as the product of the area of the concrete $A_c$ and the average compressive stress, which is equal to $f_c/2$ (Fig. 11.34). Equating the components of the internal couple in Fig. 11.34 gives

$$C = T$$

$$\frac{f_c}{2} A_c = A_{ps} f_{se}$$

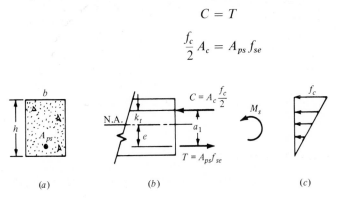

$(a)$ $(b)$ $(c)$

**Figure 11.34** Service loads act: $(a)$ cross section, $(b)$ internal couple, $(c)$ concrete stresses.

Solving for $A_c$ and setting $f_c$ equal to the maximum allowable compressive stress, $0.45f'_c$, gives

$$A_c = \frac{A_{ps}f_{se}}{0.225f'_c} \qquad (11.28)$$

where $A_{ps}$ is established by Eq. (11.27) and the designer specifies $f_{se}$ and $f'_c$.

Once $A_c$ has been computed, the dimensions of the cross section can be drawn and a more accurate estimate of the girder dead weight made. If the initial estimate of girder dead weight used to establish $M_s$ deviates significantly from the weight based on the value of $A_c$ given by Eq. (11.28), the entire computation should be repeated using the more accurate estimate of dead weight.

**Position of the tendon centroid**  The position of the tendon can be established from a consideration of the state of stress on the cross section at transfer (Fig. 11.35). If a triangular distribution of stress is assumed, the resultant $C$ of the compressive stresses must be located at the bottom kern point of the section. Since the internal couple must be equal to the moment $M_g$ due to the dead weight of the girder, the arm $a_2$ between $T$ and $C$ must equal $M_g/T$, where $T$ equals $A_{ps}f_{si}$. As indicated in Fig. 11.35, the distance $e$ of the centroid of the prestress steel from the centroid of the gross section is

$$e = k_b + \frac{M_g}{A_{ps}f_{si}} \qquad (11.29)$$

where $f_{si}$ is the stress in the steel immediately after transfer, and $k_b$ is the distance between the neutral axis and the bottom kern point

If the value of $e$ produced by Eq. (11.29) positions the tendon below the bottom surface of the cross section, the tendon is brought back into the beam and located as close to the bottom surface as the cover requirement of the ACI Code permits.

With the cross section established and the centroid of the prestressed steel located, the arm $a_1$ of the internal couple associated with a triangular stress distribution at service loads can be recomputed as the distance $e + k_t$. If the new value of $a_1$ differs significantly from the value initially assumed to solve Eq. (11.27) for $A_{ps}$, the computations should be repeated using the second value of $a_1$ to

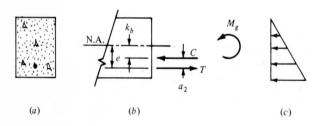

(a)  (b)  (c)

**Figure 11.35** Transfer: (a) cross section, (b) internal couple, (c) concrete stresses.

calculate improved values of $A_{ps}$ and $A_c$. On the other hand, if the computed value of $a_1$ compares closely with the initially assumed value, the service-load stresses in the concrete and the member's ultimate strength are checked to determine whether the proportions of the cross section satisfy code requirements or need to be modified further.

**Example 11.9: Design of a prestressed concrete beam for bending**   In accordance with ACI specifications, design the simply supported posttensioned beam in Fig. 11.36 with a rectangular cross section to span 60 ft (18.3 m). The beam will carry a uniform live load of 600 lb/ft in addition to its dead weight, and the tendon is to be bonded to concrete by grouting after stressing. $f'_c = 5$ kips/in$^2$, $f'_{ci} = 4$ kips/in$^2$, $f_{si} = 175$ kips/in$^2$, $f_{se} = 150$ kips/in$^2$, and $f_{pu} = 250$ kips/in$^2$. Base the design on the midspan section.

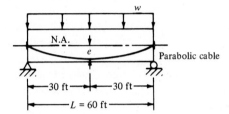

**Figure 11.36**

SOLUTION  Estimate depth

$$\text{Try } h = \frac{L}{25} = \frac{60(12)}{25} = 28.8 \text{ in} \qquad \text{use 30 in}$$

$$\text{Also try } b = 12 \text{ in.}$$

Design loads and moments

$$w_g = \frac{30(12)(0.15)}{144} = 0.375 \text{ kips/ft} \qquad w_t = 0.375 + 0.60 = 0.975 \text{ kips/ft}$$

$$M_g = \frac{w_g L^2}{8} = 168.75 \text{ ft} \cdot \text{kips} \qquad M_s = \frac{w_t L^2}{8} = 438.75 \text{ ft} \cdot \text{kips}$$

Estimate $A_{ps}$, basing the analysis on the moment at midspan created by service loads. Estimate that the arm $a = 0.5h = 15$ in (Fig. 11.37)

$$Ta = M_s$$

$$T(15 \text{ in}) = 438.75(12) \text{ in} \cdot \text{kips}$$

$$T = 351 \text{ kips}$$

$$A_{ps} = \frac{T}{f_{se}} = \frac{351 \text{ kips}}{150 \text{ kips/in}^2} = 2.34 \text{ in}^2$$

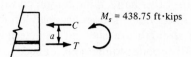

$M_s = 438.75 \text{ ft·kips}$

**Figure 11.37** Service loads acts.

Recompute $A_c$ using Eq. (11.28)

$$A_c = \frac{A_{ps} f_{se}}{0.225 f_c'} = \frac{2.34(150)}{0.225(5)}$$

$$A_c = 312 \text{ in}^2$$

Therefore 12 in by 30 in ($A_c = 360 \text{ in}^2$) seems adequate.

Establish the eccentricity $e$ of the cable assuming a triangular distribution of stress at transfer (Fig. 11.38)

$$T = f_{si} A_{ps} = 175 \text{ kips/in}^2 (2.34 \text{ in}^2) = 409.5 \text{ kips}$$

$$Ta = M_g = 168.75 \text{ ft · kips}$$

$$a = \frac{168.75(12)}{409.5 \text{ kips}} = 4.95 \quad 5 \text{ in}$$

$$e = k_b + a = \tfrac{30}{6} + 5 = 10 \text{ in}$$

You can see the above expression is equivalent to Eq. (11.29).

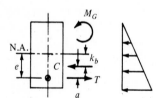

**Figure 11.38** Transfer stage.

Check stresses at midspan, at transfer $M_g = 168.75 \text{ ft · kips}$ (Fig. 11.39).

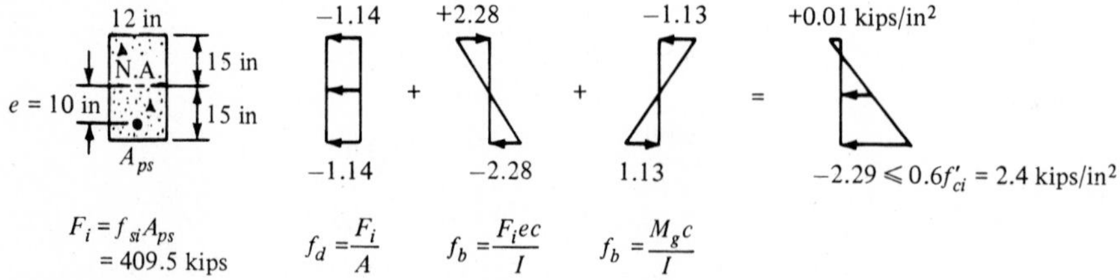

```
   12 in         −1.14    +2.28           −1.13        +0.01 kips/in²
  ┌─────┐
  │ N.A.│ ┤15 in
e = 10 in       │        +            +            =
  │     │ ┤15 in
  └─────┘
   A_ps          −1.14    −2.28          1.13       −2.29 ≤ 0.6f'_ci = 2.4 kips/in²
```

$$F_i = f_{si} A_{ps}$$
$$= 409.5 \text{ kips}$$

$$f_a = \frac{F_i}{A} \qquad f_b = \frac{F_i e c}{I} \qquad f_b = \frac{M_g c}{I}$$

**Figure 11.39**

Check stress with service loads acting after losses have occurred in tendon (Fig. 11.40).

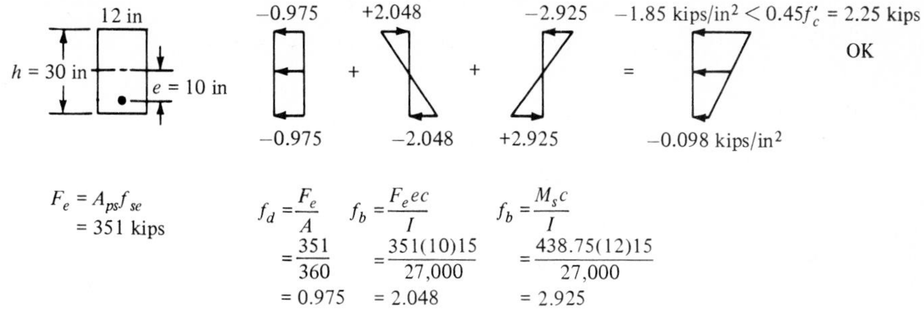

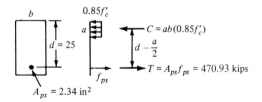

$$F_e = A_{ps}f_{se}$$
$$= 351 \text{ kips}$$

$$f_d = \frac{F_e}{A} \qquad f_b = \frac{F_e ec}{I} \qquad f_b = \frac{M_s c}{I}$$
$$= \frac{351}{360} \qquad = \frac{351(10)15}{27,000} \qquad = \frac{438.75(12)15}{27,000}$$
$$= 0.975 \qquad = 2.048 \qquad = 2.925$$

**Figure 11.40**

Compute $M_u$

$$w_u = 1.4(0.375) + 1.7(0.60) = 1.545 \text{ kips/ft}$$

$$M_u = \frac{w_u L^2}{8} = 695.25 \text{ ft} \cdot \text{kips}$$

Compute $\phi M_n$, the available design strength (Fig. 11.41)

$$f_{ps} = f_{pu}\left(1 - 0.5\rho_p \frac{f_{pu}}{f_c'}\right) = 250\left[1 - 0.5 \frac{2.34}{12(25)} \frac{250}{5}\right] = 201.25 \text{ kips/in}^2$$

**Figure 11.41**

Solve for $a$; equate $T = C$; $T = A_{ps}f_{ps} = 470.93$ kips

$$470.93 \text{ kips} = a(12)(4.25 \text{ kips/in}^2)$$

$$a = 9.2 \text{ in} \qquad a_{max} \leq 0.353d = 8.83 \text{ in}$$

therefore overreinforced

Use Eq. (11.25) to compute $M_n$;

$$M_n = 0.25f_c' bd^2 = 0.25(5)(12)(25^2) = 9753.75 \text{ in} \cdot \text{kips} = 812.81 \text{ ft} \cdot \text{kips}$$

$$\phi M_n = 0.9(812.81 \text{ ft} \cdot \text{kips}) = 731.53 \text{ ft} \cdot \text{kips} > M_u \qquad \text{OK}$$

Verify that $\phi M_n > 1.2M_{cr}$ [use Eq. (11.17) for $M_{cr}$]

$$M_{cr} = \frac{I}{y_t}\left(7.5\sqrt{f_c'} + \frac{F_e}{A} + \frac{F_e e y_t}{I}\right)$$

$$= \frac{27,000}{15}\left[7.5\sqrt{5000} + \frac{351,000}{360} + \frac{351,000(10)(15)}{27,000}\right]$$

$$= 6395.09 \text{ in} \cdot \text{lb} = 532.92 \text{ ft} \cdot \text{kips}$$

$$\phi M_n > 1.2M_{cr}$$

$$731.53 > 639.5 \quad \text{OK}$$

## 11.13 DESIGN FOR SHEAR

The design of shear reinforcement for prestressed beams is almost identical to that of nonprestressed reinforced concrete beams except for different design equations to evaluate the nominal shear strength $V_c$ of the concrete and small changes in the requirements for minimum area of shear reinforcement and maximum stirrup spacing. These small differences are caused by the longitudinal compressive stresses induced by the prestress tendons. By reducing the diagonal tension created by shear these compressive stresses raise the shear capacity of the concrete cross section. Also, as shown in Fig. 4.6g, the presence of longitudinal compressive stress reduces the slope of the plane on which the principal tensile stresses develop. As a result, the maximum stirrup spacing can be increased without reducing the the number of stirrups crossing each potential diagonal crack.

### Summary of the Design Procedure

When the shear $V_u$ produced by factored loads exceeds $\phi V_c/2$, shear reinforcement must be provided in prestressed beams. As with nonprestressed concrete members, slabs, footings, and shallow beams (as defined in ACI Code §11.5.5.1) are excluded. If $V_u$ is less than $\phi V_c$, the area and spacing of shear reinforcement are controlled by the provisions for minimum steel. If $V_u$ exceeds $\phi V_c$, the computation of the area of shear reinforcement is based on the requirement that the shear force $V_u$ produced by factored service loads must not exceed the nominal design strength $V_n$ of the cross section. This relationship can be stated as

$$V_u \le \phi V_n$$

or
$$V_u \le \phi(V_c + V_s) \tag{11.30}$$

where $V_c$ = nominal shear capacity of concrete
$V_s$ = nominal shear capacity of reinforcement = $A_v f_y d/s$
$\phi$ = capacity reduction factor = 0.85

To ensure that shear failures occur in a ductile manner by yielding of the shear reinforcement, ACI Code §11.5.6.8 specifies that $V_s$ must not exceed $8\sqrt{f'_c}b_w d$ $(0.66\sqrt{f'_c}b_w d)$. In addition, the maximum spacing of stirrups must not exceed three-fourths of the overall depth $h$ or 24 in (609.6 mm). If the shear $V_u$ on the cross section is relatively large, i.e., if $V_s$ exceeds $4\sqrt{f'_c}b_w d$ $(0.33\sqrt{f'_c}b_w d)$, the maximum permitted spacing is reduced by 50 percent and must not exceed $\frac{3}{8}h$ or 12 in (305 mm), whichever is smaller.

To ensure that stirrups have sufficient tensile strength to carry the diagonal tension in the concrete without rupturing when a diagonal-tension crack opens, ACI Code §11.5.5.3 requires that the minimum area of stirrup steel $A_v$ must not be less than

$$A_v = \begin{cases} \dfrac{50b_w s}{f_y} & \text{USCU} \\[2ex] \dfrac{0.34b_w s}{f_y} & \text{SI} \end{cases} \tag{11.31}$$

where $b_w$ = width of web

$\quad\quad s$ = stirrup spacing

$\quad\quad f_y$ = yield point of stirrup steel [must not exceed 60 kips/in$^2$ (413.7 MPa)]

If the effective prestress exceeds 40 percent of the ultimate tensile strength $f_{pu}$ of the tendon, experimental studies show that the compressive stresses induced in the concrete consistently raise the shear capacity of the member. For this design condition, the minimum area of shear reinforcement can be computed by

$$A_v = \frac{A_{ps}}{80}\frac{f_{pu}}{f_y}\frac{s}{d}\sqrt{\frac{d}{b_w}} \tag{11.32}$$

When the support reactions introduce compression into the ends of the member, sections of the beam located between the face of the support and a distance $h/2$ out from the face can be designed for the shear $V_u$ at a distance $h/2$ from the face of the support where $h$ is the overall depth of the section.

Unless otherwise stated, in all design equations for prestressed beams the effective depth $d$ equals the distance from the compression surface to the centroid of the longitudinal tension reinforcement but need not be taken less than $0.8h$. The use of $0.8h$ for $d$ regardless of the position of the tendon recognizes that because of the compression produced by prestress most of the cross section is uncracked and available to carry shear.

## Shear Strength of Prestressed Concrete

Several empirical expressions, which predict the nominal shear strength $V_c$ of concrete, have been developed from experimental studies of prestressed concrete members. For prestressed beams in which the tendons are stressed to at least 40 percent of the ultimate tensile strength $f_{pu}$, the nominal shear capacity of the concrete cross section can be conservatively predicted by

$$V_c = \begin{cases} \left(0.6\sqrt{f_c'} + 700\,\dfrac{V_u d}{M_u}\right)b_w d & \text{USCU} \\[3ex] \left(0.05\sqrt{f_c'} + 4.8\,\dfrac{V_u d}{M_u}\right)b_w d & \text{SI} \end{cases} \tag{11.33}$$

where $V_u$ and $M_u$ are the values of shear and moment produced by factored loads at the section under investigation. In the term $V_u d/M_u$, $d$ represents the distance from the compression surface to the centroid of the prestressed steel whereas the $d$ at the end of the equation is the larger of the distance from the compression surface to the centroid of the tendon or $0.8h$. Since Eq. (11.33) is empirical, it does not give valid results when either $V_u$ or $M_u$ is small. Therefore, the quantity $V_u d/M_u$ must not be taken greater than 1. In addition, $V_c$ predicted by Eq. (11.33) need not be taken less than $2\sqrt{f_c'}b_w d$ $(0.17\sqrt{f_c'}b_w d)$, nor can $V_c$ be taken greater than $5\sqrt{f_c'}b_w d$ $(0.42\sqrt{f_c'}b_w d)$.

If the size of the stirrups remains constant along the entire length of the beam, the designer will space the stirrups closest where the shear is highest. As the shear $\phi V_s$ that must be carried by the reinforcement decreases, the designer can space the stirrups farther apart. In a typical nonprestressed beam with a uniform load and simple supports, the closest spacing between stirrups occurs near the supports where the shear is highest, and the spacing between stirrups will increase steadily with increasing distance from the support as the shear that must be carried by the reinforcement decreases. In contrast, for prestressed beams, the maximum shear

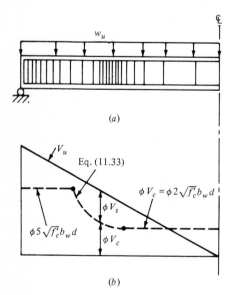

**Figure 11.42** Variation of stirrup spacing: (a) shear reinforcement, (b) shear curve showing the variation of concrete shear capacity.

reinforcement may be required near the quarter points rather than adjacent to the supports when Eq. (11.33) is used. As a result the stirrup spacing for a simply supported beam with a uniform load will vary as shown in Fig. 11.42; i.e., the spacing between stirrups is small at the supports, increases with distance out, decreases near the quarter point, and then increases again toward the center of the span. This variation of spacing can be explained by superimposing a plot of the concrete shear capacity $\phi V_c$ on the shear curve of the beam. As shown in Fig. 11.42, the difference between the ordinate of the shear curve $V_u$ and the ordinate of the concrete's shear capacity $\phi V_c$ represents the shear force that must be carried by the stirrups. Where $\phi V_s$ is large, the stirrup spacing must be small; conversely, as $\phi V_s$ decreases, the stirrups spacing may be increased. The use of Eq. (11.33) to compute the required spacing of stirrups is illustrated in Example 11.10.

**Example 11.10:**  Using Eq. (11.33) to evaluate $V_c$, determine for the beam in Fig. 11.43 the required spacing of no. 3 stirrups ($A_v = 0.22$ in²) at a section 6 ft out from support $A$. Given: $f'_c = 5.5$ kips/in², $A_{ps} = 2.4$ in², $f_y = 60$ kips/in², and $f_{pu} = 250$ kips/in². Loads shown on the beam are factored service loads.

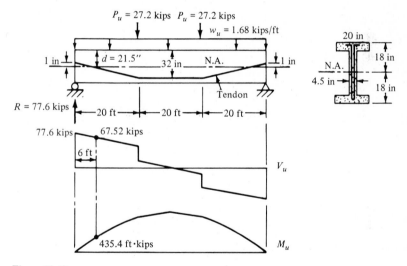

**Figure 11.43**

SOLUTION  At a distance 6 ft from the support $d = 21.5$ in, $d = 0.8h = 0.8(36) = 28.8$ in, and

$$V_c = \left(0.6\sqrt{f'_c} + 700\,\frac{V_u d}{M_u}\right)b_w d \tag{11.33}$$

where

$$\frac{V_u d}{M_u} = \frac{67.52(21.5)}{435.4(12)} = 0.28 < 1 \qquad \text{OK}$$

$$V_c = [0.6\sqrt{5500} + 700(0.28)]\,\frac{4.5(28.8)}{1000} = 31.17 \text{ kips}$$

$V_c$ is not to exceed $5\sqrt{f'_c}b_w d = 48.06$ kips and need not be less than $2\sqrt{f'_c}b_w d = 19.22$ kips; therefore use $V_c = 31.17$ kips. Compute the spacing of no. 3 stirrups, $A_v = 0.22$ in$^2$

$$V_u = \phi(V_c + V_s)$$

$$67.52 = 0.85(31.17 + V_s)$$

$$V_s = 48.3 \text{ kips}$$

$$s = \frac{A_v f_y d}{V_s} = \frac{0.22(60)(28.8)}{48.3 \text{ kips}} = 7.9 \text{ in} \qquad \text{use } 7.5 \text{ in}$$

For maximum spacing compare $V_s$ with $4\sqrt{f'_c}b_w d = 38.4$ kips. Since $V_s > 4\sqrt{f'_c}b_w d$,

$$s \leq 12 \text{ in} \qquad s \leq \tfrac{3}{8}h = \tfrac{3}{8}(36) = 13.5 \text{ in} \qquad \text{use } s = 7.5 \text{ in}$$

Check minimum $A_v$. Since $f_{se} > 0.4f_{pu} = 0.4(250 \text{ kips/in}^2) = 100 \text{ kips/in}^2$, use

$$A_{v,min} = \frac{A_{ps} f_{pu}}{80} \frac{s}{f_y} \frac{d}{d} \sqrt{\frac{d}{b_w}} = \frac{2.4}{80} \frac{250}{60} \frac{7.5}{28.8} \sqrt{\frac{28.8}{4.5}} = 0.082 \text{ in}^2$$

$$A_{v,sup} = 0.22 \text{ in}^2 > 0.082 \text{ in}^2 \qquad \text{OK}$$

Although Eq. (11.33) is simple to use, it may produce extremely conservative values of shear strength, particularly when applied to the design of composite I sections that support moving loads. If the designer wishes to take advantage of the additional shear strength of the concrete, or if the tendons are stressed below $0.4f_{pu}$, the shear strength $V_c$ of the concrete can be evaluated at a section by computing the flexural-shear strength $V_{ci}$ and the web-shear strength $V_{cw}$. $V_c$ is then taken as the smaller of $V_{ci}$ and $V_{cw}$. As specified in ACI Code §11.4.2.1, $V_{ci}$ is computed from

$$V_{ci} = 0.6\sqrt{f'_c}b_w d + V_d + \frac{V_i M_{cr}}{M_{max}} \qquad (11.34)$$

but $V_{ci}$ need not be less than $1.7\sqrt{f'_c}b_w d$, where $V_d$ equals the shear force produced at the section due to the unfactored weight of the girder and any permanently attached dead load that may be on the girder at transfer. $V_i$ and $M_{max}$ equal the shear and the moment at the section produced by the externally applied factored loads, which are positioned to maximize the moment at the section under investigation. Finally the cracking moment of the section is given by

$$M_{cr} = \frac{I}{y_t} (6\sqrt{f'_c} + f_{pe} - f_d) \qquad (11.35)$$

where $f_d$ is the flexural stress produced by unfactored dead load acting on the outside fiber of the section on the side that will be placed in tension by the externally applied loads (normally the live load). In composite construction, computation of $f_d$ is based on the noncomposite portion of the cross section. $f_{pe}$ is the compression stress on the outside fiber due to the effective prestress force after losses, $I$ is the moment of inertia of the section resisting the externally applied loads, and $y_t$ is the distance from the centroid to the outside fiber of the cross section placed in tension by the externally applied loads.

In Eq. (11.34) the last two terms represent the total shear force on the section when flexural cracking of the section impends; a tensile stress equal to the modulus of rupture develops on the outside tension fiber. Once flexural cracking impends, experimental studies indicate an additional increment of shear equal to $0.6\sqrt{f'_c}b_w d$, the first term in Eq. (11.34), will produce a diagonal shear failure. The equation assumes, of course, that the shear at the section due to unfactored dead load is less than the shear capacity of the cross section.

The equation for web-shear cracking, which normally controls the shear strength near the supports in beams with narrow webs, is given by ACI equation 11.3 as

$$V_{cw} = (3.5\sqrt{f'_c} + 0.3f_{pe})b_w d + V_p \qquad (11.36)$$

where $f_{pe}$ represents the compressive stress induced in the concrete by the effective prestress force, after losses, at the centroid of the section resisting the externally applied loads or at the junction of the web and the flange when the centroid lies within the flange. These two points represent the location of the maximum principal tensile stress. $V_p$ equals the vertical component of the effective prestress force.

Equation (11.36) is based on the assumption that a web-shear failure occurs when the maximum principal tensile stress reaches a value of approximately $4\sqrt{f'_c}$. As an alternative to Eq. (11.36), $V_{cw}$ can be evaluated as the shear force that produces a diagonal tension stress of $4\sqrt{f'_c}$ at the centroid of the section or at the intersection of the web and the flange when the centroid lies in the flange. The latter case is most likely to occur in composite construction when a thick slab is bonded to to the top flange of a precast section. The equation for principal stresses in terms of the shear and direct stress on a vertical section is

$$f_t = \sqrt{v_{cw}^2 + \left(\frac{f_c}{2}\right)^2} - \frac{f_c}{2} \qquad (11.37)$$

where $f_t$, the principal tensile stress, is set equal to $4\sqrt{f'_c}$ and $f_c$ is the direct stress at the centroid of the section or at the junction of the web and the flange when the centroid lies in the web. If the centroid lies in the web, the direct stress at the centroid is equal to the effective prestress force $F$ divided by the area of the section. Equation (11.37) is solved for $v_{cw}$. Then $V_{cw} = v_{cw}b_w d$.

In Eqs. (11.34) and (11.36) $d$ is equal to the distance between the compression surface and the centroid of the tendon but need not be taken less than $0.8h$.

**Example 11.11:** If the beam in Example 11.10 supports two concentrated service live loads of 16 kips at the third points in addition to its own dead weight, determine the shear capacity $V_c$ of the cross section at a point 6 ft from the left support using Eqs. (11.34) thru (11.37) for $V_{ci}$ and $V_{cw}$ (Fig. 11.44); $I = 48,000$ in$^4$, $A = 300$ in$^2$, $f'_c = 5500$ lbs/in$^2$, $A_{ps} = 2.4$ in$^2$, $f_{se} = 140$ kips/in$^2$ and $f_{pu} = 250$ kips/in$^2$.

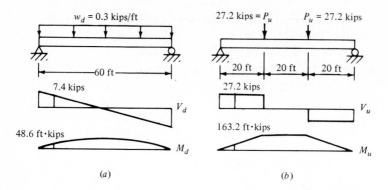

**Figure 11.44** Shear and moment curves: (*a*) unfactored dead load, (*b*) factored live load.

SOLUTION Compute $V_{ci}$. At 6 ft from the support

$$e = 21.5 - 18 = 3.5 \text{ in} \qquad F = A_{ps}f_{se} = 336 \text{ kips}$$

$$f_d = \frac{M_d c_b}{I} = \frac{48.6 \text{ ft} \cdot \text{kips}(12,000)(18 \text{ in})}{48,000} = 218.7 \text{ lb/in}^2$$

$$f_{pe} = \frac{F}{A} + \frac{Fec_b}{I} = \frac{336}{300} + \frac{336(3.5)(18)}{48,000} = 1.561 \text{ kips/in}^2 = 1561 \text{ lb/in}^2$$

Solve for $M_{cr}$ with Eq. (11.35)

$$M_{cr} = \frac{I}{y_t}(6\sqrt{f_c'} + f_{pe} - f_d) = \frac{48,000}{18}(445 + 1561 - 218.7)$$

$$= 4,766,133 \text{ in} \cdot \text{lb} = 397.2 \text{ ft} \cdot \text{kips}$$

$$V_{ci} = 0.6\sqrt{f_c'}b_w d + V_d + \frac{V_i M_{cr}}{M_{max}} \qquad (11.34)$$

$$= \frac{0.6\sqrt{5500}(4.5)(28.8)}{1000} + 7.4 \text{ kips} + \frac{(27.2 \text{ kips})(397.2)}{163.2}$$

$$V_{ci} = 79.17 \text{ kips}$$

Compute

$$V_{cw} = (3.5\sqrt{f_c'} + 0.3f_{pc})b_w d + V_p \qquad \text{Eq. (11.36)}$$

$$= \left(\frac{259.6}{1000} + 0.3\frac{336}{300}\right)[4.5(28.8)] + \frac{(15 \text{ in})(336)}{20(12)}$$

$$= 98.19 \text{ kips}$$

Or $V_{cw}$ may be taken as the shear force that produces a principal tension stress of $4\sqrt{f_c'} = 296.6$ lb/in² at the centroid of the web. Setting $f_t = 296.6$ lb/in² in Eq. (11.37) gives

$$f_t = \sqrt{(v_{cw})^2 + \left(\frac{f_c}{2}\right)^2} - \frac{f_c}{2} \quad \text{where } f_c = \frac{F}{A} = \frac{336 \text{ kips}}{300} = 1.12 \text{ kips/in}^2$$

$$296.6 = \sqrt{(v_{cw})^2 + \left(\frac{1120}{2}\right)^2} - \frac{1120}{2}$$

$$v_{cw} = 648.2 \text{ lb/in}^2$$

$$V_{cw} = v_{cw}b_w d = \frac{648.2}{1000}(4.5)(28.8) = 84.01 \text{ kips}$$

Use $V_c = V_{ci} = 79.17$ kips. Notice that $V_c$ computed from the equations for $V_{ci}$ and $V_{cw}$ produces a value of shear capacity that is $2\frac{1}{2}$ times larger than that computed by the simplified expression (11.33) in Example 11.10.

## 11.14 DEFLECTIONS

The use of high-strength materials in prestressed concrete construction leads to the design of relatively shallow beams with a large capacity for load. A well-designed prestressed concrete beam generally has a smaller depth than a reinforced concrete beam spanning the same distance. For example, the PCI Handbook[4] indicates that simply supported prestressed floor beams will typically be designed with a span-to-depth ratio of 26 while the ACI Code limits simply supported reinforced concrete beams to a span-to-depth ratio of 16, a difference of 38 percent.

Although shallow prestressed beams may have adequate bending strength to span large distances economically, they tend to be flexible and undergo large deflections under design loads. To ensure that prestressed members will have adequate bending stiffness to prevent damage to attached nonstructural elements (partitions, piping, etc.) the designer must check deflections making certain that they are within acceptable limits.

The deflections produced by prestress alone are normally larger than, and opposite in direction to, the deflections created by dead load. Therefore, when the live load is absent, the prestress force, acting continuously, produces a net upward deflection, i.e., camber, of the beam. If the live load acts infrequently or is applied only for short periods of time, creep can cause the initial camber to double or to triple over a period of several years. If the camber is large, the upward deflection and the curvature of the beams may produce any of the following undesirable conditions: unacceptable warping of floors, damage to roofing, rupture of the joints of rigid pipe, or cracking of partitions constructed of brittle materials.

To ensure that structures remain serviceable over their lifetime, both short- and long-term deflections resulting from various combinations of load and prestress should be investigated as part of a complete beam design. ACI Code §9.5.4.3 requires that deflections of prestressed beams not exceed the limiting values of deflection specified in Table 3.2. The use of minimum depths to control deflections does not apply to prestressed concrete beams because the coefficients specified in Table 3.1 are based on the behavior of nonprestressed beams only.

Deflections in prestressed beams are computed by conventional methods, such as virtual work, moment-area, or standard deflection equations. When using standard equations to compute immediate deflections due to prestress, the designer can treat the concrete as an elastic member loaded by the tendon reactions; i.e., the tendon can be removed and the forces it exerts on the concrete beam can be treated as loads (see Sec. 11.6).

If the moment-area method is used, the deflections produced by prestress can be computed by using the moment curve due to prestress. For a determinate beam or structure, the magnitude of this moment at any section is equal to the product of the tendon force and its eccentricity measured from the centroid of the gross section. The two methods of computing deflections are illustrated in Example 11.12. As it shows, lateral deflections are produced both by transverse tendon reactions and by the moments at the ends of the beam due to the eccentricity of the tendon at those points. The 200-kip axial compression force exerted by the prestress produces longitudinal shortening only and does not contribute to the lateral deflections.

Example 11.12 compares the magnitude of the immediate and the long-term deflections of a beam that supports a live load acting intermittently for short periods of time. Under these conditions long-term deflections are determined by dead load and prestress.

As long as a beam remains uncracked, i.e., the maximum tensile stress does not exceed $6\sqrt{f_c'}$ lb/in$^2$ ($0.50\sqrt{f_c'}$ for SI units) (a conservative estimate of the modulus of rupture), the moment of inertia used in deflection computations can be based on the properties of the gross section. When the maximum tensile stress, based on the properties of the gross section, lies between $6\sqrt{f_c'}$ and $12\sqrt{f_c'}$ lb/in$^2$ ($0.5\sqrt{f_c'}$ and $1.00\sqrt{f_c'}$, for SI units) cracking of the cross section is assumed; the effective value of moment of inertia should be based on the properties of both the cracked and uncracked cross sections.

**Example 11.12: Computation of deflections** Assuming an uncracked section and elastic behavior, compute for the beam in Fig. 11.45 the instantaneous vertical deflection at midspan due to prestress only using (a) deflections equations and (b) the moment-area method. The tendon force equals 200 kips, $E_c = 4000$ kips/in$^2$, and $I_g = 6667$ in$^4$.

SOLUTION (a) Compute the components of the tendon force; slope of tendon $= \frac{1}{30}$ (Fig. 11.45b)

$$F_v = \tfrac{1}{30}(200) = 6.67 \text{ kips} \qquad F_h = 200 \text{ kips}$$

On a free-body diagram of the concrete show the tendon reactions (Fig. 11.45c). At each end of the beam, the eccentric horizontal component of the tendon force produces an axial compression of 200 kips and a moment of 400 in · kips (see Fig. 3.10)

$$\Delta = \frac{PL^3}{48EI} - \frac{ML^2}{8EI} = \frac{13.33(30^3)(1728)}{48(4 \times 10^3)(6667)} - \frac{400[30(12)]^2}{8(4 \times 10^3)(6667)}$$

$$= 0.48 \text{ in} - 0.24 \text{ in} = 0.24 \text{ in}\uparrow$$

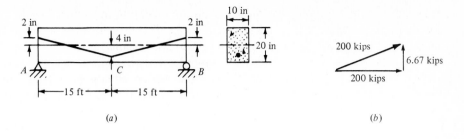

(a)                                                                                       (b)

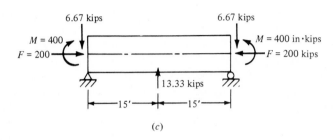

(c)

**Figure 11.45** (a) Beam, (b) components of tendon force, (c) forces exerted on the concrete by the tendon.

(b) Since the beam is loaded symmetrically (Fig. 11.46), the tangent drawn to the elastic curve at midspan is horizontal; therefore, the tangential deviation of point A from the tangent to the elastic curve at point C equals the midspan deflection

$$\Delta_c = d\frac{A}{C} = \tfrac{1}{2}(10)\frac{800}{EI}(11.67)(144) - \tfrac{1}{2}(5)\frac{400}{EI}(1.67)(144)$$

$$= 0.24 \text{ in}\uparrow$$

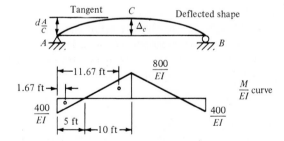

**Figure 11.46**

**Example 11.13: Influence of creep on deflection** The beam in Fig. 11.47 is prestressed by a straight cable with a constant eccentricity of 4 in. The initial tendon force equals 233 kips. After losses the effective prestress drops to 200 kips. Compute (a) the initial and (b) long-term deflections at midspan due to prestress and dead load; $E_c = 4$ kips/in$^2$, $I_g = 6667$ in$^4$, $C_c = 2$, $w_d = 0.21$ kips/ft. See Fig. 3.10 for deflection equations.

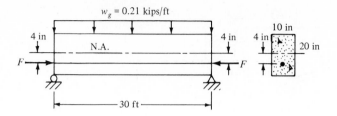

**Figure 11.47**

SOLUTION (*a*) The immediate deflection at midspan after prestress applied is

$$\Delta_p = \frac{ML^2}{8EI} = \frac{233(4)[30(12)]^2}{8(4 \times 10^3)(6667)} = 0.566 \text{ in}\uparrow$$

$$\Delta_d = \frac{5wL^4}{384EI} = \frac{5(0.21)(30^4)(1728)}{384(4 \times 10^3)(6667)} = 0.14 \text{ in}\downarrow$$

$$\Delta_{p+d} = 0.566 - 0.14 = 0.426 \text{ in}\uparrow$$

(*b*) The long-term deflection at midspan with creep coefficient $C_c = 2$ is

$$\Delta_p = \tfrac{200}{233}(0.556)(1 + C_c) = 1.43 \text{ in}\uparrow$$

$$\Delta_d = 0.140(1 + C_c) = 0.42 \text{ in}\downarrow$$

$$\Delta_{p+d} = 1.43 - 0.42 = 1.01 \text{ in}\uparrow$$

where $C_c$ is defined by Eq. (11.48)

## 11.15 END-BLOCK STRESSES

The end anchorages of tendons and the reactions at supports apply large concentrated forces to the ends of prestressed beams (Fig. 11.48). Unless properly
positioned and sized reinforcement is provided in these regions of the beam, these
forces may damage the ends of the member by crushing, cracking, or spalling the
concrete.[6] Experimental studies show that the end region in which the high local
stresses occur, called the *end block*, extends from the outside face into the beam a
distance approximately equal to the overall depth *h* of the member. The end block
acts rather like a thick bearing plate to distribute the concentrated tendon reactions
over the full area of the cross section. As indicated in Fig. 11.48, at section *A-A*,
a distance *h* from the end of the beam, the local stress concentrations have dis-
appeared and the longitudinal stresses can be calculated by superimposing the
axial and the bending stresses computed by $P/A$ and $Mc/I$.

In the end block, the variation and intensity of the stresses are very difficult
to determine. Holes created by ducts, local cracking produced by tendon forces,
the inelastic behavior of the concrete, and the stresses due to the support reactions
(typically neglected) make any type of rational stress analysis problematic.
Therefore, the design of reinforcement for the end block,[5] based primarily on the

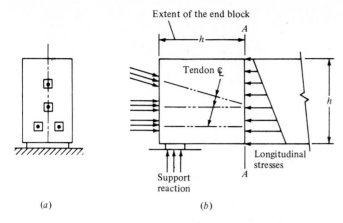

**Figure 11.48** End block: (a) end view, showing the tendon anchors; (b) side view, showing the reactions of the tendons and the support.

results of experimental studies, is usually carried out either by using empirical expressions for the internal forces or by cutting sections through the end block and using the equations of statics to determine the approximate values of the internal forces. If the analysis indicates that stresses are excessive (the case when tendon forces are large or the beam web thin), it may be necessary to thicken the web in the end block.

The location of potential cracks in the end block can be explained qualitatively by examining the forces on the concrete directly under the bearing plate. This analysis, similar to that of a footing bearing against a homogeneous soil, assumes that the base plate acts as a rigid body on an isotropic material. Because of friction between the bearing plate and the concrete a triangular element of concrete acts with the bearing plate and penetrates the end of the beam like a wedge (Fig. 11.49a). As shown in Fig. 11.49b, normal and shear forces develop on the interior faces of the wedge to resist penetration of the wedge into the end of the beam. The resultant of these forces is represented by the compression force $R$. As shown in Fig. 11.49d, between points $c$ and $f$ and points $c$ and $g$ the compression forces create two diagonal bands of compression that transfer the tendon reaction into the interior of the beam. While the longitudinal component of the diagonal force creates compression in the beam, the transverse component, acting outward, stretches the beam transversely and creates a tension field between points $f$ and $g$. Since the concrete has a low tensile strength, longitudinal cracking will develop near the center of the end block that may split the beam (Fig. 11.49e) unless the two sides are tied together by transverse reinforcement. The approximate variation of the transverse stresses along the longitudinal axis at middepth is shown in Fig. 11.49d.[6]

Resistance to displacement of the wedge into the beam is also provided by the areas directly above and below the wedge (Fig. 11.49a). The movement of the wedge inward displaces these areas laterally along the curved failure surfaces $c$-$e$ and $c$-$d$.

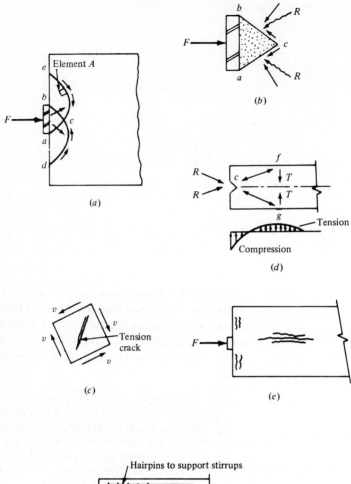

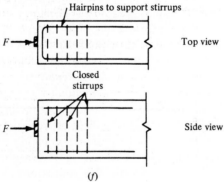

**Figure 11.49** Stresses in the end block: (*a*) end block, (*b*) forces on the concrete wedge under bearing plate, (*c*) shear stresses on element *A*, (*d*) transverse stresses at middepth, (*e*) tension cracks, (*f*)end-block reinforcement.

Resistance to this movement is provided primarily by shear stresses that develop along the failure surface. Any normal stresses that act along the failure plane are neglected to simplify the discussion. Since concrete has a low tensile strength, the diagonal-tension stresses associated with the shear can lead to cracking and spalling of the concrete adjacent to the end surface (Fig. 11.49e).

To carry the tensile stresses created in the end block by the tendon reactions, closely spaced stirrups together with horizontal bars are placed in the end block, as shown in Fig. 11.49f. Steel spirals would also be effective reinforcement. Additional steel to prevent a shear-friction failure of the end of the beam by the support reaction may also be required. The design of this steel is discussed in Sec. 4.7.

Although this discussion has centered on the loads applied by tendon anchors of posttensioned beams, pretensioned tendons, which depend on bond to provide anchorage at the ends of the beam, also create a similar state of stress in the end block.

Since this chapter is intended as an introduction to prestressed design, no detailed analysis of the end block will be made; more information can be found in Refs. 6 and 7.

## 11.16 LOSS OF PRESTRESS

Unless an accidental overload is applied to a member, the maximum stress in a prestress tendon occurs at the transfer stage when the tendon is initially stressed. Following the transfer of force from the jack to the member, a loss in tendon stress occurs ranging from 12 to 20 percent of the initial stress. That part of the loss in stress produced by slip at the anchor and by elastic shortening of the concrete takes place immediately. Other losses due to creep and shrinkage of the concrete as well as relaxation of the steel are time-dependent and occur at a diminishing rate over several years. To ensure that the required level of prestress will be present after losses, their value must be estimated and added to the required value of prestress in order to establish the value of the jacking force. If the computed losses are excessive, they can be reduced by modifying the construction procedure.

Since the ultimate flexural strength of a member is primarily a function of the area of the tendon rather than the magnitude of the effective prestress force, loss in prestress for the normal beam does not significantly affect the ultimate bending strength of a member as long as the strain at failure is sufficient to develop the full design strength of the tendon.

If the magnitude of the losses is underestimated and less prestress remains than anticipated, cracking under service loads increases. The additional cracking, which reduces the flexural stiffness, results in increased deflections. Cracking also reduces fatigue strength and shear strength. On the other hand, if losses are overestimated and a greater prestressing force remains after losses, camber and axial shortening increase while the ductility of members that fail in flexure is reduced (Fig. 11.31).

For most prestressed concrete members constructed of normal-weight concrete, subjected to average conditions of exposure, and constructed by standard methods the ACI Commentary indicates that a satisfactory design will result if losses due to creep, shrinkage, relaxation of the steel, and elastic shortening are taken as

$$\text{Losses} = \begin{cases} 35 \text{ kips/in}^2 \text{ (241 MPa)} & \text{for pretensioning} \\ 25 \text{ kips/in}^2 \text{ (172 MPa)} & \text{for posttensioning} \end{cases}$$

For posttensioned members an additional allowance for losses in prestress due to friction and anchorage slip must be added to the above losses in stress. As a practical matter both creep and shrinkage are difficult to compute accurately since both are strongly influenced by the relative humidity (which may fluctuate widely in an unpredictable manner), the type of aggregate, and the volume-to-surface ratio. Moreover, a precise determination of stress losses in prestressed concrete members is complex because the rate of loss due to one factor is continually being altered by changes in stress due to other factors. For example, the loss in stress due to creep of the concrete depends on the magnitude of the tendon force, but as creep occurs, the member shortens and the tendon force lessens. The reduction in tendon force in turn reduces the rate and magnitude of the creep. Although procedures for the most precise computations of losses require that the interdependence of losses be considered, time-dependent losses are usually treated as if they took place independently, an assumption that will be made in this text.

In posttensioned construction the force in the tendon also varies because of the friction that develops between the tendon and the concrete when the tendon is tensioned. Friction develops due to both unavoidable crookedness of the ducts (the wobble effect) and curvature of the tendons. Because of friction the force in the tendon is greatest at the jacking end and reduces with distance from the jack. If the friction losses are big because the tendon is long or the curvature large, they can be reduced by tensioning the tendon by jacks at both ends.

In the following sections each type of loss is briefly discussed. For the designer, who may occasionally be required to make a very precise analysis of losses, design aids are available[3,4] to simplify computations of losses for both normal and lightweight prestressed concrete members.

### Slip at the Anchor

At transfer in posttensioned construction when the jack stretching the tendon is released, a small amount of tendon shortening occurs because of compression of the anchorage fitting and movement of the wedges or mechanical grips that transfer the tendon force to the anchorage. The magnitude of the slippage, a function of the anchorage system, varies from 0.1 to 0.25 in and is normally specified by the manufacturer of the anchorage hardware. From Hooke's law, the loss of stress in the tendon $\Delta f_s$ due to slippage can be expressed as

$$\Delta f_s = \Delta \epsilon \, E_s = \frac{\Delta L \, E_s}{L} \tag{11.38}$$

where $\Delta L$ = end slippage
     $L$ = tendon length
     $E_s$ = modulus of elasticity of tendon

Since the loss in stress predicted by Eq. (11.38) is inversely proportional to the length of the tendon, the percentage loss in steel stress due to slippage decreases as the length of the tendon increases.

At transfer, if the tendon can be elongated by an additional increment of length equal to the anticipated slippage without being overstressed, the loss in stress due to slippage can be eliminated.

## Elastic Shortening

To establish the loss of tendon stress due to elastic shortening we shall consider the deformations of pretensioned members of constant cross section stressed by a tendon at the centroid of the concrete section (Fig. 11.50). Since the concrete and the tendon are fully bonded, after transfer they shorten by an equal amount. As a result of the shortening the initial force in the tendon drops from $F_o$ to $F_i$. Since the forces in the steel and concrete are in equilibrium, the force in the concrete must also be equal to $F_i$. The compatibility of deformations between the steel and concrete can be expressed as

$$\epsilon_c = \Delta\epsilon_s \tag{11.39}$$

where $\epsilon_c$ equals the strain in the concrete and $\Delta\epsilon_s$ equals the reduction in tendon strain produced by elastic shortening. Expressing the strain in terms of stresses gives

$$\frac{f_c}{E_c} = \frac{\Delta f_s}{E_s} \tag{11.40}$$

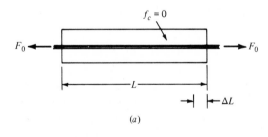

(a)

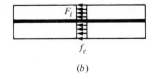

(b)

**Figure 11.50** Elastic shortening of a pretensioned member: (a) before transfer; (b) after transfer, elastic shortening occurs.

where $f_c$ is the stress in the concrete after transfer and $\Delta f_s$ is the loss of stress as a result of elastic shortening. Solving Eq. (11.40) for $\Delta f_s$ leads to

$$\Delta f_s = \frac{E_s}{E_c} f_c = n f_c \qquad (11.41)$$

Expressing $\Delta f_s$ as $f_{si} - f_s$, where $f_{si}$ and $f_s$ are the tendon stresses before and after transfer, respectively, yields

$$f_{si} - f_s = n f_c \qquad (11.42)$$

Writing Eq. (11.42) in terms of forces leads to

$$\frac{F_o}{A_{ps}} - \frac{F_i}{A_{ps}} = \frac{n F_i}{A_c} \qquad (11.43)$$

Solving (11.43) for $F_o$ after multiplying all terms by $A_{ps}$ gives

$$F_o = \frac{F_i}{A_c}(A_c + n A_{ps}) \qquad (11.44)$$

When we note that $F_i/A_c = f_c$, Eq. (11.44) can be used to express $f_c$ as

$$f_c = \frac{F_o}{A_c + n A_{ps}} = \frac{F_o}{A_t} \qquad (11.45)$$

where $A_t$, defined as the *transformed area*, equals $A_c + n A_{ps}$. If Eq. (11.45) is substituted into Eq. (11.41), an expression for the loss in prestress in terms of the initial tendon force results

$$\Delta f_s = \frac{n F_o}{A_t} \qquad (11.46)$$

Since the area of prestress steel is usually small, $A_t$ can be set equal to $A_g$ with little change in the final results. $A_g$ equals the gross area.

In posttensioned construction, elastic shortening can be neglected if all tendons are stressed and anchored in a single operation. When this procedure is followed, the jack that elongates the tendon compresses the concrete and elastic shortening takes place before the tendon is anchored.

If tendons are tensioned separately, the loss in stress caused by elastic shortening in a particular tendon is due only to those tendons which are stressed after the particular tendon has been anchored.

**Example 11.14: Loss of prestress due to axial shortening** Before transfer the tendon in a pretensioned beam is stressed to 180 kips/in². If the tendon is located at the centroid of the section (Fig. 11.51), determine the stress in both the steel and concrete immediately after transfer due to elastic shortening. Stone concrete is used; $A_{ps} = 1.2$ in², $f'_c = 4$ kips/in², and $E_s = 29,000$ kips/in².

10 in

20 in

**Figure 11.51**

SOLUTION Exact analysis gives

$$E_c = 57{,}000\sqrt{f'_c} = 3600 \text{ kips/in}^2$$

$$n = \frac{E_s}{E_c} = \frac{29{,}000}{3600} = 8.06 \qquad \text{use 8}$$

$$\Delta f_s = n\frac{F_o}{A} = 8\frac{(180 \text{ kips/in}^2)(1.2 \text{ in}^2)}{198.8 + 8(1.5)} = 8.2 \text{ kips/in}^2$$

$$f_c = \frac{F_i}{A_c} = \frac{(180 - 8.2)(1.2)}{198.8 \text{ in}^2} = 1.037 \text{ kips/in}^2$$

For an approximate computation for $\Delta f_s$ use the gross area instead of the transformed area

$$\Delta f_s = n\frac{F_o}{A_g} = 8\frac{180(1.2)}{200 \text{ in}^2} = 8.64 \text{ kips/in}^2$$

Example 11.14 illustrates that the state of stress in a prestressed member due to axial shortening can be closely approximated by basing the computations on the initial tendon force $F_o$ and the gross area. These simplifications will be used in all calculations for elastic shortening in the text.

## Shrinkage

The magnitude of the losses in tendon stress due to shrinkage in a prestressed beam depends on many factors, e.g., the amount of mixing water, the relative humidity, the length of cure, the type of aggregate, the strength of the concrete, and the shape of the cross section. The expression used to estimate the losses in tendon stress due to shrinkage is based on the recommendations in Ref. 8. This empirical expression applies to interior members as well as those exposed to the weather. An average relative humidity (RH) of 70 percent is assumed. From an analysis of test data, the loss in steel stress from the end of the curing period can be predicted by

$$\Delta f_s = \text{USH} \times \text{SSF} \times \text{AUS} \tag{11.47}$$

where USH, the ultimate loss in steel stress due to shrinkage in pounds per square inch is given by

$$\text{USH} = \begin{cases} 27{,}000 - 3000\dfrac{E_c}{10^6} \text{ but not less than} \\ 12{,}000 \text{ lb/in}^2 \qquad\qquad\qquad\qquad \text{normal-weight concrete} \\ 41{,}000 - 10{,}000\dfrac{E_c}{10^6} \text{ but not less than} \\ 12{,}000 \text{ lb/in}^2 \qquad\qquad\qquad\qquad \text{lightweight concrete} \end{cases}$$

**Table 11.7 Shrinkage factors for various volume-to-surface ratios[8]**

| Volume to surface ratio, in | Shrinkage factor SSF |
|:---:|:---:|
| 1 | 1.04 |
| 2 | 0.96 |
| 3 | 0.86 |
| 4 | 0.77 |
| 5 | 0.69 |
| 6 | 0.60 |

**Table 11.8 Shrinkage coefficients for various curing times[8]**

| Time after end of curing, d | Portion of ultimate shrinkage AUS |
|:---:|:---:|
| 1 | 0.08 |
| 3 | 0.15 |
| 5 | 0.20 |
| 7 | 0.22 |
| 10 | 0.27 |
| 20 | 0.36 |
| 30 | 0.42 |
| 60 | 0.55 |
| 90 | 0.62 |
| 180 | 0.68 |
| 365 | 0.86 |
| End of service life | 1.00 |

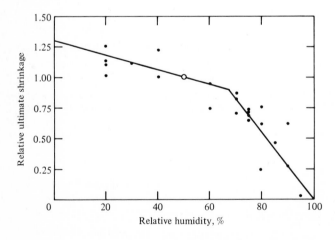

**Figure 11.52** Relative ultimate shrinkage versus relative humidity.[8]

where $E_c = 33w^{3/2}\sqrt{f'_c}$    lb/in². Values for SSF, the factor accounting for the influence of the size and the shape of the cross section, are given in Table 11.7. AUS, the factor denoting the portion of the ultimate shrinkage that has occurred from the end of the curing period, is listed in Table 11.8. If the average relative humidity is not 70 percent, Fig. 11.52 can be used to adjust the results of Eq. (11.47).

**Example 11.15:** Determine the loss in prestress due to the shrinkage of the member in Example 11.14 after 1 year with Eq. (11.47). Assume that the member is tensioned 5 days after the end of the curing period and that the relative humidity is 70 percent, $E_c = 57,000\sqrt{4000} = 3.6 \times 10^6$ lb/in².

SOLUTION By Eq. (11.47)

$$\Delta f_s = \text{USH} \times \text{SSF} \times \text{AUS}$$

$$\text{USH} = 27,000 - \frac{3000E_c}{10^6}$$

$$= 27,000 - \frac{3000(3.6 \times 10^6)}{10^6} = 16,200 \text{ lb/in}^2 > 12,000 \text{ lb/in}^2$$

For SSF, compute the volume-to-surface ratio, which is identical to the area-to-perimeter ratio

$$\frac{\text{Area}}{\text{perimeter}} = \frac{10(20)}{2(10 + 20)} = 3.33$$

By interpolation from Table 11.7 read SSF = 0.83. For AUS, from Table 11.8 read 0.2 after 5 days and 0.86 after 1 year

$$\Delta f_s = (16.2 \text{ kips/in}^2)(0.83)(0.86 - 0.2) = 8.874 \text{ kips/in}^2$$

## Creep

Loss in steel stress due to creep of the concrete, which is a direct function of the net concrete compressive stresses at the centroid of the tendon, is also influenced by many of the same factors that affect shrinkage strains, e.g., the type of aggregate, method of curing, length of cure, shape of member, age after curing when prestress is applied, and average relative humidity. Creep strains are often expressed in terms of the elastic strain by means of an experimentally determined creep coefficient $C_c$, defined by

$$C_c = \frac{\epsilon_{cu}}{\epsilon_{ci}} \tag{11.48}$$

where $\epsilon_{cu}$ is the ultimate value of strain due to creep and $\epsilon_{ci}$ is the initial elastic strain. Values of creep coefficients as a function of the average relative humidity are given in Table 11.9 .For the average prestressed member, which is made of a

## Table 11.9 Creep coefficients[9]

| Concrete strength $f'_c$ kips/ft$^2$ | Average relative humidity, % | | |
|---|---|---|---|
| | 100 | 70 | 50 |
| Average 2–4 | 1–2 | 1.5–3 | 2–4 |
| High, 4–6 | 0.7–1.5 | 1.0–2.5 | 1.5–3.5 |

high-strength concrete, a value of $C_c = 2$ is recommended. The ultimate loss in stress due to creep can be expressed as

$$\Delta f_s = E_s \epsilon_{cu} = E_s C_c \epsilon_{ci} = E_s C_c \frac{f_c}{E_c} = C_c n f_c \qquad (11.49)$$

### Relaxation

When a high-strength steel is stressed well into the elastic range, to a value of stress approaching the yield point, both plastic and elastic deformations occur. If, immediately after tensioning, a highly stressed steel tendon is connected to nearly rigid supports, such as the ends of a prestressed concrete beam, a reduction in steel stress results from the plastic deformations. This time-dependent loss of stress without change in length is called *relaxation*. The loss in stress due to relaxation, which occurs most rapidly during the first 2 hours after tensioning, is nearly complete after 2 weeks.

If the stress in the tendon is raised above the design stress for several minutes and then reduced to the required value at transfer, the loss in stress due to relaxation after the tendon is anchored can be reduced significantly—on the order of 70 percent.

For most steels a short period of overstressing will limit the loss of stress due to relaxation to approximately 2 or 3 percent of the initial stress at the time of transfer.

### Friction

As a tendon is stressed in a posttensioned member, frictional forces develop along the length of the tendon wherever contact occurs between the tendon and the walls of the duct. If the tendon is coated with mastic and wrapped with paper, the friction develops between the tendon and the surrounding concrete. The friction is a function of the crookedness of the duct (the *wobble effect*), the curvature of the tendon axis, and the roughness of the tendon surface. As a result of the friction, the force in the tendon reduces with distance from the jack. If a particular value of force is required at a certain section, the friction force between that section and the jack must be estimated and added to the required tendon force to establish the jacking force.

**Wobble** When ducts are positioned in forms, some degree of misalignment and crookedness of the duct is unavoidable because of workmanship. Figure 11.53 shows on an exaggerated scale the waves or undulations that develop in the duct for a straight tendon. These deviations occur both in elevation and in plan. As a tendon attempts to straighten during tensioning, it is forced against the sides of the duct at the crest of each wave. At each point of contact a normal force, which is proportional to the tendon force, develops between the tendon and the surrounding material. Because of the normal force, frictional forces develop at the points of contact to resist the movement of the tendon as it is drawn over a point of contact (Fig. 11.53b).

If the variation in cable tension is neglected and the cable force is taken equal to the tension at the end of the cable away from the jack, the loss in force due to friction can be expressed as

$$\Delta F_{fr} = K P_x L \tag{11.50}$$

where $P_x$ = tension in cable at interior section a distance $L$ from location of jack
$L$ = length of cable over which friction loss occurs
$K$ = experimentally determined wobble coefficient

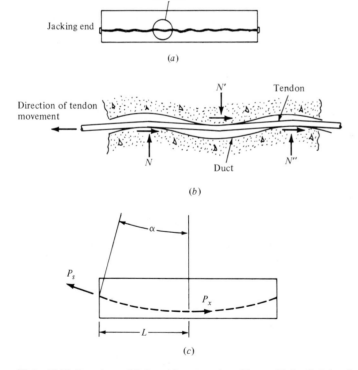

**Figure 11.53** Loss due to friction: (a) post tensioned beam, (b) detail $A$ showing unintentional crookedness (wobble) of duct, (c) friction losses in a curved tendon.

The wobble coefficient gives the friction loss per unit length of tendon per unit tendon force, in units of the reciprocal of length. Wobble coefficients for each type of tendon system are supplied by the manufacturer, and typical values of wobble coefficients are listed in the ACI Commentary and Ref. 4.

The force $P_s$ applied at the jack can be related to the internal tendon force by adding $\Delta F_{fr}$ given by Eq. (11.50) to the force $P_x$ to give

$$P_s = P_x + KP_xL$$

After factoring out $P_x$ this becomes

$$P_s = P_x(1 + KL) \tag{11.51}$$

The use of Eq. (11.51) is illustrated in Example 11.16.

**Example 11.16:** If a tension force of 200 kips is required at section 2 in the member in Fig. 11.54, what value of force should be applied at the jack? $K = 0.001 \text{ ft}^{-1}$.

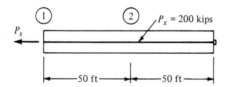

Figure 11.54

SOLUTION Although the force in the tendon varies with distance from the jacking end because of friction loss, the force in the tendon will initially be assumed constant and equal to $P_x$. The loss between the jack and the midspan section is

$$\Delta F_{fr} = KP_xL = 0.001(200 \text{ kips})(50 \text{ ft}) = 10 \text{ kips}$$

Therefore apply 210 kips at the jack. Since the force in the tendon varies from 210 kips at the jack to 200 kips at midspan, the average tendon force is 205 kips. If this value is used instead of 200 kips to compute the friction loss, $\Delta F_{fr} = 10.25$ kips, which is not significantly different from the first value.

**Curvature** When a curved tendon is used to prestress a beam, additional normal force develops between the tendon and the concrete because of the curvature of the tendon axis (Fig. 11.14). For this case ACI Code §18.6.2.2 specifies that the loss in cable tension due to friction can be computed by

$$P_s = P_x e^{KL + \mu\alpha} \tag{11.52}$$

When the curvature and the friction are moderate, that is, $KL + \mu\alpha$ is less than 0.3, the friction can be computed as

$$P_s = P_x(1 + KL + \mu\alpha) \tag{11.53}$$

where $\mu$ is the coefficient of friction associated with the curvature and $\alpha$ is the change in angle in radians of the prestressing tendon between the jack and the section at which $P_x$ is computed (Fig. 11.53c).

**Example 11.17:** The posttensioned beam in Fig. 11.55 is stressed by a parabolic tendon. If a 200 kip force is applied by the jack to the left end of the tendon at point $A$, what is the force in the tendon $P_x$ at point $B$? $K = 0.0003$ ft$^{-1}$, $\mu = 0.25$.

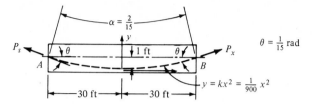

**Figure 11.55**

SOLUTION  Differentiating the equation of the parabola gives the expression for slope as $dy/dx = 2kx$. At point $B$, $x = 30$ ft and the slope of the tendon equals $\frac{1}{15}$ rad. Because of symmetry the total angle change between $A$ and $B$ is $\frac{2}{15}$ rad. Using Eq. (11.52), we get

$$P_s = P_x e^{\mu\alpha + kL}$$

$$200 \text{ kips} = P_x \exp[0.25(\tfrac{2}{15}) + 0.0003(60)]$$

$$P_x = 190 \text{ kips}$$

Since $\mu\alpha + kL = 0.033 + 0.018 = 0.051 < 0.3$, Eq. (11.53) can also be used to produce almost the identical result.

$$P_s = P_x(1 + \mu\alpha + KL)$$

$$200 \text{ kips} = P_x(1 + 0.033 + 0.018)$$

$$P_x = 190.3 \text{ kips}$$

## 11.17 PRESTRESSING COLUMNS

Since columns are normally in compression, creating additional compression by prestressing would only reduce the column's capacity for load; therefore, columns are rarely prestressed unless buckling strongly influences the mode of failure. Prestressing may be advantageous for long slender columns that carry large bending moment and small axial load. Using prestress to eliminate cracking means that the entire cross section (rather than the smaller cracked section) is available to resist bending. If the increased bending stiffness of an uncracked prestressed member results in a sizable reduction in secondary moment created by the P-delta effect, the capacity of the column for axial load may be increased.

Prestressing can also be used to prevent damage to slender precast columns during erection. For this application, the prestressing force must be large enough to neutralize the tensile bending stresses produced by the member's dead weight.

Prestressing is also applied to concrete piles so that the pile can be driven into the ground without cracking. During the driving operation, each blow of the pile driver's hammer causes a stress wave to propagate through the pile. As the wave travels up and down the pile, transient tensile stresses are induced throughout the length of the member. Moderate prestressing, on the order of 700 lb/in$^2$, will normally prevent damage to the pile.

## PROBLEMS

**11.1** The tendon in the cantilever beam in Fig. P11.1 carries a force of 540 kips. Assume elastic behavior and no cracking.

(a) Determine the forces exerted by the tendon on the concrete.

(b) Draw the shear and moment curves for the forces induced in the concrete by prestress alone.

(c) On a vertical section just to the left of point b determine the magnitude and position of the resultant force on the concrete due to prestress only. Establish the distribution of longitudinal stresses on the cross section.

(d) Neglecting the dead weight of the member, compute the reactions and draw the shear and moment curves for the beam due to prestress and the three concentrated loads shown dashed.

(e) On a vertical section just to the left of point b determine the magnitude and position of the resultant force on the concrete. Also determine the distribution of the longitudinal stresses produced by the resultant concrete force.

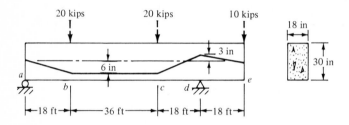

**Figure P11.1**

**11.2** The beam in Fig. P11.2 is stressed with parabolic tendons. The tendon slope is zero at the exterior ends and at midspan.

(a) Determine the forces exerted on the concrete by the tendon.

(b) Draw the shear and moment curves for the forces created in the concrete by the tendon.

(c) If the beam is constructed of normal-weight concrete $\gamma_c = 150$ lb/ft$^3$, draw the shear and moment curves for the forces created by dead load and prestress.

(d) On a vertical section at midspan locate the position of the resultant internal force due to prestress and the girder dead weight. Determine the longitudinal distribution of stresses.

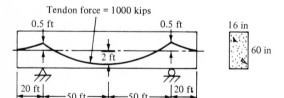

**Figure P11.2**

**11.3** Using both the internal-couple method and the method of superposition, determine the stresses on the cross section at midspan in the beam in Fig. P11.3 due to a uniform service load of 2.4 kips/ft, $A_{ps} = 3.5$ in$^2$; $f_{se} = 150$ kips in$^2$.

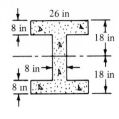

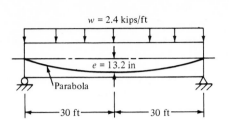

**Figure P11.3**

**11.4** Investigate the concrete stresses in the beam of Fig. P11.4 at transfer and with the total service load applied. Check stresses at midspan and at the ends. At transfer assume that only the dead weight of the beam and prestress act. Do the stresses satisfy the requirements of the ACI Code? $f_{ci} = 4$ kips/in$^2$, $f'_c = 5$ kips/in$^2$, and $\gamma_c = 150$ ln/ft$^3$. The tendon force is $F_i = 240$ kips at transfer and $F_e = 200$ kips after losses.

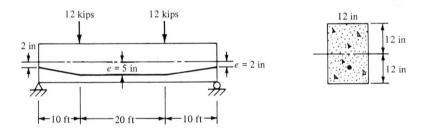

**Figure P11.4**

**11.5** (a) Considering only elastic behavior, determine the value of the uniform service load that can be applied to the beam in Fig. P11.5 if the compressive stress is not to exceed 2250 lb/in$^2$ and the tensile stress is not to exceed 200 lb/in$^2$. Base computations on the stresses at midspan.

(b) Determine the stresses at the ends of the member.

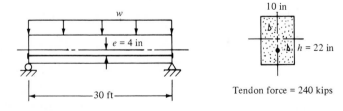

**Figure P11.5**

**11.6** Locate the kern points of the cross sections in Fig. P11.6.

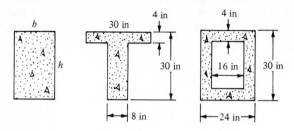

**Figure P11.6**

**11.7** Determine $M_n$, the nominal moment capacity of the cross section in Fig. P11.7 if $f'_c = 5$ kips/in², $f_{pu} = 270$ kips/in², bonded cable is used, and $f_{se} = 150$ kips/in² for (a) $A_{ps} = 2$ in² and (b) $A_{ps} = 3.2$ in².

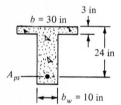

**Figure P11.7**

**11.8** In addition to its own dead weight, the simply supported uniformly loaded beam in Fig. P11.8 carries a superimposed dead load of 500 lb/ft and a live load of 1 kip/ft. The cable is bonded. $f_{pu} = 250$ kips/in², $f_{se} = 150$ kips/in², $f'_c = 6$ kips/in², and $A_{ps} = 3$ in². With regard to the midspan section:
   (a) What value of uniform load will produce initial cracking of the cross section?
   (b) Is the ultimate flexural strength of the cross section adequate for the design loads?
   (c) Is the beam over- or underreinforced?

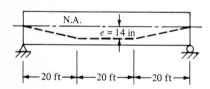

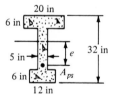

**Figure P11.8**

**11.9** Design a simply supported posttensioned beam with a rectangular cross section to support a concentrated load at midspan of 100 kips (Fig. P11.9). The depth is to be approximately 3 times the width. $f'_c = 5$ kips/in², $f'_{ci} = 4$ kips/in², $f_{pu} = 250$ kips/in², $f_{si} = 175$ kips/in², and total losses $\Delta f_s = 30$ kips/in². Assume a bonded cable. Base design on the analysis of the midspan section for flexure.

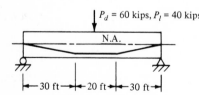

**Figure P11.9**

**11.10** The beam in Fig. P11.10 carries a uniform service live load of 0.8 kip/ft as well as its own weight. Determine the stirrup spacing throughout the length of the beam using the simplified expression for evaluating $V_c$ [Eq. (11.33)]. Use no. 3 stirrups; $f_y = 60$ kips/in$^2$, $b_w = 5$ in, $A_{ps} = 1.6$ in$^2$, $f_{pe} = 150$ kips/in$^2$, $f_{pu} = 250$ kips/in$^2$, $A_v = 0.22$ in$^2$, and $f'_c = 5.5$ kips/in$^2$.

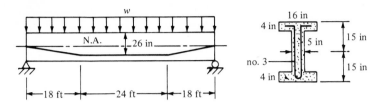

**Figure P11.10**

**11.11** The beam in Fig. P11.11 carries a moving live load as shown. Determine the stirrup spacing required at section 1 using the ACI equations for $V_{ci}$ and $V_{cw}$ to evaluate $V_c$. Use no. 3 stirrups; $A_v = 0.22$ in$^2$, $f'_c = 6$ kips/in$^2$, $f_{se} = 150$ kips/in$^2$, $f_{pu} = 250$ kips/in$^2$, and $f_y = 50$ kips/in$^2$.

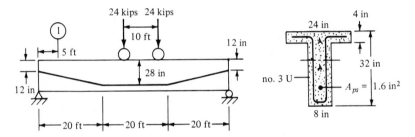

**Figure P11.11**

# REFERENCES

1. D. Billington: Historical Perspective on Prestressed Concrete, *J. PCI*, vol. 21, no. 5, p. 60, September-October 1976.
2. W. B. Bennett: Manufacture and Production of Prestressed Concrete, *J. PCI*, vol. 21, no. 5, p. 192, September-October 1976.
3. P. W. Abeles: Design of Partially Prestressed Concrete Beams, *ACI Proc.* vol. 64, no. 10, p. 671, October 1967.
4. "PCI Design Handbook," Prestressed Concrete Institute, Chicago, 1971.
5. T. Y. Lin: "Design of Prestressed Concrete Structures," 2d ed., p. 161, Wiley, New York, 1963.
6. F. Leonhardt: "Prestressed Concrete Design and Construction," 2d ed., chap. 9, Ernst, Berlin, 1964.
7. W. C. Stone: Design Criteria for Post-Tensioned Anchorage Zone Tensile Stresses, PhD dissertation, University of Texas, May 1980.
8. PCI Committee on Prestress Losses: Recommendations for Estimating Prestress Losses, *J. PCI*, vol. 20, no. 4, pp. 44–75, July-August 1975.
9. Subcommittee 5, ACI Committee 435: Deflection of Prestressed Concrete Members, *ACI Proc.*, vol. 60, no. 12, p. 1697, December 1963.

# CONVERSION FACTORS

| | | | |
|---|---|---|---|
| Length | 1 in = 25.4 mm | Area | $1\ \text{ft}^2 = 0.0929\ \text{m}^2$ |
| | 1 ft = 0.3048 m | | $1\ \text{in}^2 = 645.2\ \text{mm}^2$ |

Moment of
inertia $\quad 1\ \text{in}^4 = 0.4162 \times 10^6\ \text{mm}^4$

Pressure
$1\ \text{lb/in}^2 = 6.895\ \text{kPa (N/m}^2)$

| | | |
|---|---|---|
| Force | 1 lb = 4.448 N | |
| | 1 kip = 4.448 kN | $1\ \text{lb/ft}^2 = 47.880\ \text{Pa}$ |
| | 1 kip/ft = 14.594 kN/$m$ | $1\ \text{kip/in}^2 = 6.895\ \text{MPa}$ |
| | | $1\ \text{kip/ft}^2 = 47.880\ \text{kPa}$ |

| | | | |
|---|---|---|---|
| Mass | 1 lb = 0.454 kg | Moment | $1\ \text{ft} \cdot \text{kip} = 1.356\ \text{kN} \cdot \text{m}$ |
| Weight | $1\ \text{lb/ft}^3 = 16.018\ \text{kg/m}^3$ | | |

*Prefixes most often used with SI units*

| | | |
|---|---|---|
| $10^9$ | giga | G |
| $10^6$ | mega | M |
| $10^3$ | kilo | k |
| $10^{-3}$ | milli | m |

# BAR AREA AND SPACING IN SLABS

**Table B.1  Area of groups of standard bars, in$^2$**

| Bar no. | Number of bars | | | | | | |
|---|---|---|---|---|---|---|---|
| | 2 | 3 | 4 | 5 | 6 | 7 | 8 |
| 4 | 0.39 | 0.58 | 0.78 | 0.98 | 1.18 | 1.37 | 1.57 |
| 5 | 0.61 | 0.91 | 1.23 | 1.53 | 1.84 | 2.15 | 2.45 |
| 6 | 0.88 | 1.32 | 1.77 | 2.21 | 2.65 | 3.09 | 3.53 |
| 7 | 1.20 | 1.80 | 2.41 | 3.01 | 3.61 | 4.21 | 4.81 |
| 8 | 1.57 | 2.35 | 3.14 | 3.93 | 4.71 | 5.50 | 6.28 |
| 9 | 2.00 | 3.00 | 4.00 | 5.00 | 6.00 | 7.00 | 8.00 |
| 10 | 2.53 | 3.79 | 5.06 | 6.33 | 7.59 | 8.86 | 10.12 |
| 11 | 3.12 | 4.68 | 6.25 | 7.81 | 9.37 | 10.94 | 12.50 |
| 14S | 4.50 | 6.75 | 9.00 | 11.25 | 13.50 | 15.75 | 18.00 |
| 18S | 8.00 | 12.00 | 16.00 | 20.00 | 24.00 | 28.00 | 32.00 |

**Table B.2  Spacing of bars for slab reinforcement as a function of steel area, square inches of steel per foot of slab**

| Spacing, in | Bar no. | | | | | | | | |
|---|---|---|---|---|---|---|---|---|---|
| | 3 | 4 | 5 | 6 | 7 | 8 | 9 | 10 | 11 |
| 3 | 0.44 | 0.78 | 1.23 | 1.77 | 2.40 | 3.14 | 4.00 | 5.06 | 6.25 |
| $3\frac{1}{2}$ | 0.38 | 0.67 | 1.05 | 1.51 | 2.06 | 2.69 | 3.43 | 4.34 | 5.36 |
| 4 | 0.33 | 0.59 | 0.92 | 1.32 | 1.80 | 2.36 | 3.00 | 3.80 | 4.68 |
| $4\frac{1}{2}$ | 0.29 | 0.52 | 0.82 | 1.18 | 1.60 | 2.09 | 2.67 | 3.37 | 4.17 |
| 5 | 0.26 | 0.47 | 0.74 | 1.06 | 1.44 | 1.88 | 2.40 | 3.04 | 3.75 |
| $5\frac{1}{2}$ | 0.24 | 0.43 | 0.67 | 0.96 | 1.31 | 1.71 | 2.18 | 2.76 | 3.41 |
| 6 | 0.22 | 0.39 | 0.61 | 0.88 | 1.20 | 1.57 | 2.00 | 2.53 | 3.12 |
| $6\frac{1}{2}$ | 0.20 | 0.36 | 0.57 | 0.82 | 1.11 | 1.45 | 1.85 | 2.34 | 2.89 |
| 7 | 0.19 | 0.34 | 0.53 | 0.76 | 1.03 | 1.35 | 1.71 | 2.17 | 2.68 |
| $7\frac{1}{2}$ | 0.18 | 0.31 | 0.49 | 0.71 | 0.96 | 1.26 | 1.60 | 2.02 | 2.50 |
| 8 | 0.17 | 0.29 | 0.46 | 0.66 | 0.90 | 1.18 | 1.50 | 1.89 | 2.34 |
| 9 | 0.15 | 0.26 | 0.41 | 0.59 | 0.80 | 1.05 | 1.33 | 1.69 | 2.08 |
| 10 | 0.13 | 0.24 | 0.37 | 0.53 | 0.72 | 0.94 | 1.20 | 1.52 | 1.87 |
| 12 | 0.11 | 0.20 | 0.31 | 0.44 | 0.60 | 0.78 | 1.00 | 1.27 | 1.56 |

# NOTATION

| | |
|---|---|
| $a$ | depth of equivalent stress block; arm of internal couple |
| $a_1$ | arm of internal couple in prestressed beam when service loads act |
| $a_2$ | arm of internal couple in prestressed beam at transfer |
| $A$ | area of the gross section; parameter equal to area of tension concrete divided by number of bars in Eq. (3.20) |
| $A_b$ | area of individual bar |
| $A_c$ | area of concrete in equivalent stress block; core area of spiral column; area of critical section for punching shear |
| $A_g$ | gross area of cross section |
| $A_l$ | total area of longitudinal torsion steel |
| $A_n$ | one of several areas formed by subdivision of larger area |
| $A_{ps}$ | area of prestressed steel |
| $A_s$ | area of flexural steel |
| $A_s'$ | area of compression steel |
| $A_{sb}$ | area of balanced steel |
| $A_{st}$ | total area of column steel |
| $A_t$ | area of one leg of closed stirrup used to carry torsion |
| $A_v$ | total area of shear reinforcement supplied by stirrup |
| $A_{vf}$ | area of shear-friction steel |
| $b$ | width of compression zone at extreme fiber; width of member |
| $b_0$ | perimeter of critical section for punching shear |
| $b_{\text{eff}}$ | effective width of compression zone of T beam |

$b_w$      width of beam web

$c$      distance of neutral axis from compression surface as flexural failure impends; distance from the extreme fibers, top and bottom, to neutral axis in elastic equation for flexural stress

$c_1$      width of column, equivalent column, capital, or bracket measured in direction of span for which moments are being determined (see Tables 10.9 to 10.10)

$c_2$      width of column, equivalent column, capital, or bracket measured perpendicular to direction of the span for which moments are being computed (see Tables 10.9 to 10.10)

$c_b$      distance of neutral axis from compression surface as flexural failure impends when cross section reinforced with balanced steel $A_{sb}$

$c_t$      distance from neutral axis to top fiber

$C$      resultant of longitudinal compressive stresses due to flexure or axial load and flexure; torsional factor in Eq. (10.6) measuring torsional resistance of cross section

$C_b$      clear side cover; or twice the clear distance between bars in row

$C_c$      creep coefficient

$C_f$      compression force in flanges

$C_m$      correction factor applied to the maximum end moment in column with unequal end moments [see Eq. (7.25)]

$C_s$      force in compression steel; clear depth of cover; half the clear spacing between bars in row of reinforcement: depth of cover from surface of bar to face of concrete

$C_t$      coefficient in interaction equation for shear and torsion

$C_w$      compression force in web

$d$      distance from extreme compression fiber to centroid of flexural reinforcement; diameter of a standard concrete test cylinder; distance (in transfer equation for moment of inertia) between centroid of area and axis about which moment of inertia is computed

$d'$      distance from extreme compression fiber to centroid of compression steel

$d_b$      nominal diameter of reinforcing bar or wire

$d_c$      distance from tension surface to center of bar closest to that surface [see Eq. (3.20)]

$D$      unfactored dead load or forces created by dead load

$D_f$      depth of footing below grade

$e$      eccentricity of load; eccentricity of prestressed tendon; base of natural logarithms

$E$      modulus of elasticity; load or force created by earthquake motion

$E_c$      modulus of elasticity of concrete [see Eqs. (2.1) and (2.2)]

$E_{cb}$      modulus of elasticity of concrete in beam

$E_{cs}$      modulus of elasticity of concrete in slab

$f$      flexural stress

$f_b$      stress due to moment

| | |
|---|---|
| $f_c$ | maximum flexural compressive stress in concrete produced by service loads; compressive stress in concrete |
| $f_c'$ | 28-day compressive strength of concrete, $lb/in^2$ or MPa |
| $f_{ci}'$ | compressive strength of concrete at time of initial prestresses |
| $f_{ct}$ | split-cylinder tensile strength of concrete |
| $f_d$ | flexural stress produced by unfactored dead load on extreme fiber on side of prestressed beam where external loads produce tensile stress; direct stress $P/A$ |
| $f_h$ | maximum tensile stress that can be anchored by standard hook |
| $f_{ps}$ | stress in prestressed reinforcement when flexural failure impends |
| $f_{py}$ | yield strength of prestressed steel |
| $f_r$ | modulus of rupture, flexural tensile strength of concrete |
| $f_s$ | stress in tension steel |
| $f_s'$ | stress in compression steel |
| $f_{sb}'$ | stress in compression steel when balanced failure impends |
| $f_{se}$ | effective stress in prestressed tendon after losses |
| $f_{si}$ | stress in prestressed steel immediately after transfer |
| $f_t$ | tensile stress in concrete; principal tension stress |
| $f_y$ | yield point of nonprestressed reinforcement |
| $f_2$ | lateral pressure |
| $F$ | fluid pressure or forces created by fluid pressure; coefficient listed in ACI design aid (see Table 3.4), used to select proportions of rectangular cross section |
| $F_c$ | compression force in concrete |
| $F_{cl}$ | longitudinal component of concrete force from torsion |
| $F_{cv}$ | vertical component of concrete force from torsion |
| $F_e$ | prestressed force after all losses |
| $F_i$ | prestressed force immediately after transfer, before long-term losses |
| $F_o$ | prestressed force before transfer |
| $F_r$ | shear-friction force |
| $F_s$ | tension force in steel |
| $F_s'$ | compression force in steel |
| $h$ | depth of beam or slab; wall thickness; diameter or thickness of column; vertical distance between parabolic tendon and its chord (termed cable sag) |
| $h_{core}$ | diameter of core in spiral column, measured to outside diameter of spiral |
| $h_{min}$ | minimum depth of beam or slab for ensuring that deflections will not be excessive |
| $H$ | unfactored earth pressure or forces produced by earth pressure |
| $I$ | moment of inertia of gross section about centroidal axis |
| $I_b$ | moment of inertia of the gross section of beam about centroidal axis |
| $I_c$ | moment of inertia of column |
| $I_{cr}$ | moment of inertia of cracked transformed section |

| | |
|---|---|
| $I_e$ | effective moment of inertia used in deflection computations to account for flexural cracking [see Eq. (3.16)] |
| $I_g$ | moment of inertia of gross cross section about the centroidal axis; moment of inertia of girder |
| $I_s$ | moment of inertia of gross area of slab about centroidal axis |
| $I_{sb}$ | moment of inertia of slab and beam cross section, slab extending between panel centerlines |
| $I_{se}$ | moment of inertia of reinforcement about centroidal axis of cross section |
| $j$ | factor giving arm of internal couple for service-load moment as function of effective depth $d$ |
| $J_c$ | polar moment of inertia of critical section for punching shear |
| $k$ | effective length factor for columns; coefficient giving position of neutral axis as function of $d$ for service-load stresses; distance to kern point of cross section from neutral axis; constant in Eq. (5.1) |
| $k_b$ | distance from neutral axis to bottom kern point |
| $k_t$ | distance from neutral axis to top kern point |
| $K$ | wobble coefficient, friction force per pound of tendon force per foot of tendon units of $ft^{-1}$ |
| $K_b$ | flexural stiffness of beam |
| $K_c$ | flexural stiffness of column |
| $K_{ec}$ | flexural stiffness of equivalent column |
| $K_s$ | flexural stiffness of slab |
| $K_t$ | torsional stiffness of arm of equivalent column |
| $K_u$ | parameter in ACI design aids (see Table 3.3) |
| $l$ | span length of beam or one-way slab (for continuous structures measured between centerlines for determination of moments; for discontinuous structures clear span plus depth of member but not greater than center-to-center distance of supports) |
| $l_1$ | span length, center to center of supports, of two-way slab in direction moments are being computed |
| $l_2$ | span length, center to center of supports, in direction perpendicular to $l_1$ |
| $l_a$ | additional length of embedment beyond the centerline of support or point of inflection |
| $l_c$ | length of column between slab centerlines |
| $l_d$ | development length, minimum length required to anchor bar, usually stressed to $f_y$ |
| $l_e$ | length of straight bar with same anchorage capacity as hook |
| $l_g$ | length of girder, center to center of supports |
| $l_n$ | clear span for positive moment and shear, average of adjacent clear spans for negative moment |
| $l_u$ | unsupported length of column |
| $L$ | span length |
| $L_0$ | initial length of member |

| | |
|---|---|
| $m$ | ratio of volume of longitudinal torsional reinforcement to volume of stirrup steel |
| $m_x$ | moment per unit length of slab |
| $M$ | bending moment |
| $M_0$ | simple beam moment for panel of a two-way slab [see Eq. (10.12)] |
| $M_1$ | smaller end moment at end of column produced by factored load, positive for single-curvature bending and negative for double-curvature bending |
| $M_2$ | larger value of column end moment produced by factored loads, always positive |
| $M_a$ | maximum moment produced by service loads |
| $M_b$ | moment acting with axial force $P_b$ to produce balanced failure in column |
| $M_c$ | magnified moment in slender column, includes additional moment produced by eccentricity of the axial load; positive design moment in a two-way slab |
| $M_{cr}$ | cracking moment; minimum moment required to produce initial flexural cracking |
| $M_i$ | internal moment |
| $M_m$ | modified moment required to evaluate $V_c$ in presence of axial load |
| $M_{max}$ | maximum value of moment; in equation for $V_{ci}$ moment at section produced by externally applied factored loads |
| $M_n$ | nominal flexural strength; value of moment associated with failure |
| $M_s$ | service-load moment; in two-way slab negative design moments at supports |
| $M_u$ | moment due to factored loads |
| $n$ | modular ratio; number of stirrrups |
| $N$ | normal force on shear-friction failure surface |
| $N_u$ | factored value of axial load, positive for compression and negative for tension; used in equations for $V_c$ |
| $p$ | soil pressure due to unfactored loads; pressure |
| $p_{min}$ | minimum value of soil pressure |
| $p_{max}$ | maximum value of soil pressure |
| $p_u$ | soil pressure due to factored loads |
| $P$ | unfactored concentrated load applied to beam or column |
| $P_0$ | axial strength of a concentrically loaded column |
| $P_b$ | value of axial load producing balance failure in column |
| $P_c$ | Euler buckling load |
| $P_n$ | nominal axial strength of column for particular value of eccentricity; concentrated load producing failure of beam or column |
| $P_s$ | force in tendon adjacent to jack |
| $P_u$ | factored load; axial force in a column produced by factored load |
| $P_x$ | tendon force a distance $x$ from jack |
| $q$ | tension reinforcement index ($\rho f_y/f_c'$); soil bearing capacity |
| $q_a$ | allowable bearing capacity of soil considering all loads above base of foundation |

| $q_n$ | allowable bearing capacity of soil for pressures in excess of those created by weight of soil above foundation base |
|---|---|
| $Q$ | concentrated service load; in equation for shear stress moment of area of cross section above level at which shear stress is to be computed |
| $r$ | radius of gyration of gross area; radius of circle |
| $R$ | resultant of force system; reaction at a support (may be subscripted to indicate the location of support) |
| $s$ | spacing between bars, stirrups, ties, coils of spiral |
| $T$ | torque; tension force; forces created by temperature, creep, shrinkage, or support movement |
| $T_1$ | in trial method, approximation value of tension force based on estimating arm of internal couple (if estimate is repeated to improve the value of $T_1$, subscript 2 is used to denote improved value) |
| $T_c$ | nominal torsional strength provided by concrete; tension force carried by concrete |
| $T_{cv}$ | torsional capacity of concrete in presence of shear |
| $T_n$ | nominal torsional strength provided by both concrete and torsional reinforcement |
| $T_s$ | tension force in steel; torsional strength supplied by torsional reinforcement |
| $T_u$ | torque on vertical section produced by factored loads |
| $u$ | bond stresses; parameter in moment-magnification equations to account for the moment produced by eccentricity of axial load |
| $u_u$ | nominal bond strength in terms of stress |
| $U$ | design strength required to support the force produced by factored loads; bond force per unit length |
| $U_u$ | nominal bond strength, force per unit length |
| $v$ | shear stress produced by service loads |
| $v_c$ | shear strength of concrete, force per unit area |
| $v_{cw}$ | web shear strength in terms of stress $= V_{cw}/b_w d$ |
| $v_t$ | shear stress produced by a torque; total shear stress |
| $v_u$ | nominal shear stress produced by factored loads |
| $V$ | shear force produced by service loads |
| $V_a$ | shear strength produced by aggregate interlock |
| $V_c$ | shear strength of concrete in units of force; shear force carried by compression zone of cracked section |
| $V_{ci}$ | flexural shear strength; shear force associated with flexural shear failure in prestressed beam |
| $V_{ct}$ | shear strength of concrete in presence of torsion |
| $V_{cw}$ | web shear strength: shear force associated with web shear failure in prestressed beam |
| $V_d$ | dowel shear |
| $V_i$ | shear at section produced by externally applied factored loads, which are positioned to maximize moment at section under study |
| $V_n$ | nominal shear strength of a cross section $= V_c + V_s$ |

| | |
|---|---|
| $V_p$ | vertical component of force in a sloping tendon after losses |
| $V_s$ | shear strength of reinforcement |
| $V_u$ | shear force due to factored loads |
| $w$ | width of crack at tension surface; unfactored uniform load |
| $w_c$ | unit weight of concrete, $lb/ft^3$ or $kg/m^3$ |
| $w_d$ | unfactored dead load, force per unit length or force per unit area |
| $w_l$ | unfactored live load, force per unit length or force per unit area |
| $w_p$ | distributed load exerted on concrete by curved prestressed tendon |
| $w_s$ | unit weight of soil, $lb/ft^3$ or $kg/m^3$ |
| $w_u$ | factored load per unit length or per unit area |
| $W$ | wind load or force created by wind; uniformly distributed load |
| $W_u$ | uniformly distributed factored load, force per unit length |
| $x$ | length of the short side of rectangular area; horizontal distance |
| $x_1$ | length of short side of rectangular closed stirrup |
| $y$ | length of long side of rectangular area; vertical distance |
| $y_1$ | length of long side of closed rectangular stirrup |
| $y_t$ | distance from centroid of gross area to extreme tension fiber |
| $\bar{y}$ | distance between centroid of gross area and reference axis |
| $\bar{y}_n$ | distance between centroid of area $A_n$ and reference axis |
| $Y$ | distance from neutral axis to extreme compression fiber, distance between a reference axis and centroid of an area |
| $z$ | parameter measuring width of flexural cracks |
| $Z$ | section modulus |
| $Z_b$ | section modulus bottom |
| $Z_t$ | section modulus top |
| $\alpha$ | ratio of stiffness of beam along column centerline to stiffness of slab section contained between panel centerlines in direction in which moments are computed; angle between inclined stirrups and longitudinal axis of beam; change in slope in radians of prestressed tendon between jack and section at which prestressed force, adjusted for friction loss, is computed |
| $\alpha_1$ | value of $\alpha$ in direction $l_1$ |
| $\alpha_2$ | value of $\alpha$ in direction $l_2$ |
| $\alpha_c$ | ratio of stiffness of columns above and below slab to sum of stiffness of beams and slabs framing into joint $= \Sigma K_c / \Sigma (K_s + K_b)$ |
| $\alpha_{ec}$ | ratio of equivalent column stiffness [see Eq. (10.4)] to sum of stiffnesses of beams and slabs framing into joint $= K_{ec} / \Sigma (K_S + K_b)$ |
| $\alpha_m$ | average value of $\alpha$ for panel with edge beams |
| $\alpha_{min}$ | smallest value of $\alpha_c$ required to prevent excessive increase in positive slab moment because of support rotation (see Table 10.6) |
| $\alpha_t$ | slope of torque-versus-reinforcement-index curve [see Eq. (5.10)] |
| $\beta$ | ratio of distance from the neutral axis of cracked section to tension surface divided by distance from neutral axis to centroid of flexural steel; in biaxial bending, measure of curvature of three-dimensional interaction surface; ratio of long to short sides of rectangular footing; ratio of long to short clear spans in two way slab |

| | |
|---|---|
| $\beta_1$ | ratio of depth of stress block $a$ to the distance between neutral axis and extreme compression fibre $c$ [see Eqs. (3.22$a$) and (3.22$b$)] |
| $\beta_a$ | ratio of unfactored dead load to unfactored live load |
| $\beta_b$ | ratio of area of reinforcement terminated to total area of reinforcement |
| $\beta_c$ | ratio of long to short sides of a column or a loaded area [see Eq. (4.24)] |
| $\beta_d$ | ratio of maximum factored dead load moment to maximum factored total load moment |
| $\beta_s$ | in two-way slab, ratio of length of continuous edges to perimeter of panel |
| $\beta_t$ | in exterior panel of two-way slab, ratio of edge-member torsional stiffness to flexural stiffness of slab in direction in which moments are computed $= E_{cb}C/2E_{cs}I_s$ |
| $\gamma$ | in ACI design aids ratio of distance between rows of reinforcement on opposite sides of column to depth of column in direction of bending |
| $\gamma_v$ | fraction of unbalanced slab moment transferred into columns by shear couple |
| $\delta$ | magnification factor [see Eq. (7.24)] |
| $\delta_s$ | factor to adjust positive moment in slab with flexible columns [see Eq. (10.15)] |
| $\Delta$ | small displacement; when $\Delta$ precedes a variable, it indicates a small change in the variable |
| $\Delta_a$ | additional deflection occurring over time due to creep and shrinkage |
| $\Delta_i$ | initial deflection |
| $\epsilon$ | strain |
| $\epsilon_c$ | strain in concrete |
| $\epsilon_{ci}$ | initial elastic strain in concrete |
| $\epsilon_{cu}$ | ultimate creep strain |
| $\epsilon_s$ | strain in steel |
| $\epsilon_s'$ | strain in compression steel |
| $\epsilon_{sb}'$ | strain in compression steel; section reinforced with balanced steel area |
| $\epsilon_y$ | yield-point strain |
| $\theta$ | slope angle |
| $\mu$ | coefficient of friction; coefficient of friction due to curvature |
| $\xi$ | parameter for evaluating capacity of standard hook (see Table 6.2) |
| $\pi$ | 3.14159 |
| $\rho$ | area of flexural steel divided by effective area of concrete $= A_s/b_w d$ |
| $\rho'$ | compression-steel ratio $= A_s'/b_w d$ |
| $\rho_b$ | balanced-steel ratio $= A_{sb}/b_w d$ |
| $\rho_p$ | prestressed-steel ratio $= A_{ps}/b_w d$ |
| $\rho_s$ | ratio of volume of spiral steel to volume of core, measured out to out of spiral |
| $\phi$ | strength-reduction factor |
| $\omega$ | $\rho f_y/f_c'$ |
| $\omega_p$ | $\rho_p f_{ps}/f_c'$ |

# ANSWERS TO SELECTED PROBLEMS

## Chapter 1

**1.1** At midspan required $\phi M_n = +86.84$ ft · kips and $-285.52$ ft · kips; at supports required $\phi M_n = -344.52$ ft · kips

## Chapter 3

**3.1** (a) 491 lb/ft, (b) 391 lb/ft
**3.2** $M_{cr} = 20$ ft · kips
**3.4** $P = 2.38$ kips
**3.5** (a) $I_{cr} = 5206$ in$^4$, (b) 3522 in$^4$, (c) 7336 in$^4$, (d) 17,483 in$^4$
**3.6** $h_{min} = 10.96$ in
**3.10** Instant deflection, total load $= 0.153$ in
**3.11** (a) $z = 199$, (b) $z = 171$, (c) $z = 157$
**3.13** (1) with $f'_c = 2500$ lb/in$^2$, $M_n = 1838$ in · kips, $\varepsilon_s = 0.008$
**3.14** $M_u = \phi M_n = 106.1$ ft · kips, $\varepsilon_s = 0.0204$
**3.16** $\phi M_n = 115.17$ ft · kips
**3.19** $A_{s, min} = 0.583$ in$^2$, $\phi M_n = 45$ ft · kips
**3.20** $\phi M_n = 109.6$ ft · kips
**3.21** $A_s = 4.92$ in$^2$, $A'_s = 1.78$ in$^2$
**3.23** $A_{sb} = 3.45$ in$^2$
**3.24** $A_s = 3.88$ in$^2$, $A_{sb} = 14.73$ in$^2$
**3.25** $\phi M_n = 300$ ft · kips
**3.27** $A_{s, min} = 0.864$ in$^2$ controls

## Chapter 4

**4.1** $\phi V_c = 8.87$ kips
**4.2** $A_{v, min}$ controls, $s = d/2 = 7.75$ in, use 7.5 in
**4.3** Stirrups not required $\phi V_c = 36.7$ kips

**4.7** At section $A$, $s = 10$ in; at section B use $d/2$
**4.8** At section 1, $s = 7.5$ in if short equation for $V_c$ used; $s = 10$ in if long form used
**4.11** $w_u = 0.562$ kips/ft$^2$

# Chapter 5

**5.1** 11,520 in$^3$
**5.2** Yes, 120 in · kips exceeds $0.85(0.5\sqrt{f_c'}\Sigma x^2 y) = 80.64$ in · kips
**5.5** 14- by 22-in section not adequate, use $b = 17$ in or more
**5.7** Use no. 4 stirrups at 3.85 in; $A_l = 1.5$ in$^2$

# Chapter 6

**6.1** 42 in $< l_d = 62.4$ in, anchorage no good
**6.2** 5 ft-10 in $< l_d = 87.4$ in. Maximum diameter bar permitted is no. 9
**6.3** $l_d = 58$ in, 54 in $< 58$ in; not properly anchored
**6.4** Anchorage Ok
**6.5** $l_{min} = 46.6$ in
**6.6** Minimum lap length $= 25.3$ in

# Chapter 7

**7.2** $kl_u/r = 57.2 > 22$, long column
**7.3** Case of braced frame: column $ab$, $kl_u/r = 25.8 < 38$, column short; column $cd$, $kl_u/r = 24.9 < 31$, column short
**7.5** $P_u = 830$ kips
**7.6** Plastic centroid 7.5 in from left edge
**7.11** One possible solution: 12 in square, reinforced with four no. 10
**7.13** One possible solution: 15-in-diameter column with six no. 10, no. 3 spiral with 1.5-in pitch
**7.15** One possible solution: 10- by 18-in section with four no. 11
**7.16** One possible solution: 16-in-diameter column with seven no. 9
**7.18** $P_n \approx 498$ kips
**7.19** $P_n \approx 277.7$ kips
**7.20** No
**7.22** $M_c = 50.7$ ft · kips, required $A_{st} = 2.86$ in$^2$
**7.23** Yes, 5.06 in$^2 > 3.84$ in$^2$
**7.26** $P_{u, max} = 261.9$ kips

# Chapter 8

**8.1** $p_{max} = 12.06$ kips/ft$^2$, $p_{min} = 8.32$ kips/ft$^2$
**8.2** One solution: 14 in deep, 5.5 ft wide, reinforced with no. 5 at 8 in
**8.3** One solution: 8 ft 2 in square, 28 in deep, reinforced with eight no. 9 bars each way
**8.6** $P = 181.6$ kips, punching shear controls

# Chapter 10

**10.2** Joint satisfactory for shear

# Chapter 11

**11.1**  Prestress plus load, top stress $= 1.2$ kips/in$^2$ compression; bottom stress $= 0.8$ kips/in$^2$ compression

**11.3**  Top stress $= 2.13$ kips/in$^2$ compression; bottom stress $= 0.31$ kips/in$^2$ tension

**11.4**  1.68 kips/in$^2$ compression at top, 0.29 kips/in$^2$ tension bottom

**11.5**  $w = 1.41$ kips/ft

**11.6**  (a) $k_t = k_b = h/6$, (b) $k_t = 4.78$ in, $k_b = 7.69$ in, (c) $k_b = k_t = 7.21$ in

**11.7**  (1) 908.1 ft · kips, (2) 1078.1 ft · kips

**11.8**  $w_{cr} = 1.93$ kips/ft, $\phi M_n = 1223.5$ ft · kips

**11.9**  One solution: $b = 20$ in, $h = 60$ in, $A_{ps} = 8.3$ in$^2$, $d = 48.75$ in

**11.11**  $V_{ci} = 54.32$ kips controls, space no. 3 stirrups at 7 in

# INDEX

# INDEX

Aggregates:
  influence on air content, 22
  influence on creep, 29
  influence on modulus of rupture, 26
  influence on shrinkage, 32
  maximum size, 19
  types, 19
American Concrete Institute (ACI) Building Code:
  scope, 3–4
  system of units, 2
American Institute of Steel Construction (AISC), 3, 56
Amplification factor (*see* Moment magnification)
Anchorage, 210, 215, 224–228
  (*See also* Bond)
Arch action, 114

Beam-columns (*see* Columns)
Beams:
  area of steel: influence on failure mode, 71, 73
  maximum for nonrectangular sections, 104
  maximum for rectangular cross section, 85
  maximum for T beams, 121
  minimum for rectangular cross section, 80
  minimum for T beams, 121
  required for flexure, 90
  selection by trial method, 98–103

Beams (*Cont.*):
  balanced steel: nonrectangular cross sections, 103–105
  rectangular cross sections, 83–85
  compression steel: design procedure, 110
  influence on behavior, 107–110
  continuous: analysis by ACI coefficients, 372–375
  pattern loading for maximum moment, 371–372
  crack width, 68–69
  ACI procedure to control, 69
  cracking moment, 43–45
  equation to predict, 44
  depth to control deflection, 61–66
  design procedure, 89
  doubly reinforced, 53, 110–113
  effective flange width, 119
  failure: overreinforced, 71–73
  underreinforced, 74–79
  moment redistribution, 380–382
  neutral axis, service loads, 48–50
  with balanced steel, 83
  at failure, 74–76
  prestressed (*see* Prestressed beams)
  proportioning, 88
  use of design aids, 93–97
  required clearances for reinforcement, 91
  shallow, 41, 56, 88
  shear strength, 147–148
  stress block, 76–77
  T-beam design, 121–122

535